PLANTATIONS, PROLETARIANS AND PEASANTS IN COLONIAL ASIA

PLANTATIONS, PROLETARIANS AND PEASANTS IN COLONIAL ASIA

Edited by
E. Valentine Daniel, Henry Bernstein
and Tom Brass

Routledge
Taylor & Francis Group
LONDON AND NEW YORK

First published 1992 in Great Britain by
Frank Cass and Co.Ltd.

Published 2019 by Routledge
2 Park Square, Milton Park, Abingdon, Oxon, OX14 4RN
52 Vanderbilt Avenue, New York, NY 10017

Routledge is an imprint of the Taylor & Francis Group, an informa business

British Library Cataloguing in Publication Data

Plantations, Proletarians and Peasants in
Colonial Asia. — (Special Issues of
"Journal of Peasant Studies" Series, ISSN
0306–6150; Vol. 19, Nos. 3 & 4)
I. Daniel, E. Valentine II. Series
305.5
ISBN 0–7146–3467–0

Library of Congress Cataloging-in-Publication Data

Plantations, proletarians, and peasants in colonial Asia / edited by
E. Valentine Daniel, Henry Bernstein, and Tom Brass.
p. cm.
Articles first appeared in the Journal of peasant studies, vol.
19, nos. 3 & 4, 1992.
Includes bibliographical references.
ISBN 0–7146–3467–0
1. Plantations—South Asia—History—Congresses. 2. Plantation workers—South Asia—History—Congresses. 3. Peasantry—South Asia–
–History—Congresses. 4. Colonies—South Asia—Congresses.
5. Tropical crops—Economic aspects—South Asia—History–
–Congresses. 6. Indentured servants—South Asia—History–
–Congresses. I. Daniel, E. Valentine. II. Bernstein, Henry.
III. Brass, Tom, 1946– .
HD1471.S58P55 1992
306.3′49—dc20 92–30826
CIP

This group of studies first appeared in a Special Issue on 'Plantations, Proletarians and Peasants in Colonial Asia' of *The Journal of Peasant Studies*, Vol. 19, Nos. 3 & 4, 1992,

Typeset by Regent Typesetting, London

ISBN 13: 978-0-7146-3467-8 (hbk)

Contents

This volume originated in a conference on 'Capitalist Plantations in Colonial Asia', held at the Centre for Asian Studies of the University of Amsterdam and Free University of Amsterdam in September 1990. The conference was organised by Jan Breman and Val Daniel, and sponsored by the Joint Committees on South and Southeast Asia of the American Council of Learned Societies and Social Science Research Council, the World Institute for Development Economics Research of the United Nations University, and the Royal Netherlands Academy of Sciences. Presented here is a selection of the papers from the conference, edited by Val Daniel with the assistance of Laurie Lindsay for the SSRC, and by Henry Bernstein and Tom Brass for *The Journal of Peasant Studies*.

HB
TB

Introduction: Proletarianisation and Deproletarianisation on the Colonial Plantation

TOM BRASS and HENRY BERNSTEIN

I

Introduction

The contributions to this collection focus on the production of rubber, sugar, tea, and several less strategic plantation crops, in colonial Indochina, Java, Malaya, the Philippines, India, Ceylon, Mauritius and Fiji (although geographically anomalous, both the latter are included because of the centrality to their sugar plantations of indentured labour from India). The country or region, period, crop, and principal themes covered in each article are summarised in Table 1, which indicates that the major concerns of the authors are the origins and recruitment of plantation labour, the labour processes into which the plantation workforce was deployed, and the labour regimes governing this. Several contributions also address the ideological representation of the colonial 'coolie' (Kelly, Breman and Daniel) and the colonial plantation as alien 'enclave' (Meyer), while McCoy traces the exceptional formation and development of an indigenous planter class which gave 'Philippines plantations a depth of social and political penetration not found elsewhere in tropical Asia'.

What is missing from this collection, however, are similarly detailed and considered accounts of the forms and effects of workers' struggles on and against plantations, despite the useful glimpses of 'resistance' provided by Murray, Ramasamy, McCoy, Behal and Mohapatra, and Kelly. Accordingly, the importance of different forms of class struggle over the commodification, decommodification, and recommodification of labour power on the colonial plantation is a central theme in our theoretical discussion below.

Tom Brass is at the Faculty of Social and Political Sciences, University of Cambridge, Free School Lane, Cambridge CB2 3RQ, UK, and Henry Bernstein is at the Institute for Development Policy and Management, University of Manchester, Oxford Road, Manchester M13 9QS, UK.

TABLE 1

Author	*Country/region*	*Period*	*Crop*	*Main Themes*
Murray	Indochina	1910–40	Rubber	Labour recruitment, plantation labour regimes, worker resistance
Knight	Java (Pekalongan–Tegal)	1830–1930	Sugar	Relations plantation/peasant economy, origins of plantation labour
Ramasamy	Malaya	late 19th century–1941	Rubber	Chinese and Indian labour recruitment, labour control, worker resistance
McCoy	Philippines (Negros)	*c.* 1850–present	Sugar	Formation of an indigenous planter class, and its political strategies
Behal and Mohapatra	India (Assam)	1840–1908	Tea	Rise and fall of indentured labour, labour recruitment, labour process, role of the colonial state
Das Gupta	India (Bengal, Assam, South India)	19th century–1920s	Various	Labour recruitment and reproduction, labour regimes
Meyer	Ceylon	19th century–1930s	Tea, and others	Plantation/village interlinkages, dualist representations, their sources and political implication
Warner	India–Mauritius	1830s–1860s	Sugar	Recruitment of indentured labour, and use of returnees
Kelly	Fiji	1880s–1920	Sugar	Indentured labour and its discourses
Breman and Daniel	South and Southeast Asia	late 19th century–early 20th century	Plantations and other sectors	Colonial capitalism and its representations of 'the coolie'

—

A further point of note concerning the various contributions is the period they address, from about the last third of the nineteenth century through the capitalist depression of the 1930s to the beginning of the Second World War. While recognising the salient legacy of earlier colonial plantations, both the slave plantations of the Americas and the earlier nineteenth-century plantations in Asia (Knight on Java, Behal and Mohapatra on Assam, Das Gupta on Bengal, Meyer on Ceylon), what is the significance of the 1870–1940 period? First, it was the apogee of European colonialism in Asia (and Africa), and, second, its onset coincides with the beginning of *imperialism* in the sense designated by Lenin: the 'highest stage of capitalism' marked by the transformation of competitive into monopoly capital, the hegemonic role of finance capital, and the internationalisation of capital driven by the need for super-profits. Again, several contributors (Murray, McCoy, Behal and Mohapatra) provide glimpses of the effects of the new imperialist world economy, in terms of the concentration and organisation of plantation capital, the pressures of overproduction and price fluctuations on the world market, and how these were transmitted to the plantation labour process in the form of workforce restructuring and technical change.

For Lenin, of course, the colonial expansion of the late nineteenth century, above all in Africa and parts of Southeast Asia, was an integral part of the emergence of capitalist imperialism, which, however, is not coterminous with nor reducible to colonial rule, as Lenin [1964: 263–4] illustrated in the cases of Argentina (post-colonial) and Portugal (itself a minor colonial power in Africa but dominated by British imperialism). The analysis of imperialism reminds us both that the great expansion of plantation export agriculture in the late nineteenth century, and the unfree labour regimes it employed, was not restricted to colonial territories, and that unfree labour was also central to sectors other than plantation agriculture. On the first point, for example, Latin America's massive export boom from the 1870s manifested 'a virtually unique combination in the nineteenth century of political independence and primary commodity led incorporation into the international capitalist economy' [*Archetti et al.*, 1987: xvi], fuelled by a range of forms of indentured, bonded and other unfree labour [*Brass*, 1990a]. On the second point, for example, from the late nineteenth century onwards capital in Southern Africa exhibited an almost insatiable demand for labour for mining as well as farming (see, *inter alia*, van Onselen [1976]; Levy [1982]; First [1983]; Kimble [1985]; Miles [1987]). In fact, some of the mechanisms and practices of unfree labour in South Africa to this day provide useful points of comparison with those described in this volume in the context of plantations in colonial Asia.

To summarise the three major features of the period of global capitalism with which the contributions to this volume are centrally concerned. First, the formation of a world market shaped by the needs of increasingly rapid capitalist industrialisation in the course of the nineteenth century generated an accelerated demand for agricultural commodities of all kinds, with a particular burst of expansion from the 1870s onwards (and despite the depression of 1873–96, which Lenin identified as the catalytic moment in the emergence of imperialism). As well as a new scale of international demand for grains, sugar, tea, coffee, tobacco and bananas, there was an even greater expansion of demand for crops associated with industrial processing and manufacturing, such as vegetable oils, cotton, sisal, jute and rubber, the last an exemplary raw material of the 'Second Industrial Revolution' of the late nineteenth century (Murray in this volume; see also Barraclough [1964: Ch. 2] and Hobsbawm [1987: Ch. 3]). Second, the movement of these commodities in ever increasing quantitites was made possible both by technical changes that dramatically lowered the costs of long distance bulk transport, and by the opening of strategic new transport routes.

Third, how was the supply of these commodities ensured? In his outline of plantation agriculture, Courtenay [1965: Ch. 3] speaks of the 'rise of the industrial plantation' in the nineteenth century, and notes in particular the pace of its expansion from around the 1870s. The 'industrial plantation' replaced previous types in the 'traditional' areas in the Americas (a transformation Courtenay associates with the change in the Caribbean and Brazil from chattel slavery to indenture and other forms of contract labour), and in Asia was connected both with change in and massive expansion of existing plantation areas such as Java (Knight), Negros (McCoy), Assam (Behal and Mohapatra) and Ceylon (Meyer), and with the rapid development of new plantion 'frontier' zones in Indochina (Murray), Malaya (Sundaram [1986: Ch. 7]), and Sumatra (Stoler [1985], Breman [1989]).

What makes the 'industrial plantation' in the epoch of imperialism distinctive is the *connection* between its organisation of production (labour processes, technologies, management structures and imperatives), its character as (part of) corporate capitalist enterprise, and the new, more highly integrated, conditions of its world market location, involving linkages with finance capital, shipping, industrial processing and manufacturing – aspects of that late nineteenth century development aptly described as a 'worldwide shift towards agribusiness' (Stoler [1985: 17]; see also Beckford [1972]). The 'industrial plantation' thus represented new forms of capital, and indeed a new form of capitalism (imperialism). Lenin's view that the dynamic of the internationalisation of capital is the pursuit of super-profits returns us, then, to questions concerning the conditions and mechanisms of the exploitation of labour, in the plantation, in other colonial capitalist sectors, and also (by implication) in the post-colonial imperialist periphery.

Given the continuing debate about the place of the plantation in the colonial economy, together with the latter's capitalist/non-capitalist nature, the most problematic issues concern the theorisation of the social relations of production structuring the plantation economy, and in particular their free/unfree nature.[1] This in turn entails confronting two interrelated revisionist characterisations of the plantation workforce: in economic terms as effected by neo-classical theory, and (more insidiously) in what might be called 'culturalist' terms, as effected by moral economy, survival strategies and resistance theory. In their different ways, all these (re-) interpretations of the plantation system in general, and in particular its meaning for the workforce, involve either the denial or the dilution of its unfree element.

By contrast, here it is intended to examine the dynamic of freedom/unfreedom as this informs the recruitment/reproduction of plantation labour, and its proletarian/peasant origins, together with the multiple trajectories/combinations which have been outcomes of class formation/struggle in this context (depeasantisation, peasantisation, proletarianisation, deproletarianisation, reproletarianisation).[2] As it structures the reproduction of different variants of labour-power employed on the plantation, class struggle is waged both from below (in the form of experiencing/attempting a process of becoming, in the case of the Caribbean a crucial development for a post-emancipation workforce composed of ex-slaves) and also from above (in the double form of conflict among the different components of capital over access to and control of labour-power, and by capital to displace/pre-empt the emergence of a specifically proletarian consciousness).

The Problematic of Free Wage Labour in Capitalism

The necessary starting point for a consideration of the trajectories followed by plantation labour is Marx's emphasis on the freedom of wage labour in the relationship between labour and capital, the social relation of production that constitutes the distinctive character of capitalism. As is well known, the designation of the freedom of wage labour has a double aspect: labouring subjects ('direct producers') are 'freed' from access to the means of production that secure their reproduction, and consequently they are (and must be) free to exchange their labour-power with capital for wages with which to purchase subsistence.[3]

Generally speaking, both aspects of wage labour capture the difference between capitalism and pre-capitalist modes of production. The first aspect signals the moment of *dispossession* of pre-capitalist producers, part of the process of primitive accumulation. Dispossession is the condition of the formation of a class of free wage labour which is the second aspect, or moment of *proletarianisation*. The contrast here is that, whereas in pre-capitalist modes of production labour is exploited by means of extra-economic compulsion, under capitalism proletarians are owners of the commodity labour-power which they exchange with capital under the 'dull compulsion of economic forces'.

Having outlined these familiar positions, it is useful to interrogate them a little further. First, with regard to the moment of dispossession, it is necessary to recognise that not all social subjects in all pre-capitalist formations necessarily had property or usufruct rights in land and/or other means of production, let alone that all pre-capitalist formations guaranteed the means of subsistence. Indeed, as several contributors to this volume show, strongly differentiated pre-capitalist formations in Asia contained landless labourers 'available' for recruitment by colonial capitalism.[4] In short, dispossession of pre-capitalist producers was not always necessary to the (initial) formation of a class of capitalist wage labour.[5]

This brings us to the second, and more general, moment of proletarianisation. This too has an important double aspect, albeit often overlooked or confused, that is conveyed theoretically in the distinction between *labour-power* and *labour*, and more concretely in the difference between *labour market* and *labour process*. Labour-power is the capacity to work that is the *property* of workers (their mental and physical energies): the commodity they exchange with capital for wages. Labour is the use value of labour-power: the expenditure of the mental and physical energies of workers in a production process controlled by capitalists, the products of which are the property of the latter and not the former. Correspondingly, the labour market designates the site in which labour-power is exchanged, and the labour process the site in which labour is exploited by the imperatives of capitalist production, accumulation, and profit.[6] Evidently, both labour markets and labour processes within capitalism exhibit a wide range of variation in their specific conditions, how their contradictions are manifested, and the forms of class struggle to which they give rise.

What is the connection between on the one hand the theoretical rationale of the freedom of wage labour and on the other the pervasive incidence of *unfree* wage labour throughout the history of capitalism, which is the central theme of this volume? We suggest that assaults by capital (actual or potential, preemptive or reactive) on the freedom of wage labour – the ability of workers to enter *and withdraw* from particular labour markets and labour processes – are a *general* feature of capitalism.[7] That is, the capitalist class as a whole (or its representatives) in some cases, or particular groups/types of capitalists and/or individual capitalists in other cases, seek to define and redefine the labouring subject so as to limit the exercise of his/her freedom as the owner of the commodity labour-power.

Such actions and struggles on the part of capital (with their economic, political and ideological expressions) aim to bring about *deproletarianisation*, in the sense of diminishing or eliminating altogether the freedom of wage labour as defined above: the ability of owners of the commodity labour-power to exchange it as they choose (on which more, see below). In other words, it is precisely by means of deproletarianisation that capital is able to effect a *double* dispossession of its workforce: both from the means of labour, and also from the means of commodifying labour-power itself. Moreover, the ability of capital to engineer unfreedom in labour markets – to impose restrictions in the process of *recruiting* workers – has implications for the freedom of capital in *using* workers in labour processes from which they are unable to withdraw. Evidently certain historical conditions, such as those of colonial imperialism, provide capital with reasons and opportunities for waging this particular kind of class struggle against labour. In this connection we should also note that the trajectories of wage labour are often marked by competition between branches/groups of capitalists and/or individual capitalists for access to and control over labour-power.

On the side of workers, this suggests that their class struggle against capital is commonly a struggle to assert, reassert or extend their freedom to dispose of their own labour-power: either separately from or combined with class struggle to destroy capitalism, therefore, such actions always constitute a potential/actual threat to existing property relations. Accordingly, when workers seek to defend and exercise their freedom as owners of labour-power, they are also required to contest on an individual or collective basis the attempts by capital to impose/reimpose and reproduce unfreedom both in the labour market and the labour process. That the experience of the proletarian condition does not automatically generate an immediate, unambiguous or exclusive class consciousness on the part of workers, therefore, may in given circumstances be a consequence of the fact that the aim of deproletarianisation is precisely to prevent, deflect or distort the development of just such a consciousness of class.[8] For this reason, the experience of *becoming* a worker in the employ of capital, or the change from class-in-itself to class-for-itself, has crucial implications for the self-perception and the perception-of-others on the part of the labouring subject, as well as the forms taken by any resulting political action. Viewed thus, unfree relations of production are an integral part of both class struggle and capitalist accumulation in the context of plantation agriculture.

II

(Re-) Visions of the Past: Paradise Lost?

In contrast to marxist interpretations of unfreedom as an exploitative relationship which permits the extraction of surplus labour, the revisionist project eliminates the element of coercion from most or all agrarian relationships, and subsequently reclassifies them as free wage labour. Accordingly, revisionist contributions to debates about the plantation labour theorise what Marxists term unfreedom as a form of equal exchange, or a benign (and thus tension-free) arrangement to the benefit of all parties to the relation.[9] Eschewing objective definitions, these texts adopt subjective criteria when conceptualising freedom/unfreedom, and consequently maintain that what outside observers wrongly interpret as unfree relations derives from the voluntary entry by rural labour into 'reciprocal' and essentially desirable working arrangements which provide the labouring subject with economic security in the form of a 'subsistence guarantee' or 'employment insurance'.[10] Accordingly, revisionists theorise the debt component of relations such as indenture, peonage, contract or bonded labour as evidence not of the coercive power exercised by employers or planters but much rather of the enhanced bargaining power exercised by workers.[11]

Significantly, this revisionist view concerning the social relations of production on plantations is epistemologically no different from (and in some instances is explicitly determined by) neo-classical economic theory.[12] Within such an economic framework, silence on the part of the choice-making subject is interpreted as assent: accordingly, a result of a methodological inability and/or theoretical unwillingness to address non-economic aspects of unfreedom, and the consequent failure to problematise coercion, is the epistemological impermissibility of unfreedom within a neo-classical economic framework. Having banished coercion, neoclassical economics is left with a form of market essentialism, a reductionist approach whereby anything and everything involving the planter/workforce relationship is *ipso facto* proof of the harmonious operation of a free (or 'perfect') market in which choice-making individual labourers express subjective preferences.[13]

The fact that within this framework value is subjective and not objective, and consequently all parties to market exchanges receive the value of their own individual economic contribution and nothing more, necessarily precludes analytical categories such as exploitation, class formation/reproduction, and class struggle. Since neo-classical economic theory cannot accept that in some circumstances employers prefer to recruit unfree labour, and thus may actually strive to bring about an imperfect market, it correspondingly fails to recognise the attempt by either party to change the relationship from what it is to what in the opinion of the respective protagonists it ought to be. This in turn negates the neo-classical view of the market as an arena of harmony, and undermines not only the notion of an equilibrium structuring the exchange between employer and labourer, but also the assumption that planters are interested in the operation of a free market and strive towards its realisation.

Unfreedom may indeed be profitable, yet attempts by capital to solve its crises by increasing the level of exploitation – and hence oppression – cannot but generate conflict. Accordingly, any debate about the profitability/non-profitability of a labour process structured by unfreedom can only explain its economic 'rationality' from the viewpoint of the employer. Since it is unable to accommodate the notion of class struggle, neoclassical theory cannot conceive that what it categorises as an economically 'rational' system (profitable, etc.) may still be swept away because of opposition from labour.

Paradoxically, even some variants of Marxism also overlook the fact that unfreedom is constucted and undermined by class struggle, and thus – like neo-classical economics – focus solely or largely on whether or not unfreedom is profitable and/or compatible with capitalism.[14] However, unfreedom may be overturned as a result of a specifically capitalist contradiction between the relations and forces of production.[15] Thus a plantation unit or system employing unfree labour 'efficiently', and which may indeed be profitable in the short term, will in the long term not be able to realise surplus-value due to a crisis of overproduction. It is important to note, however, that this is not a problem for capital as long as it remains possible on the one hand to continuously lower production costs through restructuring, and on the other hand to find expanding markets for its commodities.

Ironically, attempts by neo-classical theory to combat negative portrayals of black slaves on the cotton plantations of the antebellum American South through arguments about 'positive' aspects of black *economic* participation/autonomy in the plantation system (material well-being, self-improvement, etc.) find parallels in attempts by left scholars similarly to challenge negative stereotypes by invoking positive images of a resilient black *culture* produced by plantation slaves themselves.[16] The difficulty faced by both variants is that they license a discursive slide whereby the defence of blacks (implicity or explicitly) may ultimately – as the case of Fogel and Engerman demonstrates – be transformed into a *celebration* of the plantation, and in particular its mode of unfreedom.[17] This in turn opens up a theoretical space for anti-universalising/decentred postmodern analysis: the latter objects to (and thus denies the efficacy/existence of) unfreedom on the grounds that, in the overarching metanarratives of colonial and capitalist discourse, free labour is privileged as the totalising agent of universal progress.[18] Like neo-classical economic theory, which precludes a connection between capitalism and unfreedom by redefining the latter as free wage labour, postmodernism dismisses bondage as a figment of western discourse, thereby banishing it from indigenous accumulation and simultaneously reifying unfreedom as a cultural 'other'.[19]

Much the same is true of the frequently invoked 'survival strategies' framework.[20] Insofar as it entails an 'adaptive' approach by an individual choice-making subject, therefore, the concept 'survival' is similarly compatible with neo-classical economic theory. Within such a framework, moreover, there is also an analogous progression in textual interpretation whereby 'survival' shifts rapidly from being an end in itself to being a *positive* countervailing aspect of the colonial plantation, a position which then links up with the neo-classical claim that indentured workers actively *chose* such employment, which consequently could have been neither exploitative nor oppressive.[21] Significantly, the 'survival strategies' framework is in many ways similar to the notion of 'self-help' which structures much of the currently fashionable theory of peasant 'resistance'.[22] Where resistance occurred on the plantation system, the argument goes, then the unfree relations of production structuring its labour process were consequently rendered unviable; from this position it is once again a short step to the neo-classical economic argument that, as workers could have made it non-operational had they chosen so to do, the continued existence on the plantation of unfreedom was *ipso facto* chosen by the workers themselves, hence non-coercive/non-exploitative and to their advantage.[23]

III

Turning from revisionist interpretations to concepts/explanations of unfreedom utilised in this volume, the general approach of Kelly, Murray, and Das Gupta to the role of coercion in the construction/reproduction of the plantation workforce in colonial Asia is that, while unfree labour was indeed a characteristic of the system, it was also economically irrational.[23a] However, rather than irrationality, which conveys the impression that, from the viewpoint of capital, better (or crisis-free) economic paths existed but were not followed, it is more productive to investigate contradiction (capital followed the only path open to it, which necessarily generated more acute forms of class struggle). That is, unfree labour could be seen as an economically rational response by capitalist plantations to the attempts by workers to assert independence in one of two ways: repeasantisation or reproletarianisation.

TABLE 2

POSSIBLE LABOUR TRAJECTORIES OF THE PLANTATION WORKFORCE[a]

Recruitment Property Relation[b]		*Plantation Property Relation*[c]		*Post-Plantation Property Relation*[d]
Depeasantisation	→	Proletarianisation	→	Repeasantisation
Depeasantisation	→	Deproletarianisation	→	Repeasantisation
Depeasantisation	→	Repeasantisation	→	Proletarianisation
Proletarianisation	→	Deproletarianisation	→	Reproletarianisation
Proletarianisation	→	Peasantization	→	Reproletarianisation
Proletarianisation	→	Deproletarianisation	→	Peasantisation
Deproletarianisation	→	Deproletarianisation	→	Peasantisation
Deproletarianisation	→	Deproletarianisation	→	Proletarianisation
Deproletarianisation	→	Peasantisation	→	Proletarianisation
Deproletarianisation	→	Proletarianisation	→	Peasantisation
Deproletarianisation	→	Deproletarianisation	→	Deproletarianisation

Notes: [a] Only some of these trajectories are covered in the text. [b] The relation of production structuring the recruitment of migrant labour for the plantation. [c] The property relation structuring the reproduction of migrant labour-power on the plantation itself. [d] The property relation structuring the employment of such workers once they cease to be permanent employees of the plantation.

Trajectories of Plantation Labour

Table 2 schematically illustrates a number of trajectories of the ways in which a workforce was recruited for, reproduced on, and subsequently employed by the colonial plantation in Asia. These trajectories follow a number of differently combined property relations. For example, local smallholders could be depeasantised in the process of recruitment, and settled on the plantation either as free wage labour (in which case the transformation corresponds to one of full proletarianisation, or depeasantisation → proletarianisation), as unfree workers (depeasantisation → deproletarianisation), or as smallholding tenants (depeasantisation → repeasantisation). Alternatively, an already existing proletariat might be recruited and settled on the plantation as tenants (proletarianisation → peasantisation) or as unfree labour (proletarianisation → deproletarianisation). Yet another path is followed where, once on the plantation, tenants and/or unfree workers are subsequently transformed into an off-plantation proletariat (peasantisation or deproletarianisation → reproletarianisation).

The importance of these combinations lies in their implications for the process of both the economic formation and politico-ideological consciousness of class on the part of the plantation workforce, including the emergence of a petty-bourgeois producer/proprietor disposition as distinct from that of the landless agricultural labourer. For this reason, it is crucial to distinguish the origins of the plantation workforce, in terms of peasant or proletarian, and similarly to distinguish repeasantisation or deproletarianisation as the outcome of recruitment.

Origins of Plantation Labour I: Recruitment

A particular target of the revisionist critique of unfreedom concerns the process of recruitment by contractors of migrants for work on the plantations: specifically, the extent to which it was structured by deception and thus depended on the ignorance of the labouring subject. Revisionists such as Bauer [1979] object that, while duplicity may indeed structure recruitment in the short term, it is difficult to accept the premiss that the reproduction of unfree relations over time depends simply on an unsustainable view that its subjects continue indefinitely in a state of ignorance with regard to issues such as working conditions, wage levels, etc., which obtain in the labour process to which they migrate.[24] As the contribution by Carter points out with regard to Indian indentured labour recruited for work on the sugar estates of Mauritius, information concerning working conditions is rapidly communicated through the social networks of the migrants themselves.

The difficulty lies with the conclusion that revisionism draws from questioning the efficacy of 'duplicity'. Since for revisionists the mirror image of unfreedom structured by non-class-specific 'duplicity' is, and can only be, an equally non-class-specific form of freedom, an abstracted (dehistoricised/depoliticised) 'knowledge' about working conditions on plantations necessarily implies a corresponding acceptance by the labouring subject of the indenture relationship that makes migration possible. Revisionism then goes on to deny the element of unfreedom on the grounds that, if peonage, bondage, contract or indenture does not entail trickery, such relations must accordingly be the result of choice, are therefore voluntary and consequently free. However, because it fails to address the issue of class, and thus the presence of an incipient or developed proletarian consciousness, the fact that the non-existence of 'duplicity' leads not to an endorsement (or, in neo-classical economics, 'choice') of plantation unfreedom but much rather to a rejection of it on the basis of a class-specific alternative political consciousness is a question that revisionism cannot pose.

A related argument concerns the role of famine as a 'push' factor in indentured migration. Here, too, revisionists maintain that famine induced emigration constitutes evidence of the 'optimising peasant migrant', and thus provides additional support for a positive view of colonial indentured labour as a non-coercive outcome of rational economic choice.[25] What such arguments ignore, however, is that the unfree component of indenture is defined not so much by the manner of entry into the relationship as by the (in)ability to terminate the contract once it has been established. While recruitment may be voluntary, in the sense that the labourer willingly offers him/herself for work, it does not follow that the production relation which results is correspondingly 'free' in terms of the worker's capacity to withdraw from or re-enter the labour market or the labour process. Accordingly, neither the absence of 'duplicity' nor the presence of scarcity make the subsequent production relation free, since the latter is not characterised by its mode of recruitment.

Origins of Plantation Labour II: Depeasantisation

Generally speaking, the process of agrarian transformation linked to the presence in a particular context of the plantation economy adheres to one of three patterns, in each of which the workforce combines a greater or lesser requirement to sell its labour-power under conditions of unfreedom with a specific set of property relations.[26] The first involves a linkage between indebtedness and either the partial or full separation of peasant proprietors from their means of production: the accompanying process of agrarian change is usually presented in terms of *depeasantisation* leading to *proletarianisation* or *semi-proletarianisation* a transformation associated historically with imperialism, and in particular the manner whereby large-scale export-oriented colonial plantations acquired control over the labour-power of indigenous smallholders by depriving them of the ability to obtain all their subsistence requirements from their own land.

There are two problems with this view. The first concerns the reason advanced for the necessity of unfree plantation labour: that it is associated historically with the presence in such contexts of labour shortages. The view that unfreedom derives from labour scarcity is linked to the work of Nieboer [1910], who argued that as inhabitants of non-capitalist societies usually prefer to work for themselves rather than for others, actual or potential employers of labour-power must rely on unfree workers so long as unappropriated land (or 'open resources') is still available.[27] Once land becomes private property (or a 'closed resource'), unfree labour ceases to be necessary since the sale of labour-power is now the only method of obtaining subsistence.[28] The difficulty is that this view overlooks connections between unfreedom and surplus labour (and hence restructuring and deproletarianisation), and – where it encounters the stereotype of colonial Asia as an unproblematically labour surplus region – creates a theoretical space for delinking unfreedom and the plantation economy.[29]

The second problem concerns the way in which the concepts 'proletarianisation' and 'semi-proletarianisation' are theorised. As has already been noted, in marxist political economy a proletariat refers to a work-force composed of wage labourers who are doubly free: from the ownership of means of production, and also from the control of any group of employers or any one of the latter in particular.[30] Because proletarianisation is frequently equated with landlessness and semi-proletarianisation with insufficient land for subsistence cultivation, however, all workers fully or partially separated from the means of production tend to be regarded as free wage labourers, notwithstanding that their labour-power may be under the control of a planter or labour contractor.[31] Accordingly, this process of agrarian change involves depeasantisation combined not with proletarianisation or semi-proletarianisation but much rather with *deproletarianisation.*[32]

A number of contributions to this volume hold the view that unfree relations on the colonial plantation in Asia were determined by the need to dislodge recalcitrant peasants from the land so as to make their labour-power available for the process of capitalist accumulation. For example, Murray argues that the resort by capital to extra-economic coercion was due to an underdeveloped labour market, or an absolute scarcity of plantation workers in a frontier region. This in turn is linked to the need on the part of capital to habituate its reluctant workforce to the rhythms of industrial labour, a situation in which migrants had little or no intention of remaining in the position of full-time proletarians. Das Gupta makes the same point, observing that peasants turned into garden workers were subjected to a regularity and discipline quite unlike the pre-industrial rhythm of work.

There are four difficulties with such a view. First, it opens up a theoretical space for idealist explanations rooted in the notion of a 'natural' tribal/village/peasant existence. Second, evidence suggests that in their home villages contract workers were *already* landless labourers, that is actually/potentially a proletariat. Third, evidence also suggests that workers *were* available for employment by capital, but – if permitted – chose to sell their labour-power to non-plantation enterprises (see below). And fourth, it overlooks the dual origin of unfree relations in the process of class struggle: accordingly, these can occur in both labour scarce and labour surplus areas, in the latter because workers have too little bargaining power and in the former because (from the viewpoint of capital) they have too much.[33]

Outcome I: Repeasantisation

The second form of agrarian transformation involves a linkage between unfreedom and the attachment of an individual to land, the object on the part of the planter being to appropriate either surplus labour or surplus product. The most common forms of tenure associated with the presence of such smallholding units linked economically to the plantation (regardless of proximity) are sharecropping, serfdom, or labour-service tenancy, and in terms of agrarian change corresponds either to *peasantisation* or more usually where proletarianisation has already occurred, to *repeasantisation.*[34]

The main difficulty with the repeasantisation thesis is that it is premised on an essentialist (and hence untenable) concept of peasant economy, regarded as possessing a distinctive and autonomous dynamic. This licenses a dualist analytical framework whereby the supposed economic viability of petty commodity production is equated unproblematically with the culturally-determined 'resistance' and 'survival' of a distinct mode of production outside, and against, both the plantation and capitalism. Such a view fails to take into account the extent to which smallholding linked to colonial plantations was internally differentiated in socio-economic terms: that is, between on the one hand the better-off peasants, composed of proprietors and/or tenants whose main income derived from the sale of the product of labour rather than labour-power itself, and on the other those 'peasants' the majority of whose income came from the sale of labour-power and not its product, and who were as a result never anything other than labourers with temporary/conditional access to land rather than peasants who sold their labour-power.[35] In other words, the extent to which repeasantisation was not so much a unilinear once-and-for-all transition as a multiple and continous process of transformation, that itself hid an accompanying and underlying trend towards proletarianisation/deproletarianisation.[36]

According to Kelly, in the case of Fiji the inherent economic rationality of smallholding production when compared with large-scale plantations led to the displacement of the latter by the former, or a process of substantial repeasantisation. Similarly, in the case of tea plantations in Assam, Das Gupta argues that by 1906 the leasing of small plots to the plantation workforce was commonplace, and Carter also suggests that sugar plantations in Mauritius parcelled out land to their Indian workforce from the late 1870s. Interestingly, this point overlaps to some extent with a variant of the revisionist position on the non-coercive nature of plantation recruitment (see above), the argument being that workers were attracted by the possibility of landholding on the plantation, and that this in some sense mitigated against the negative depiction of unfreedom (that is, there were compensating factors). Kelly, for example, notes that the majority of indentured workers who migrated from India to Fiji never returned because in Fiji they obtained access to land.

A related problem concerns the claim by Das Gupta that capitalist plantations were able to externalize part of the cost of the reproduction of labour-power by passing this on to the non-capitalist 'subsistence' sector, a view that fits in with a perception of migrants as basically landholders experiencing depeasantisation or repeasantisation rather than landless agricultural workers being deproletarianised. The problematic nature of this dualistic approach emerges from the article by Behal and Mahopatra, who show that high mortality and low fertility among tea plantation coolies in Assam was due precisely to the non-existence of a 'subsistence' sector combined with below-subsistence level wages, and that by Knight, who suggests that an economically autonomous smallholding peasantry which controlled its land (and thus its subsistence base) was already on the decline in rural Java during the mid-nineteenth century (see also Knight [1988]).

The contributions by Breman and Daniel, dealing with the construction in colonial and nationalist discourse of the meaning attributed to 'coolie', and by Knight on the sugar industry in Java, in fact challenge the whole repeasantisation thesis, together with its accompanying argument that capitalism perceives this as an acceptable method of obtaining a settled/permanent source of cheap labour-power. According to Breman and Daniel, however, the latter corresponded historically in Asia to what they term a 'proto-proletariat', composed of cultivators with little or no land, artisans and declassed artisans, service workers, and 'lumpens'. Precisely in what way these categories were different from a proletariat remains unclear. Similarly, because he associates indebtedness solely with the exercise of planter control over smallholders, Knight proceeds to question the efficacy/existence of coercive institutional mechanisms – such as debt bondage – by which a repeasantised agricultural sector is subordinated to, and thus compatible with, the economic interests of sugar plantations. However, debt bondage is compatible with planter control not just over a workforce composed of landholding peasant family farms but also – and perhaps more significantly – over a workforce consisting of landless agricultural labourers. In short, the object of unfree agrarian relations was not to force a longstanding or newly reconstituted peasantry to become workers (depeasantisation/repeasantisation → proletarianization) so much as to stop precisely just such a transition from taking place; that is, to prevent/preclude a *de facto* proletariat from becoming, continuing or acting as such.[37]

Outcome II: Deproletarianisation

Unlike the preceding two variants of agrarian change, both of which usually involve plantation workers either owning means of production or not yet completely separated from them, this form entails the imposition/reimposition of unfree production relations on fully proletarianised workers whose sole property is their labour-power, or the decommodification of landless labour: in short, *proletarianisation* combined with *deproletarianisation*. The use of indenture or debt bondage against free workers already separated from means of production suggests that reasons for the existence of unfree relations are unconnected with the need of planters to compel unwilling and/or wholly self-sufficient peasant proprietors to sell their labour-power, and again raises the issue of the connection between unfreedom and worker availability. In such circumstances, the concept 'labour shortage' posesses a specific politico-ideological meaning. The term is applied by planters not to an absolute unavailability of labour-power (additional workers are needed, yet none exist) but to situations where market forces or political consciousness permit free workers to act as (and reap the benefits from being) proletarians.[38]

Generally speaking, deproletarianisation (or the economic and politico-ideological decommodification of labour-power) corresponds to workforce restructuring by means of introducing or reintroducing unfree relations, a process of class composition/recomposition which accompanies the struggle between capital and labour. In contexts/periods where/when further accumulation is blocked by overproduction, economic crisis forces capital to restructure its labour process either by replacing free workers with unfree equivalents or by converting the former into the latter. The economic advantage of deproletarianisation is that such restructuring enables planters first to lower the cost of local workers by importing unfree, more easily regulated, and thus cheaper outside labour, and then to lower the cost of the latter if/when the original external/local wage differential has been eroded.[39] In this way it is possible either to maintain wages at existing (low) levels or even to decrease pay and conditions of both components of the plantation workforce, thereby restoring/enhancing profitability and with it the accumulation process on (or linked to) the plantation system.

In ideological terms, the object of the deproletarianisation/ decommodification of distinct forms of labour-power employed on the plantation is similarly either to prevent the emergence of a specifically proletarian consciousness or to curtail the latter where it already exists. Hence the utilisation of unfree components from the industrial reserve army of labour not as an addition to the existing (free) workforce but rather as a substitute for – and thus competitors with – the latter has dire consequences for the development of working class political consciousness, in agriculture no less than manufacturing.[40] There are numerous instances of racist responses on the part of an existing plantation workforce displaced by the nationally/ethnically/regionally specific labour-power of cheap/unfree migrants recruited by planters engaged in the restructuring of the labour process.[41]

Where an initially progressive proletarian class struggle shows signs of being/becoming effective, the attempt by capital to demobilise it by means of workforce restructuring may convert what is an actually or potentially revolutionary situation into a politically reactionary combination of nationalism and racism. Accordingly, in such circumstances the form taken by class struggle waged from above in turn affects the form taken by class struggle waged from below. Although it may continue to reproduce itself in economic terms, therefore, and thus constitute a (segmented) class-in-itself, working class recomposition takes the all-important form of class-for-itself only where/when such politico-ideological division is transcended. In the case of Sri Lanka, for example, Breman and Daniel (this volume) point to the existence of just such a division between Tamil coolies and attempts by Sinhalese nationalists to repatriate/resettle them elsewhere. In a similar vein, the contribution by Ramasamy indicates how ethnically-specific forms of control exercised by headmen over Chinese labour employed on the plantations of pre-war Malaysia served to counter and undermine the emergence of collective action in this context.

Evidence contained in most of the contributions to this volume supports the view that what planters were really engaged in was deproletarianisation. In the case of Fiji, for example, the indenture system was introduced in 1879 only after the inhabitants from other Pacific islands would no longer agree to work in its sugar plantations. Kelly observes that those secured in this manner were looking for waged work and were already members of a labouring class when recruited for indenture. Although not interpreted by Kelly as such, the categorisation by planters of their unfree workforce in ethnic terms, and the consequent politico-ideological identification of the coolie as the 'other', was precisely to subvert and/or deflect attention from (and thus deny) the continuing attempt by Indian indentured workers to assert/reassert their identity as a *proletariat,* as sellers of labour-power. Similarly, Carter notes that Indians recruited for indentured labour on the sugar estates of Mauritius were already seeking work away from home when recruited; moreover, when a malaria epidemic resulted in labour shortages, anti-vagrancy legislation was introduced in order to prevent time-expired immigrant workers from selling their labour-power. More generally, Breman and Daniel provide numerous instances of the existence throughout Asia of landless subjects willing and able to sell their labour-power.

As is clear from the contributions by Behal and Mohapatra, Murray, and Knight, much the same is true of those recruited for work in the tea plantations of Assam, the rubber plantations in colonial Indochina, and the sugar factories of Java, where the main object of the unfreedom inherent in institutional forms such as debt bondage and penal contracts was to *deproletarianise* and not depeasantise the plantation workforce. Far from being stereotypical petty commodity producers that capital would have to force off the land, therefore, it is clear from the contribution by Murray that as early as the 1920s work-seekers in the region were attracted to the industrial enterprises and in transport, commercial, and trading fields. In other words, these subjects were trying to become or remain proletarians, and sell their labour-power on the open market.

This very point is underlined by Knight, who indicates that by the 1920s and 1930s a *de facto* proletariat already existed in rural Java, where sugar production was carried out with a workforce composed of landless locals and temporary migrants who found other employment as (and were dependent on the income from) wage labour in the off-peak season. The same applies in the case of Santal tribals migrating from Chotanagpur to Assam, who – as emerges clearly from the paper by Behal and Mohapatra – exhibited a marked *proletarian* disposition, as indicated by their preference for better-paid wage labouring jobs on road construction in Chittagong, as construction workers on the Assam-Bengal Railway, and as coolies on those plantations where wages were higher and employment conditions better.

Similarly, villagers in colonial Ceylon are described by Meyer as being highly wage conscious subjects who refused to accept wage retention and unfair treatment meted out on the plantation, which suggests that planters turned to (unfree) Tamil coolies from India precisely because local Sinhalese were not merely conscious of the worth of their own labour-power but also showed a willingness to withhold it until employers recognised this.[42] In other words, planters in this context were also engaged in an analogous process of restructuring, in the course of which the two ethnically-specific components of the workforce ('foreign' Tamil coolies, local Sinhalese villagers) were played off against each other. This was the dynamic that informed the restructuring of the 1930s and 1940s: migration of Tamils from India stopped, the industrial reserve army of labour contracted, the negotiating power of the plantation workforce was accordingly enhanced, the latter formed trade unions and withdraw labour, planters turned to villagers and used them as strike-breaking workers, which in turn generated intra-workforce antagonism that took the form of ethnic conflict. As soon as economic recovery began, and villagers themselves started to exercise their bargaining power again, employers replaced them once more with Tamil coolies from India.[43]

In the context of falling tea, sugar, or rubber prices due to capitalist overproduction yet increasing competition between employers (planters and others) for first-time migrants, migrants generally, and time-expired landless coolies, the object of legislation introduced by the colonial state in India or Indochina was to make continued accumulation possible by a combination of deregulated recruitment and a more regulated workforce. In this way it prevented the operation of *an already formed* labour market, and thus lowered the cost of labour-power, a situation in which plantation workers were denied the ability to sell their only commodity to the highest bidder. There was no need, therefore, for capital to pay adequate wages or invest in improving working conditions, since in such circumstances the owner of labour-power was unable to negotiate over its exchange and use in terms of price, intensity and context, or to withdraw from the labour process if any/all of these were not to his/her liking.[44]

Deproletarianisation as a planter response to worker organisation/militancy is further supported by its link to strike action. Thus, as Murray shows, contract workers recruited for rubber plantations in Indochina were permitted neither to engage in collective bargaining nor to withdraw their labour-power. In the case of sugar plantations in Fiji, Kelly observes that a strike by indentured workers in 1886 led directly to the passage of new labour laws. Significantly, the object of subsequent strikes carried out by Indian plantation workers in 1920 and 1921 was for complete wage parity with European labourers, that is nothing less than the full recommodification of erstwhile indentured labour-power. In the case of the tea plantations in Assam, however, neither Das Gupta nor Behal and Mohapatra mention that penal contracts were introduced specifically to curtail the bargaining power workers were beginning to exercise; thus the

Workmen's Breach of Contract Act, 1859, was itself a reaction by planters and the colonial state to strikes by plantation labour for higher wages during the 1840s and 1850s (see Guha [1981: 77–78]).

That labour control, in the specific form of workforce restructuring, rather than an absolute labour shortage was the reason for planter recourse to unfree relations is suggested by Knight's observation that free local workers, unwilling to labour in existing conditions, were replaced with bonded (and therefore cheaper) external migrants especially recruited by sugar agribusiness capital in Java during the capitalist crisis of the late 1920s. In this respect, planter restructuring of the workforce as part of the class struggle in Assam, Indochina, Java and Fiji was remarkably similar to what happened in Guyana; it was only when recently emancipated slaves began to act as free workers (or a proletariat), and undertook successful strike action during the 1840s, that sugar planters began to import large numbers of unfree indentured workers from India [*Rodney*, 1981].

Outcome III: Reproletarianisation

Just as there is disagreement over the reasons for the inception and reproduction of indenture and/or contract in terms of their positive/negative characterisation, so the end of these relations, and their replacement by free wage labour (or reproletarianisation), pose similar theoretical problems. About the latter two opposing views exist: the demise of unfreedom is attributed to pressure exercised either from above or from below.

Accordingly, one view is that reproletarianisation happens automatically as a result of spontaneous pressure from below, or when the 'masses' so wish. As has been argued above, such a view informs much 'resistance' and postmodern theory as well as neo-classical economic analysis: unfreedom is reconstituted in positive terms, respectively either as 'resistance', as an expression of 'cultural' autonomy, or as the outcome of 'preference' expressed by choice-making individual workers. What exists, therefore, coincides with (and is nothing other than a reflection of) social or individual self-empowerment.

There are three variants of the argument that the demise of unfreedom and consequent reproletarianisation was the result of pressure from above. The first is that abolition of indenture was the outcome of the purely humanitarian motives of a benign ruling class.[45] The second variant is that indenture was abandoned by the ruling class not for humanitarian motives but because it ceased to be profitable, a view which structures the contribution to this volume by Behal and Mohapatra.[46] The third variant maintains that, although indenture was actually an economically viable relation of production on the plantation (thereby denying an impetus to reproletarianisation), its continued existence nevertheless either constituted an obstacle to, or was incompatible with, the development of the wider capitalist system.[47] In so far as no reference is made to what those below thought about it, or how they contributed to its demise/abandonment (if indeed that is what happened), all these variants might be termed the 'reverse resistance' position, or the view that the abolition of indenture and other forms of unfree relations on the plantation system was determined solely by ruling class action (reproletarianisation effected by the colonial state).

Related issues here concern the extent to which unfree labour practices continued beyond their legislative abolition (thereby rendering problematic any claims regarding reproletarianisation) and, in cases where reproletarianisation succeeded in eliminating unfreedom, whether it signalled a decline in the power exercised by planters over their workforce. On the first point, it has been argued that unfree production relations on tea plantations in South India continued at least until 1929, while on Assam plantations they did not actually end until 1951.[48] With regard to the second point, even in cases where the economic power of planters was challenged by legislative action, they quickly resorted to other forms of control compatible with a deproletarianisation → reproletarianisation transformation. For example, when the power they exercised over their Tamil coolies was marginally curtailed by the Indian Emigration Act of 1922, tea planters in Ceylon turned explicitly to caste distinctions, encouraging/reinforcing the caste identity of the worker (by paying special attention to religious observance and social custom) as a method of continuing their control. Significantly, for planters the attraction of caste was not its element of hierarchy but the primacy and innateness of intra-workforce 'cultural' division; in short, a system in which everyone knew their place, as ordained (and thus rendered immutable) by 'nature'. The role of caste was accordingly the reproduction of an inherent (or 'natural') concept of otherness *within* the ranks of the plantation workforce that maintained a sense of unity (and thus work discipline) while simultaneously hindering any attempts to combine effectively against the planter class.[49]

The contributions to this volume which address the issue of the demise of unfree migration to the colonial plantation, implicitly accepting thereby the accomplishment of the deproletarianisation → reproletarianisation transformation, attribute this to the rising cost of indentured and/or contract labour. Hence the uneconomic nature of unfree labour on the sugar plantations in colonial Fiji is linked by Kelly to its demise, and

Behal and Mohapatra similarly argue that the high cost to the planter of labour recruitment by means of the *arkatti* system led to the demise of unfree migration to Assam. It is necessary to ask, however, whether planters did not in fact attempt to wrest control over migrants from the contractors either by operationalising the debt mechanism at the point of production or by financing/setting up their own network of labour contractors. According to Das Gupta and Knight, this is precisely what occurred from 1915 onwards in Assam and during the 1920s in Java.[50] Furthermore, in those instances where migration was and continued to be expensive at the point of recruitment, it would have been possible for planters subsequently to recoup this expenditure over time, in the form of better control, higher labour intensity, prevention/pre-emption of capitalist competition for workers, lower wage rates and reduced labour-related costs.

In order to sustain the claim regarding the expense of unfree labour, and by implication the relative cheapness of its free equivalent, it would be necessary to calculate the real cost of unfree migration in relation to the profitability not just of recruitment but of tea production using free wage labour throughout the period in question. And this raises another crucial point: if the high cost of recruitment was due to profitability of contracting, then clearly someone other than the planter or migrant was making a lot of money. This in turn prompts two questions. First, if Behal and Mohapatra are right, then Das Gupta's claim that plantations had no multiplier effect on the surrounding economy is problematic. And second, it would be important to know how this money was used, and its effects on the economy of the villages where the contractors resided: in short, whether or not the profits of labour contracting were ploughed back into land purchases, as Carter argues happened in the case of returnee-recruiters in Mauritius, and whether or not this licensed a process of primitive accumulation and the emergence of a rich peasantry that combined labour contracting with expanding its cultivation.[51]

Conclusion

In overall terms, deproletarianisation as an outcome of class struggle between capital and labour on Asian plantations during the colonial era raises some important issues concerning class formation in these contexts: specifically, in terms of parallels with current developments in the global economy and possible alternative trajectories under imperialism. To begin with, current arguments about the new international division of labour find echoes in, and simultaneously emphasise the contemporary relevance of, debate about the developmental role of the colonial plantation. There is a parallel between the need and ability of capital invested in both colonial plantations and contemporary export processing zones to force down wages, since both produce for export and do not require a local market for their commodities. For capital the economic logic is in both instances the same: since there no need for local consumers, for competitive purposes wages can be decreased by whatever means possible, especially by deskilling (on which see Breman and Daniel in this volume) and by decommodification of labour-power, thereby eliminating or undermining trade union activity. Thus payment made to the work-force in both the colonial plantation and the world market factory could/can be maintained or lowered without risk to the overall realisation of surplus-value.[52]

Second, deproletarianisation as the outcome of class struggle suggests that a proletarian class consciousness may have been present much earlier than is generally supposed in colonial contexts, and also that it may have developed among those more usually characterised as allegedly backward 'peasants'. Because of this, not only was the state required to assume its regulatory functions on behalf of capital in such direct ways (as the contribution by Behal and Mohapatra implies), but capital itself was forced thereby to engage in far more coercive (and perhaps more typical) forms of struggle on the 'periphery'. One might only speculate as to the general direction of Third World struggles, together with the pace and political outcome of decolonisation, had these initial attempts to develop a specifically proletarian consciousness not been blocked in this manner.[53]

NOTES

1. See, for example, Jain [1970], Florescano [1975], Duncan & Rutledge [1977], Cooper [1977, 1980], Mandle [1978], Cross and Marks [1979], Kahn [1981], Alavi *et al.* [1982], Fox-Genovese and Genovese [1983], Albert and Graves [1984, 1988], Munslow and Finch [1984], Marks and Richardson [1984], Saunders [1984]. Moreno Fraginals *et al.* [1985], Stoler [1985], Emmer [1986], Solow & Engerman [1987], Cross and Heuman [1988], Blackburn [1988], Berlin and Morgan [1991].
2. Since it conveys more accurately an open-ended outcome which in turn recognises the possibility of reversal in the course of class struggle, the term 'dynamic' is used here instead of the more usual concept 'transition' (which has about it an air of finality, a once-and-for-all accomplishment). Given the diverse chronological range of colonialism in Asia, Latin America, and the Caribbean, the attempt here to compare plantation economies in different regions is based on a broad element of periodicity. On this point Mandle [1978: 9–10] observes that 'parts of the South even after the Civil War more closely resembled societies in the Caribbean than in the northern United States'.
3. On this point, Marx [1976: 271–2] observes:

> ... labour-power can appear on the market as a commodity only if, and in so far as,

> its possessor, the individual whose labour-power it is, offers it for sale or sells it as a commodity. In order that its possessor may sell it as a commodity, he must have it at his disposal, he must be the free proprietor of his own labour-capacity, hence of his person. He and the owner of money meet in the market, and enter into relations with each other on a footing of equality as owners of commodities, with the sole difference that one is a buyer, the other a seller: ... [...] The second essential condition which allows the owner of money to find labour-power in the market as a commodity is ... that the possessor of labour-power, instead of being able to sell commodities in which his labour has been objectified, must rather be compelled to offer for sale as a commodity that very labour-power which exists only in his living body.

4. On the pre-colonial/pre-capitalist presence and extent of landless labour in South India, see Kumar [1965]. For a restatement of this position, and its application to colonial Asia generally, see Breman and Daniel (this volume).
5. 'Initial' because as capitalism develops its appetite for labour-power, its markets are fed by recruits from the ranks of small producers up to or beyond the point at which the need for labour is satisfied by the reproduction of the existing working class and reserve army of labour. This continuing supply of new workers from the ranks of small farmers, artisans and others who are unable to reproduce themselves as petty commodity producers *within capitalism* (due to competition from larger capitals, class differentiation of petty commodity production, etc.) involves different mechanisms from any initial dispossession of pre-capitalist producers in the course of primitive accumulation.
6. It is important to distinguish between a free market in labour, in which both free and unfree labour-power can circulate, and a free labour market, in which only free wage labour circulates. A result of the failure to make this distinction is market essentialism (see below), a theoretical effect of which is the conflation of two relationally distinct transactions: on the one hand a direct exchange between worker and employer (labour-power as the private property *of* the individual subject), and on the other an indirect transaction involving only employers (or the latter and contractors), one of whom transfers an unfree worker to the other (private property *in* the labour-power of the individual subject). In the antebellum American South, for example, plantation slaves were no less such for being hired out by their owners, either to other planters or to manufacturing employers in local towns [*Starobin*, 1970: 128–37].
7. On this point, see texts by, *inter alia*, Aufhauser [1973], van Onselen [1976], Corrigan [1977], Meillassoux [1981: 91ft.], Patnaik [1985], Brass [1986, 1988], Miles [1987], Cohen [1987], Zegeye and Ishemo [1989].
8. In the case of the tea gardens in Assam, for example, Guha [1981: 107] attributes the lack of political success on the part of the CPI to collusion along class lines between European planters and Indian capitalists to prevent parties of the left making any headway in mobilising plantation workers. The corresponding success among tea garden workers on the part of Congress was thus due to the fact that the latter came to an agreement with planters during the late 1940s to permit unionisation only by the the more 'moderate' INTUC specifically with the object of excluding communist influence from the plantation.
9. For examples of texts which subscribe to this view, see, *inter alia*, Bauer [1979], Miers and Kopytoff [1977], Reid [1983], Hart [1986], Knight [1986; 1988], Bardhan [1984; 1989], Nicholas [1988]. For the way in which pro-slavery ideology in the Antebellum American South projected slavery as beneficial to its subject ('positive good') and cast planter/slave relations in terms of 'reciprocity', see Tise [1987] and Glickstein [1991: 175]
10. For a critique of this position, see Brass [1986, 1990a, 1990b].
11. For an example of a revisionist text which regards unfreedom as evidence of the enhanced bargaining power of agricultural labour, see Bauer [1979: 46–7, 54].
12. It is not without significance that neoclassical economic theory, which emerged in the 1870s, was an explicitly anti-Marxist response to the development of the labour

movement. In politico-ideological terms, it constituted a project of innateness, or the dehistoricising and reconstituting as immutable of what had been regarded by classical economic theory as terrain changed by conflict; in neo-classical analysis, such terrain became an unchanging and thus a 'natural' socio-economic order.

13. For the application of a neoclassical economic framework to unfree plantation labour, see, *inter alia*, Evans [1970], Brown and Reynolds [1973], Fogel and Engerman [1974], Galenson [1981], and Engerman [1986]. For critiques from different theoretical viewpoints of the methods utilised in this process, see Gutman [1975], Sutch [1975], David *et al.* [1976], Greenberg [1977], Mandle [1978: 24–7], Fox-Genovese and Genovese [1983: 91ff.], Graves [1984a] and Brass [1991b]. The inapplicability of a market essentialist framework to the tea plantation economy of Assam is highlighted by Siddique [1990: 191], who observes:

    > ... the supply of immigrant labour for the tea industry [in Assam] was not determined by the operation of normal market forces because the planters failed to offer competitive market wages to attract labourers from distant places. Naturally, the planters had always preferred to employ those labourers who would be readily available to work at very low wages, possibly lower than the market rate of wages. Therefore, the tendency of the planters had been not to recruit labourers from the local market (where labour was generally in short supply), but to get another class of labourers who would work for a lower wage and would be bound to work in plantations as and when required under their strict control. In other words, the planters were looking for a system of indentured labour, ... the free play of the labour market in Assam was effectively thwarted through state intervention, thereby leading to the introduction of bonded labour in the tea plantations.

14. See, for example, Williams [1964] and Genovese [1966: 275ff.]. For another version of the socio-economically unspecific 'irrationality' of plantation unfreedom, see Tinker [1984: 77].
15. The element of contradiction refers here to economic processes generated by unfree production relations which in turn place limits on the unproblematic continuation of capital accumulation. Insofar as it prevents/preempts full proletarianisation, therefore, unfree labour may in certain situations constitute an obstacle to further capitalist development in one of the following three ways. First, the most profitable enterprises may be deprived of the additional quantities of labour-power necessary for expanded reproduction during periods of economic expansion. Second, the low wages paid to unfree labour may perpetuate correspondingly low levels of consumption on the part of the rural population, which in turn delineates the upper limit to the domestic market for commodities. And third, low wages consequent on unfreedom may also mean that employers have little incentive to adopt labour-displacing technology: there is accordingly no increase in the level of the productive forces, and labour productivity remains correspondingly low. It must be emphasised, however, that there are exceptions to all these contradictory processes: for example, unfreedom is not of itself and in all circumstances incompatible with advanced technology.
16. For the economic self-improvement/achievement of black slaves on the plantation in the American South, see Fogel and Engerman [1974: 108–9, 127]. For examples from the other end of the political spectrum of a 'culturalist' defence of slaves (*not* slavery), see James [1977: 240; 1980: 173–90; 1982: 11–14] and Genovese [1971: 300; 1975]. In part, the defence of slave economic life on the plantation by those on the political right and of slave culture by those on the political left were both attempts to rescue black slaves from the negative image associated with the passive/docile 'sambo' stereotype projected in the work of Elkins [1959].
17. In this connection it is perhaps salutary to note that, in its attempt to undermine the exceptionalist image of German fascism,

    > the right has rethought its analysis of National Socialism in part by borrowing, altering, and often gravely distorting the concepts, methods and conclusions of

social history. The right's discourse is filled with pleas to normalise the study of Nazism, to empathise with the little man and to recognise that many aspects of the Third Reich, including its most horrendous acts, were not unique ... It was history from below which first sought to reconstruct non-elite groups' perceptions of fascism' [*Nolan*, 1988: 52].

18. See, for example, Prakash [1990a, 1990b, 1991], whose work illustrates how a postmodern concept of unfreedom is compatible with 'resistance' theory and 'popular culture' (on which more, see footnotes 19 and 22). Maintaining that because it lacked a discursive form in the precolonial era, debt bondage in the Indian state of Bihar automatically had no relational existence, Prakash concludes [1990a: 197–8] that unfreedom is the significant and invented other of 'colonial discourse', and consequently that freedom is the correspondingly unacceptable embodiment of totalising Eurocentric notions of human destiny. Against this it is possible to make two points. First, that his symptomatically postmodern outside-of-discourse/language-there-is-nothing view is a palpably idealist position. And second, that he fails accordingly to distinguish between an *ideology* of unfreedom (which may indeed have been absent, although precisely what constitutes proof of this remains problematic) and *de facto* unfreedom, which had a material existence that predated colonialism (regardless of whether or not it was identified by its subject as such). On this point, see also footnote 23 below.
19. As has been argued elsewhere [*Brass*, 1991a], it is now possible to trace a common epistemological lineage from the moral economy argument (which includes not only Eric Wolf, Hamza Alavi, and J.C. Scott, but also E.P. Thompson and C.L.R. James) through the middle peasant thesis (of Wolf and Alavi) and 'resistance' theory (of J.C. Scott) to the postmodern/deconstructionist framework of the new social movements and subaltern studies project. In so far as it conceptualises all viewpoints as equally valid, and thus leaves the field open to the most powerful, any attempt at a non-political deconstruction is in ideological terms not just an apolitical or value-free technique – as its postmodern practioners claim – but rather a *re*politicisation of theoretical practice similar in outcome to the effect of *laissez-faire* which structures bourgeois economics.
20. See, for example, Lieten, Nieuwenhuys and Schenk-Sandbergen [1989].
21. For just such a positive theorisation of 'survival' by female indentured labour on plantations in Fiji, see Lal [1989: 179].
22. For more recent examples of the ubiquitous 'resistance' theory, see Scott [1985], Colburn [1989], and Haynes and Prakash [1991]. Notwithstanding his claim that hierarchical inversion in oral tradition was a form of 'resistance' practiced by unfree labour in the southern part of Bihar in eastern India, Prakash [1991: 170] finally concedes that, after all, it might just be the case that 'the resistance contained in oral traditions was not particularly significant; they may have made the burden of bondage a little easier to shoulder, moderated the impact of hierarchy, but did little to change the "real" condition of the Bhuinyas ... one may say that the oral traditions deluded the Bhuinyas into thinking that they were reconstituting hierarchy and labour relations when, in fact, they did nothing of the sort'. In other words, what is idealised by Prakash as a form of 'resistance' may amount to no more than a contextually/historically specific form of false consciousness.
23. A symptomatic text in this regard is that by McCreery [1990], where a 'survival strategy' framework is combined with the culturally-based 'resistance' theory of Scott in order to invert the meaning of unfreedom on coffee plantations in Guatemala during the late nineteenth century. Despite noting that '[f]orce, the coffee planters reasoned, was what the Indian understood, force would do, and force was what the Indians got', that '[l]abor recruiters and state agents in villages jailed workers, beat and defrauded them, kidnapped their wives and children, and burned their houses', that from the 1870s 'the state ... had the ability to deliver effective and immediate violence to the countryside', and that '[l]ife on the run and cut off from ... the community, ... was precarious ... [Labour contractors and village authorities], pressed to deliver labor, ... increasingly

took their search for [workers] into every corner of the municipality, making evasion more difficult' [1990: 161, 164, 166, 168], McCreery nevertheless claims that the state was powerless to enforce unfreedom on the indigenous population. For this reason, therefore, the debt bondage relation is theorised in a positive fashion as a mechanism that enabled plantation workers 'to force from their employers as much money as possible', and is accordingly subsumed by McCreery under the rubric of 'culture' which – along with folk tales, dances, religious ceremony and ritual – becomes yet another form of unproblematically successful 'resistance' on the part of indigenous rural communities [1990: 167, 169, 172]. On this point, see also note 19 above.

23a. A few of the points made below address issues raised by the contributors in their papers as originally presented at the CASA Conference.

24. This critique has been levelled against non-revisionist interpretations of the Latin American *enganche* system by Bauer [1979: 38] and Taylor [1984: 367–8, 371, 376], and in the case of Asia by Brown [1990] against the description by Breman [1989] of the horrors experienced by coolies employed on the tobacco plantations in the Dutch colony of Sumatra.

25. For an example of the theorisation of indentured migration from India to Surinam in terms of 'choice', see Emmer [1986: 187, 189]. The problematic nature of this view is underlined by the frequent and ambiguous references by Engerman [1986: 263, 268, 276, 282] to indenture as a contract which its subject 'presumably entered into'.

26. A variant of this view is the neo-classical argument made by Higgs [1977], who offers a symptomatically harmonious view of agrarian transformation in the American South. However, unlike Fogel and Engerman – for whom it is slavery itself which benefited its subject – Higgs projects the same analytical framework onto the post-abolition period, and depicts a similarly non-coercive process which might be described as proletarianisation without class struggle. According to this view, ex-slaves unproblematically abandoned the plantation economy following emancipation, a move which employers in general and planters in particular were powerless to resist. Hence the dismissal by Higgs of the efficacy of peonage (for contrasting views concerning the latter, see Daniel [1972], Ransom and Sutch [1977: 149ff.], and Dittmer [1980: 72ff.]).

27. For a reaffirmation of this view, see Dovring [1965], Domar [1970], van den Boogaart and Emmer [1986] and Engerman [1986]. As Domar [1970: 31] points out, this argument can be traced back from Nieboer through Loria's *Les Bases Economiques de la Constitution Sociale* (1893), Edward Gibbon Wakefield's *A View of the Art of Colonisation* (1834), to Adam Smith's *Wealth of Nations* (1776). Not the least of the difficulties associated with Nieboer's argument is its compatibility with the idealised Chayanovian theory of peasant economy and the theory of the articulation of modes of production. The link between the non-existence of private property in land and a corresponding difficulty in procuring free wage labour for accumulation in the colonies is also considered by Marx [1976: 934–5].

28. This view also structures the argument in two recent texts linking the presence of unfree labour in Russia and on the plantation systems of America, the Caribbean, South Africa, and Australia to the existence in all these contexts of labour shortages (Miles [1987: 205, 214], Kolchin [1987: 18, 359ff.]). The recourse by Andean landowners and Caribbean planters to a workforce subordinated by indenture, vagrancy legislation, or the *enganche* system is also attributed to labour scarcities arising from the continued access by peasants to land (Kloosterboer [1960: 197, 199–200], Bauer [1979: 56–7], LeGrand [1984: 33ff.]). The presence of unfree labour in Asia generally during the nineteenth century has been similarly ascribed to a situation where 'financial inducements alone did not suffice to disconnect the peasant from his native heath' (Lasker [1950: 210]).

29. For a discussion of unfreedom in areas of labour surplus/shortage, and the contrasting significance of control over land and labour in Asia (where a shortage of land was coupled with plentiful labour) and Africa (where land was abundant and labour scarce), see Watson [1980].

30. Hence the necessity for such workers to rely for their income on an untrammelled

ability to sell their only commodity, labour-power.

31. For examples of this argument applied to Latin America and the Caribbean, see Blanchard [1982: 120–21] and Bergad [1983a, 1983b]. Similarly, Legassick and de Clercq [1984: 143–4] maintain that contract/indentured migrant labour in South Africa constituted a proletariat because it had been separated from the means of production.
32. Even those who recognise that a proletariat is not a necessary outcome of depeasantisation nevertheless misunderstand the reason why the result is unfree rather than free relations of production. Thus Miles [1987] attributes unfreedom not so much to the need by plantations to cut costs by undermining/prempting workforce bargaining power as to resistance by smallholders to proletarianisation. Evidence suggests that struggle conducted by those experiencing proletarianisation is to low wage levels and poor working conditions rather than to proletarianisation *per se*. For example, even tribals in India who sold their labour-power on a seasonal basis would withdraw this in circumstances where pay and conditions were not to their liking: thus in the 1950s Bhil tribals from western India who migrated to work the cotton, tobacco, and ground-nut cash crops produced in Gujarat discontinued providing seasonal labour because employers worked them too hard [*Nath*, 1960: 43–4]. The occurrence of unfree relations in such circumstances would clearly be unconnected with the need on the part of employers to habituate their workforce to the rhythms of wage labour.
33. See Brass [1986: 51]. By contrast, in a recent collection of texts (Berlin and Morgan [1991]) the emphasis of the debate about the plantation workforce has shifted in the opposing direction, from slave-as-potential-proletarian to slave-as-an-actually-independent-producer. For the application of the latter view to indentured labour, see Emmer [1984: 102].
34. The labour-shortage-because-of-peasantisation argument has been advanced to account for the continuation of unfree relations generally by Engerman [1986: 279], and more particularly by Kloosterboer [1960: 193–4] to account for their continuation after the abolition of plantation slavery in British Guiana where a relatively low population density was combined with the presence of 'open resources' in land; by contrast, unfree labour was unnecessary in the smaller Caribbean islands (Antigua, St. Kitts, Barbados) with a high population density and 'closed resources'. On the general question of the organisational attempts by planters in Assam to block access by their workforce to waste or Crown Lands, see Renford [1987: 65].
35. It should be noted that such a socio-economic differentiation process should be applied to all the components (permanent/casual, local/migrant) of the plantation workforce which have access to land. Although official data – which anyway overestimated nominal wage rates – show that during 1905/06 the average monthly cash earnings for a male adult plantation labourer in Assam was no more than Rs 5/12/9 (Rupees/annas/pies) (Siddique [1990: 168, Table 5.6]), Harold Mann [1967: 72] estimated that an average labouring family (composed of an adult male, his spouse, and three children) employed on the tea plantations of Assam in May 1906 would require almost double this amount, or an income of Rs 10/10/0 per month, to obtain the minimum diet. Following strike action during the early 1920s, planters maintained that, as workers could not possibly have any genuine economic grievances because they were both well-paid and received concessions (land on which to graze livestock or to cultivate, housing, etc.), strikes could only be the result of activity by outside agitators. Regarding this claim by employers about the value of payment-in-kind, Bose [1954: 104–5; see also 79–80] has observed:

> Of all the concessions, the most widely advertised one was the grant of land for cultivation. Now, according to the Report (of the Assam Labour Enquiry Committee, 1921–22), the total area of land held by the workers either from the Government or tea-garden authorities, was 126, 951 acres. The total number of adult labourers was 571, 526. This would mean a little over one-fifth of an acre per adult worker. Taking one acre to produce about 18 maunds of paddy or 10 maunds of rice in a year, at Rs 5/- per maund, we find that the average monthly income from

> land came to about Re. 1/- per head. This does not take into account any money spent on cultivating the land. Moreover, whenever a worker worked on his land he had to absent himself from [tea-] garden work and lose his wage. There was also the rent which had to be paid for holding land (about Rs 1/9/3 per acre).

The same source concludes that 'if we calculate the money equivalent of all these concessions and add it to the cash wage, then we ... find that *the total monthly remuneration of a worker fell far short of what should have been a minimum living wage* (original emphasis)'. On the general issue of organisational attempts by tea planters in Assam to fix maximum payments and/or reduce wages, see Renford [1987: 57 footnote 94, 58, 59, 63–4, 65, 72–3]. For a similar argument in the case of South Indian tea plantations, see Raman [1991: 252ff.].

36. Evidence from the Caribbean suggests that, where/when it operated within the context of the plantation economy, petty commodity production not only remained wholly dependent on the latter but was seen by its subject as being reproduced by capital for its own benefit. The precarious and economically unviable nature of smallholding cultivation emerges clearly in the case of the Island of St. Vincent, where peasants rapidly ceased to be such in one of two ways (West India Royal Commission [1897, Appendix C – Part VIII: 113]). First, when arrowroot prices were low, planters leased out land to their workforce on a sharecropping basis: however, as the repeasantised subjects themselves pointed out, the object of this transformation was not so much to create a class of smallholders as to cultivate arrowroot more cheaply with the existing workforce ('labourers shrewdly guess, and fail not to say, that this has been done from no wish to form again a class of small growers but to lessen the loss of cultivation ...') and for planters represented the only alternative to abandoning the plantation economy altogether; when crop prices rose sufficiently, planters would invariably take back this sharecropped land to resume direct cultivation with hired labour. And second, when arrowroot prices were high, as in the 1870–85 period, not only did workers on sugar plantations who were able to do so lease in land to cultivate this crop but the plantations themselves also switched from sugar to arrowroot, with the result that even those who had managed to become independent smallholders were no longer able to compete as overproduction led to falling prices.

37. In the case of West Indian sugar plantations at the turn of the century, for example, the potential dangers posed to employers by an emerging proletarian consciousness among the workforce is highlighted in a revealing memorandum submitted to the Royal Commission by the Rev S F Branch, who warned (West India Royal Commission [1897, Appendix C – Part VII: 114]):

> As an alternative to parcelling out estates among labourers for cultivation, the scale of wages should vary with the price of produce. Although it has more or less brought home to the labourer that arrowroot now does not pay, and in some way sugar too (the high local retail prices of sugar makes him suspicious), yet the labourer argues that when these productions sold at remunerative prices there was little advance on the present barely living wage; hence there is a feeling of unfairness that as he did not share in the prosperity of his employer formerly, he should not be called on now to suffer with him when prices are low. The [plantation] labourers in St. Vincent have not yet been educated enough to combine for higher wages, but it would be wise if the employers in time, by a liberal policy, gave no occasion for combination. There are not wanting signs that if our industries revive there are some self-seeking persons who will teach the labourer to put pressure on his employer. We have an excitable, suspicious, and still ignorant labouring population, drifting away from the old clerical control, as shown in the numerous new friendly societies, all self-managed, and not as heretofore directed by ministers of religion ... The labourer, by his lay, friendly and religious societies, and his revolt from the healthy control of the Church and ministers of Wesleyan body, shows he is tasting the pleasure of thinking and determining for himself. The employer in the near future must be prepared to reckon with men who are trying to

think out the problems of life, but who are still uneducated, and may be dangerous in asserting what they consider to be their rights.

38. Attributing indentured migration generally to a 'distaste by free labourers to bear the non-pecuniary costs of production upon the plantations', Engerman [1986: 277] implies that everywhere and at all times free labour was unwilling to undertake plantation work under *any* circumstances. Such a view overlooks the fact that planters employed unfree contract/indentured labour because locals were exercising not an absolute but rather a relative unwillingness to work, and withholding their labour-power in order to secure improvements in pay and conditions for the application of this commodity on the plantation itself. In other words, the object of employing unfree labour in such circumstances was to compel free locals to accept plantation work conditions and pay levels that they would otherwise have rejected.
39. Instances abound of cost cutting achieved on the plantation system through restructuring based on deproletarianisation. One is Queensland, where legislation enacted from the 1880s onwards prevented Pacific Island labour from obtaining more lucrative employment outside the sugar plantation, thereby providing the latter with a captive workforce that not only cost less than half other forms of labour-power but also worked longer on a daily/annual basis (Parnaby [1964: 130], Graves [1984: 270], Saunders [1984: 227]). Another is Puerto Rico where, because newly emancipated ex-slaves began to act as a proletariat and bargain with their erstwhile masters for higher wages and better working conditions, the colonial state reimposed unfreedom during 1874 in the form of contracts preventing freedmen resident on sugar plantations from selling their labour-power without the consent of their employers; when in the 1890s sugar plantation workers on the island struck for higher wages and improved working conditions, employers imported contract labour from the British West Indies (Ramos Mattei [1985: 158–78; 1988: 38]). Yet others are Trinidad, where over the 1891–1909 period sugar planters also managed to hold down labour costs by importing unfree indentured migrants specifically with the object of undermining the bargaining power of local free workers (Ramesar [1984: 60, 65]), and Surinam, where the same considerations applied (Emmer [1984: 98]). Similarly, had they not been able to use unfree indentured workers as strike-breakers against a local plantation workforce of ex-slaves which had withdrawn its (free) labour in pursuit of higher wages in Jamaica during the mid-1850s, planters would have been compelled to raise pay levels for locals by 25 per cent (Tinker [1974: 217–18]). In this respect the situation in Jamaica was no different at the turn of the century, when local workers who wanted to remain free and were thus unwilling to accept the terms and conditions imposed on indentured labour (to whom they referred as 'slave coolies'), complained that the presence of the latter undermined their own bargaining power and prevented them from earning decent wages on the plantation (West India Royal Commission [1897, Appendix C – Part XIII: 306, 330, 331, 418]). For the cost differentials involved in utlising different forms of labour-power employed on sugar plantations in Hawaii during the mid-1880s, see Lind [1938: 200].
40. Opposition to the abolition of slavery among both the southern and northern white working class in antebellum America stemmed from the use of unfree black labour as strike-breakers in the manufacturing sector together with the presentation by southern planters of emancipation as an endeavour by northern capital to lower industrial wages by expanding the reserve army of labour with ex-slaves (Dew [1966: 30], Fite [1968: 187–9], Foner and Genovese [1969: 57–8]). Writing in 1909, Fite [1968: 187] makes the point that: 'In resisting increase of wages capital naturally called to its aid every possible form of cheap labour: more employment was given to women, to young boys as apprentices, to unskilled labor, [blacks], and immigrants, and more use was made of labor-saving machinery; toward these cheap rivals the unchanging attitude of the labor world was that of extreme hostility'.
41. Examples of racism as a result of restructuring include Jamaica and Guyana, where during the 1850s and 1860s the use by planters of indentured labour from India to break strikes organised by the local Creole ex-slaves attempting to prevent wage cuts led to

racial conflict between Creole ex-slaves and Indian coolies (Tinker [1974: 217]; see also note 38 above). In the 1880s, members of the Trinidad Working Men's Association advocated a reduction in the number of imported coolies, while blacks in Jamaica and the French Caribbean colonies of Guadeloupe and Martinique used their electoral power to block further immigration 'on the ground of its constituting an unfair competition to [local] labour' (West India Royal Commission [1897, Appendix C – Part IV: 296], Beachey [1957: 109], Tinker [1974: 218–19]). Similarly, throughout the latter part of the nineteenth century white workers in Australia opposed in racist terms and aimed trade union struggles against the employment of cheap migrant labour from Melanesia on the sugar plantations of Queensland (see Hunt [1978: 80ff.]). Contemporary instances of indigenous working class racism determined by restructuring include attacks by free locals against illegal Mexican immigrants to the USA during the 1950s and against unfree Haitian immigrants by free locals being displaced from employment on the sugar plantations in the Dominican Republic during the 1980s (Cohen [1987: 52–3, 129–30, 187, 193ff.], Plant [1987: 69–70]).

42. See also Meyer [1990: 170], where it is suggested that such attitudes on the part of villagers employed on plantations dated from the mid-nineteenth century.
43. On this point, see also Meyer [1990]. Such restructuring would help to explain not just the labour unrest on the plantations of Ceylon at this period as well as the conflict between villagers and coolies but also the ethnic-specific idioms in which such antagonism was couched. The importance of such a process for the dissemination among and acceptability to the agrarian working class of a (bourgeois) nationlist myth premissed on ethnic specificity and threats to the 'purity' thereof (the 'enclave' myth and its attendant dualisms), together with its effects in terms of current Tamil/Sinhalese conflicts, scarcely requires elaboration.
44. In this respect, of especial value in the contribution by Behal and Mohapatra is the juxtaposition of and contrast between planter ideology and the views of their workforce. Maintaining that coercion was necessitated by the absence of a labour market, employers stressed that, compared with their place of origin and the penal contract notwithstanding, migrants were in fact better off working on the tea plantations in Assam since they obtained benefits such as free housing, assured wages, and free medication. Significantly, this view coincides almost exactly with that of the revisionists who offset (or deny) the importance of unfree production relations by reference to their supposed role as 'employment insurance' and 'subsistence guarantee' (see above). Against this view Behal and Mohapatra cite a confidential report by a group of Santal headmen who visited the tea gardens in 1894 and interviewed the labourers, the latter complaining about low wages, poor working conditions, and the inability to avail themselves of alternative and better-paid employment.
45. This argument, applied to slavery by Coupland [1933], is advanced with regard to indenture by Engerman [1986: 284].
46. This argument, associated with Eric Williams' [1964] interpretation of the reasons for the end of slavery in the Caribbean, is also invoked by Willemina Kloosterboer to account for the decline of unfree labour during the capitalist crisis a century later. She notes [1960: 206]:

> The economic situation remains the determining factor even today ... [this] is indicated by the fact that during the slump of the 1930s when fewer labourers were required steps were taken to abolish compulsory labour, while compulsory labour sprang into life again in many places during the [second world] war ... when there was an increasing demand for manpower.

Accordingly, just as free workers are displaced by unfree equivalents when/where the latter are cheaper and/or more easily disciplined than the former, so the opposite holds true: thus it is considerations of profitability which underlie the move towards reproletarinisation. In the case of the antebellum American South, for example, Robert [1938: 208] notes that:

> Partly because of disciplinary troubles, but primarily on account of the extremely high rates asked by the owners of [leased-out black slaves from the plantations of Virginia and Northern Carolina], in the very last years of the 1850s tobacco manufacturers began to employ white girls in the making of plugs....

47. See, for example, Tinker [1974: 334ff.; 1984: 83], for whom the abolition of indentured emigration from India in 1916 was due on the one hand to competition within the sending country for wartime resources in the form of transportation and labour-power, and on the other to the fear on the part of the colonial government that its continuation would politically strengthen nationalist opposition. The latter in turn is linked by Emmer [1986: 188, 205] to the wish on the part of an increasingly important domestic capitalist class to retain control over its own sources of labour-power for agricultural and industrial accumulation within India (for a different view on this, see Engerman [1986: 277]). The acuteness of this struggle between the different components of the ruling class over access to and control of labour-power emerges clearly from the strong and successful opposition dating from the early 1880s on the part of tea planters in Assam to a proposal to repeal the 1859 Masters' and Workman's Act [*Renford*, 1987: 62–3], and also their mobilisation against the Ilbert Bill, a consequence of which would be that 'the power of district magistrates and judges to intervene on behalf of the tea garden coolie might pass into native hands' [*Renford*, 1987: 62].
48. On the historical longevity of unfree relations in the plantations of South India and Assam, see Raman [1991: 248] and Guha [1981: 106]. For similar arguments regarding the persistence of unfree agrarian relations well into this century in contexts such as the United States, India, and the Philippines, see US Congress [1970], Friedland and Nelkin [1971], Daniel [1972], Singh and Iyer [1985], and ICL Research Team [1979: 51ff.]. In this connection it should also be remembered that by far the largest number of unfree workers who suffered and died in the course of building the Burma railway for the Japanese military during the second world war were not allied prisoners of war but coolies recruited from Burma and Indonesia. [*Kinveg*, 1992].
49. The explicitly political object of reinforcing caste distinctions among the plantation workforce emerges clearly in Green [1925: iii–v]:

> In these democratic days, one hears and reads frequently of the need for breaking down caste, which is regarded by many as one of the great obstacles to progress in the development of India on modern lines. To my mind caste will always exist, and what is wanted is the breaking down of the many barriers that exist between caste and caste from an arrogant idea of superiority of one caste over another ... there is no need to abolish caste. [...] ... in the development of social India each caste had its individual job to do and was expected to do it and there is something to be learnt from that.

Observing that the coolie was becoming freer and acquiring greater independence, the same text concludes: '... as is only to be expected, he is beginning to display the common symptoms of the young and growing – a greater independence of character, and frequently an exaggerated idea of his own importance. This is an excellent thing ... but one is bound to recognise that it must have an appreciable effect upon the relations between employee and employer.' (For the extent to which both workers and trade unions on Sri Lankan plantations continue to be segmented along caste lines, see Hollup [1991]) The same political objective was voiced by representatives of the planter class in the Caribbean during the 1830s and 1840s. Hence the view (cited in Mangru [1987: 40, 43]) that:

> It is of great importance to us [planters] to endeavour to provide a portion of other labourers, whom we might use as a set-off, and, when the time for it comes, makes us, as far as possible, independent of our [local black] population; and it has occurred to us that a moderate number of Bengalese ... might be very suitable for our purpose ... [...] I think the safety of the whites depends very much upon the want of union in the different races of labourers, and I should be glad to see some

more Madeiranese, and, if possible, Chinese coming in [to British Guiana]: the Coolies too would always hold by the whites.

In a similar vein, Thompson [1947: 114] reports that in Malaya during 1946 the 'British planting community [was] well aware of this growing rift between Chinese and Indian labourers ... the chairman of the United Planters Association of Malaya (UPAM) stated that Tamil rubber tappers were making a real effort to break away from the labor unions, which had been Chinese-dominated, and that *this tendency should be deliberately encouraged by the district planters*' (emphasis added).

50. This point is also made by Bose [1954: 85] and Guha [1981: 85–6], and confirmed for an earlier period by an observation [India 1931: 376] in the Royal Commission on Labour in India, to the effect that in the Assam tea gardens 'where successive amendments of the law have removed all the statutory restraints, there is ample evidence to show that [during 1929] the old faith in restraint ... persists. The planter ... has deliberately allowed the old ideas of the penal contract to linger in the minds of his workers'.
51. This last point necessarily raises the question of the village-level effects of recruitment/outmigration. In the case of tribal migration from Chotanagpur to the Assam tea gardens, for example, it would be important to know what connection existed between emigration and the Birsa Munda movement which took place during the 1890s [*Singh*, 1983]. In other words, did factors such as the pauperisation of tribal migrants employed in Assam, the recruitment of tribal labour by Hindu contractors, the reinvestment of contracting profits in means of production, or the perception by village headmen of labour outflow and/or capital inflow as a socio-economic disintegration of the existing tribal structure – and hence a threat to their own political authority, contribute to the reassertion of ethnic identity and the hostility towards non-tribals (*dikus*) which characterised the movement. On this point generally, Omvedt [1980: 194] rightly observes that: 'There has been little study of these labour-contractors as a significant social stratum, a kind of "labour aristocracy" that often arose from the ranks of the workers themselves. There is little knowledge, also, about the exact links they had in rural areas with the landlords and headmen who "released" the labourers to them for recruitment ...'.
52. The points of similarity between export processing zones (EPZs) which characterise the new international division of labour [*Fröbel et al.*, 1980] and the plantation system are outlined below:

	EPZ	*Plantation*
Labour-power	deskilled	deskilled
Response to class struggle	decommodification	decommodification
Production relations	unfree	unfree
Productive forces	advanced	advanced
Labour process	deregulated	deregulated
Role of the state	interventionist	interventionist
Markets	international	international

53. One economic argument, made in relation to the plantation system in both the antebellum American South and the Caribbean (see Genovese [1966], Dupuy [1983]), is that historically underdevelopment necessarily results from unfree relations of production which structured plantation agriculture. In the case of the cotton South, therefore, unfreedom was an obstacle to further captial accumulation because slaves anyway had a low level of subsistence, and demand was therefore low, and what was required in terms of subsistence was anyway met from within the plantation system, which in turn meant that a further domestic outlet for Northern manufactures was blocked. The fact that a colonial plantation agriculture was unable to create an extended market for commodities meant that the indigenous bourgeoisie could not therefore complete the economic project of captialism. This issue in turn raises the

political issues associated with Trotsky's [1962] concept of permanent revolution, especially when linked to the presence in the plantation context of a workforce exhibiting an incipient proletarian consciousness (subsequently to be closed off in the course of deproletarianisation).

REFERENCES

Alavi, H. *et al.*, 1982, *Capitalism and Colonial Production*, London: Croom Helm.

Albert, B. and A. Graves (eds.), 1984, *Crisis and Change in the International Sugar Economy 1860–1914*, Norwich: ISC Press.

Albert, B. and A. Graves (eds.), 1988, *The World Sugar Economy in War and Depression 1914–40*, London: Routledge.

Archetti, E. *et al.* (eds.), 1987, *Sociology of 'Developing Societies': Latin America*, London: Macmillan.

Aufhauser, K.R., 1973, 'Slavery and Scientific Management', *Journal of Economic History*, Vol.33, No.4.

Bardhan, P., 1984, *Land, Labour and Rural Poverty*, Delhi: Oxford University Press.

Bardhan, P. (ed.), 1989, *The Economic Theory of Agrarian Institutions*, Oxford: Clarendon Press.

Barraclough, G., 1964, *An Introduction to Contemporary History*, London: C.A. Watts.

Bauer, A.J., 1979, 'Rural Workers in Spanish America: Problems of Peonage and Oppression', *Hispanic American Historical Review*, Vol.59, No.1.

Beachey, R.W., 1957, *The British West Indies Sugar Industry in the Late 19th Century*, Oxford: Basil Blackwell.

Beckford, G., 1972, *Persistent Poverty: Underdevelopment in Plantation Economies of the Third World*, New York: Oxford University Press.

Bergad, L., 1983a, 'Coffee and Rural Proletarianization in Puerto Rico, 1840–1898', *Journal of Latin American Studies*, Vol.15, Part I.

Bergad, L., 1983b, *Coffee and the Growth of Agrarian Capitalism in Nineteenth Century Puerto Rico*, Princeton, NJ: Princeton University Press.

Berlin, I. and P.D. Morgan (eds.), 1991, *The Slaves' Economy: Independent Production by Slaves in the Americas*, a Special Issue of *Slavery & Abolition*, Vol.12, No.1.

Blackburn, R., 1988, *The Overthrow of Colonial Slavery 1776–1848*, London: Verso.

Blanchard, P., 1982, *The Origins of the Peruvian Labor Movement 1833–1919*, Pittsburgh, PA: University of Pittsburgh Press.

Bose, S.K., 1954, *Capital and Labour in the Indian Tea Industry*, Bombay: All-India Trade Union Congress.

Brass, T., 1986, 'Unfree Labour and Capitalist Restructuring in the Agrarian Sector: Peru and India', *Journal of Peasant Studies*, Vol.14, No.1.

Brass, T., 1988, 'Slavery Now: Unfree Labour and Modern Capitalism', *Slavery & Abolition*, Vol.9, No.2.

Brass, T., 1990a, 'The Latin American *Enganche* System: Some Revisionist Reinterpretations Revisited', *Slavery & Abolition*, Vol.11, No.1.

Brass, T., 1990b, 'Class Struggle and the Deproletarianization of Agricultural Labour in Haryana (India)', *Journal of Peasant Studies*, Vol.18, No.1.

Brass, T., 1991a, 'Moral Economists, Subalterns, New Social Movements, and the (Re-) Emergence of a (Post-) Modernised (Middle) Peasant', *Journal of Peasant Studies*, Vol.18, No.2.

Brass, T., 1991b, 'Market Essentialism and the Impermissibility of Unfree Labour', *Slavery & Abolition*, Vol.12, No.3.

Breman, J., 1989, *Taming the Coolie Beast: Plantation Society and the Colonial Order in Southeast Asia*, Delhi: Oxford University Press.

Brown, I., 1990, Review of *Taming the Coolie Beast* by Jan Breman, *Journal of Peasant Studies*, Vol.18, No.1.

Brown, W.W. and M.O. Reynolds, 1973, 'Debt Peonage Re-examined', *Journal of Economic History*, Vol.33, No.4.

Cohen, R., 1987, *The New Helots: Migrants in the International Division of Labour*, Aldershot: Gower Publishing Co. Ltd.

Colburn, F.D. (ed.), 1989, *Everyday Forms of Peasant Resistance*, New York: M.E. Sharpe.

Cooper, F., 1977, *Plantation Slavery on the East Coast of Africa*, New Haven, CT: Yale University Press.

Cooper, F., 1980, *From Slaves to Squatters: Plantation Labor and Agriculture in Zanzibar and Coastal Kenya, 1890–1925*, New Haven, CT: Yale University Press.

Corrigan, P., 1977, 'Feudal Relics or Capitalist Monuments? Notes on the Sociology of Unfree Labour', *Sociology*, Vol.11, No.3.

Coupland, R., 1933, *The British Anti-Slavery Movement*, London: Thornton Butterworth Ltd.

Courtenay, P.P., 1965, *Plantation Agriculture*, London: G. Bell & Sons.

Cross, M. and A. Marks (eds.), 1979, *Peasants, Plantations and Rural Communities in the Caribbean*, Guildford: Department of Sociology, University of Surrey.

Cross, M. and G. Heuman (eds.), 1988, *Labour in the Caribbean: From Emancipation to Independence*, London: Macmillan.

Daniel, P., 1972, *The Shadow of Slavery: Peonage in the South 1901–1969*, Urbana IL: University of Illinois Press.

David, P.A. *et al.*, 1976, *Reckoning with Slavery: A Critical Study in the Quantitative History of American Negro Slavery*, New York: Oxford University Press.

Dew, C.B., 1966, *Ironmaker to the Confederacy: Joseph R. Anderson and the Tredegar Iron Works*, New Haven, CT: Yale University Press.

Dittmer, J., 1980, *Black Georgia in the Progressive Era 1900–1920*, Urbana., IL: University of Illinois Press.

Domar, E.D., 1970, 'The Causes of Slavery or Serfdom: A Hypothesis', *Journal of Economic History*, Vol.30, No.1.

Dovring, F., 1965, 'Bondage, Tenure, and Progress: Reflections on the Economics of Forced Labor', *Comparative Studies in Society and History*, Vol.VII, No.3.

Duncan, K. and I. Rutledge (eds.), 1977, *Land and Labour in Latin America: Essays on the Development of Agrarian Capitalism in the Nineteenth and Twentieth Centuries*, Cambridge: Cambridge University Press.

Dupuy, A., 1983, 'Slavery and Underdevelopment in the Caribbean: A Critique of the "Plantation Economy" Perspective', *Dialectical Anthropology*, Vol.7, No.3.

Elkins, S., 1959, *Slavery: A Problem in American Institutional and Intellectual Life*, Chicago, IL: University of Chicago Press.

Emmer, P.C., 1984, 'The Importation of British Indians into Surinam (Dutch Guiana), 1873–1916', in Marks and Richardson (eds.).

Emmer, P.C., 1986, 'The Meek Hindu: The Recruitment of Indian Labourers for Service Overseas, 1870–1916', in P.C. Emmer (ed.).

Emmer, P.C. (ed.), 1986, *Colonialism and Migration: Indentured Labour Before and After Slavery*, Dordrecht: Martinus Nijhoff Publishers.

Engerman, S.L., 1986, 'Servants to Slaves to Servants: Contract Labour and European Expansion', in P.C. Emmer (ed.).

Evans, R., 1970, 'Some Notes on Coerced Labor', *Journal of Economic History*, Vol.30, No.4.

First, R., 1983, *Black Gold: The Mozambican Miner, Proletarian and Peasant*, Brighton: Harvester Press.

Fite, E.D., 1968, *Social and Industrial Conditions in the North During the Civil War*, New York: Frederick Ungar Publishing Co.

Florescano, E. (ed.), 1975, *Haciendas, Latifundios y Plantaciones en América Latina*, México: Siglo Ventiuno Editores.

Fogel, R.W., and S.L. Engerman, 1974, *Time on the Cross: The Economics of American Negro Slavery*, London: Wildwood House.

Foner, L. and E.D. Genovese (eds.), 1969, *Slavery in the New World*, Englewood Cliffs, NJ: Prentice-Hall, Inc.
Fox-Genovese, E. and E.D. Genovese, 1983, *Fruits of Merchant Capital: Slavery and Bourgeois Property in the Rise and Expansion of Capitalism*, New York: Oxford University Press.
Friedland, W.H., and D. Nelkin, 1971, *Migrant: Agricultural Workers in America's Northeast*, New York: Holt, Rinehart & Winston.
Fröebel, F., *et al.*, 1980, *The New International Division of Labour*, Cambridge: Cambridge University Press.
Galenson, D., 1981, *White Servitude in Colonial America*, Cambridge: Cambridge University Press.
Genovese, E., 1966, *The Political Economy of Slavery: Studies in the Economy and Society of the Slave South*, London: MacGibbon & Kee.
Genovese, E., 1971, 'American Slaves and their History', in A.J. Lane (ed.), *The Debate Over Slavery: Stanley Elkins and his Critics*, Urbana, IL: University of Illinois Press.
Genovese, E., 1975, *Roll, Jordan, Roll: The World the Slaves Made*, London: Andre Deutsch.
Glickstein, J.A., 1991, *Concepts of Free Labor in Antebellum America*, New Haven, CT: Yale University Press.
Graves, A., 1984a, 'The Nature and Origins of Pacific Islands Labour Migration to Queensland, 1863–1906', in Marks and Richardson (eds.).
Graves, A., 1984b, 'Crisis and Change in the Queensland Sugar Industry, 1862–1906', in Albert and Graves (eds.).
Green, L.B., 1925, *The Planter's Book of Caste and Custom*, Colombo: The Times of Ceylon Co. Ltd.
Greenberg, M., 1977, 'The New Economic History and the Understanding of Slavery: A Methodological Critique', *Dialectical Anthropology*, Vol.2, No.2.
Guha, Amalendu, 1981, 'Formation of a Working Class in Assam Plantations: A Study in Retrospect', in North East India Council for Social Science Research (ed.), *Problems of Tea Industry in North East India*, Calcutta: Research India Publications.
Gutman, H.G., 1975, *Slavery and the Numbers Game: A Critique of Time on the Cross*, Urbana, IL: University of Illinois Press.
Hart, G., 1986, 'Exclusionary Labour Arrangements: Interpreting Evidence on Employment Trends in Rural Java', *Journal of Development Studies*, Vol.22, No.4.
Haynes, D. and G. Prakash (eds.), 1991, *Contesting Power: Resistance and Everyday Social Relations in South Asia*, Delhi: Oxford University Press.
Higgs, R., 1977, *Competition and Coercion: Blacks in the American Economy, 1865–1914*, Cambridge: Cambridge University Press.
Hobsbawm, E.J., 1987, *The Age of Empire 1875–1914*, London: Weidenfeld & Nicolson.
Hollup, O., 1991, 'Trade Unions and Leadership among Tamil Estate Workers in Sri Lanka', *Journal of Contemporary Asia*, Vol.21, No.2.
Hunt, D., 1978, 'Exclusivism and Unionism: Europeans in the Queensland Sugar Industry 1900–1910', in A. Curthoys and A. Markus (eds.), *Who Are Our Enemies? Racism and the Australian Working Class*, Sydney: The Australian Society for the Study of Labour History.
ICL Research Team, 1979, *The Human Cost of Bananas*, Manila: n.p.
India, 1931, *Report of the Royal Commission on Labour in India*, London: HM Stationery Office.
Jain, R.K., 1970, *South Indians on the Plantation Frontier in Malaya*, New Haven, CT: Yale University Press.
James, C.L.R., 1977, 'The Atlantic Slave Trade', in *The Future in the Present*, London: Allison & Busby.
James, C.L.R., 1980, 'The Making of the Caribbean People', in *Spheres of Existence*, London: Allison & Busby.
James, C.L.R., 1982, 'Introduction' to R. Ramdin, *From Chattel Slave to Wage Earner*, London: Martin Brian & O'Keeffe.

Kahn, J.S. 1981, 'Mercantilism and the Emergence of Servile Labour in Colonial Indonesia', in Kahn and Llobera (eds.), *The Anthropology of Precapitalist Societies*, London: Macmillan.

Kimble, J., 1985, 'Migrant Labour and Colonial Rule in Southern Africa: The Case of Colonial Basutoland, 1890–1930,' Unpublished Ph.D. Thesis, University of Essex.

Kinvig, C., 1992, *The River Kwai Railway: The Story of the Burma–Siam Railroad*, London: Brassey's (UK) Ltd.

Kloosterboer, W., 1960, *Involuntary Labour since the Abolition of Slavery*, Leiden: E.J. Brill.

Knight, A., 1986, 'Mexican Peonage: What Was It and Why Was It?', *Journal of Latin American Studies*, Vol.18, Part I.

Knight, A., 1988, 'Debt Bondage in Latin America', in L. Archer (ed.), *Slavery and Other Forms of Unfree Labour*, London: Routledge.

Knight, G.R., 1988, 'Peasant Labour and Capitalist Production in Late Colonial Indonesia: The "Campaign" at a North Java Sugar Factory, 1840–70', *Journal of Southeast Asian Studies*, Vol.XIX, No.2.

Kolchin, P., 1987, *Unfree Labor: American Slavery and Russian Serfdom*, Cambridge, MA: Harvard University Press.

Kumar, D., 1965, *Land and Caste in South India: Agricultural Labour in Madras Presidency in the Nineteenth Century*, Cambridge: Cambridge University Press.

Lal, B.V., 1989, 'Kunti's cry: Indentured Women on Fiji Plantations', in J. Krishnamurty (ed.), *Women in Colonial India: Essays on Survival, Work and the State*, Delhi: Oxford University Press.

Lasker, B., 1950, *Human Bondage in Southeast Asia*, Chapel Hill, HC: University of North Carolina Press.

Legassick, M., and F. de Clercq, 1984, 'Capitalism and Migrant Labour in Southern Africa: The Origins and Nature of the System', in Marks and Richardson (eds.).

LeGrand, C., 1984, 'Labor Acquisition and Social Conflict on the Colombian Frontier, 1850–1936', *Journal of Latin American Studies*, Vol.16, Part I.

Lenin, V.I., 1964, 'Imperialism: The Highest Stage of Capitalism', in *Collected Works*, Vol.22, Moscow: Progress Publishers.

Levy, N., 1982, *The Foundations of the South African Cheap Labour System*, London: Routledge & Kegan Paul.

Lind, A.W., 1938, *An Island Community*, Chicago, IL: University of Chicago Press.

Lieten, G.K., O. Nieuwenhuys, and L. Schenk-Sandbergen (eds.), 1989, *Women, Migrants and Tribals: Survival Strategies in Asia*, New Delhi: Manohar.

Mandle, J.R., 1978, *The Roots of Black Poverty: The Southern Plantation Economy after the Civil War*, Durham, NC: Duke University Press.

Mangru, B., 1987, *Benevolent Neutrality: Indian Government Policy and Labour Migration to British Guiana, 1854–1884*, Hertford: Hansib.

Mann, H.H., 1967, 'Note on the Diet of Tea Garden Coolies in Upper Assam and its Nutritive Value', in D. Thorner (ed.), *Harold H. Mann: The Social Framework of Agriculture*, Bombay: Vora & Co.

Marks, S. and P. Richardson (eds.), 1984, *International Labour Migration: Historical Perspectives*, London: Temple Smith.

Marx, K., 1976, *Capital*, Volume 1, Harmondsworth: Penguin Books.

McCreery, D., 1990, 'Hegemony and Repression in Rural Guatemala, 1871–1914', *Peasant Studies*, Vol.17, No.3.

Meillassoux, C., 1981, *Maidens, Meal and Money*, Cambridge: Cambridge University Press.

Meyer, E., 1990, 'Aspects of the Sinhalese – Tamil Relations in the Plantation Areas of Sri Lanka under the British Raj', *Indian Economic and Social History Review*, Vol.27, No.2.

Miles, R., 1987, *Capitalism and Unfree Labour: Anomaly or Necessity?*, London: Tavistock Publications.

Miers, S., and I. Kopytoff (eds.), 1977, *Slavery in Africa: Historical and Anthropological Perspectives*, Madison: University of Wisconsin Press.

Moreno Fraginals, M. *et al.* (eds.), 1985, *Between Slavery and Free Labor*, Baltimore, MD: T. Johns Hopkins University Press.

Munslow, B. and H. Finch (eds.), 1984, *Proletarianisation in the Third World: Studies in the Creation of a Labour Force Under Dependent Capitalism*, London: Croom Helm.

Nath, Y.V.S., 1960, *Bhils of Ratanmal*, Baroda: The Maharaja Sayajirao University of Baroda.

Nicholas, S. (ed.), 1988, *Convict Workers: Reinterpreting Australia's Past*, Cambridge: Cambridge University Press.

Nieboer, H.J., 1910, *Slavery as an Industrial System*, The Hague: Martinus Nijhoff.

Nolan, M., 1988, 'The *Historikerstreit* and Social History', *New German Critique*, No.44.

Omvedt, G., 1980, 'Migration in Colonial India: The Articulation of Feudalism and Capitalism by the Colonial State', *Journal of Peasant Studies*, Vol.7, No.2.

Patnaik, U., 1985, 'Introduction', in Patnaik and Dingwaney (eds.).

Patnaik, U., and M. Dingwaney (eds.), 1985, *Chains of Servitude: Bondage and Slavery in India*, New Delhi: Sangham Books.

Parnaby, O.W., 1964, *Britain and the Labor Trade in the Southwest Pacific*, Durham, NC: Duke University Press.

Plant, R., 1987, *Sugar and Modern Slavery: A Tale of Two Countries*, London: Zed Press.

Prakash, G., 1990a, 'Bonded Labour in South Bihar: A Contestatory History', in S. Bose (ed.), *South Asia and World Capitalism*, Delhi: Oxford University Press.

Prakash, G., 1990b, *Bonded Histories: Genealogies of Labor Servitude in Colonial India*, Cambridge: Cambridge University Press.

Prakash, G., 1991, 'Becoming a Bhuinya: Oral Traditions and Contested Domination in Eastern India', in Haynes and Prakash (eds.).

Raman, K.R., 1991, 'Labour Under Imperial Hegemony: The Case of Tea Plantations in South India, 1914–46', in S. Bhattacharaya *et al.* (eds.). *The South Indian Economy: Agrarian Change, Industrial Structure, and State Policy 1914–47*, Delhi: Oxford University Press.

Ramesar, M.D., 1984, 'Indented Labour in Trinidad 1880–1917', in K. Saunders (ed.).

Ramos Mattei, A., 1985, 'Technical Innovations and Social Change in the Sugar Industry of Puerto Rico, 1870–1880', in M. Moreno Fraginals *et al.*, (eds.).

Ramos Mattei, A., 1988, 'The Rise of a Sugar Proletariat Labor Force and the Emergence of Strikes in the Puerto Rican Sugar Industry 1873–1905', in M. Cross and G. Heuman (eds.).

Ransom, R.L., and R. Sutch, 1977, *One Kind of Freedom: The Economic Consequence of Emancipation*, Cambridge: Cambridge University Press.

Reid, A. (ed.), 1983, *Slavery, Bondage and Dependency in Southeast Asia*, St. Lucia: University of Queensland Press.

Renford, R.K., 1987, *The Non-Official British in India to 1920*, Delhi: Oxford University Press.

Robert, J.C., 1938, *The Tobacco Kingdom: Plantation, Market, and Factory in Virginia and North Carolina, 1800–1860*, Durham NC: Duke University Press.

Rodney, W., 1981, *A History of the Guyanese Working People, 1881–1905*, London: Heinemann.

Saunders, K., 1984, 'The Workers' Paradox: Indentured Labour in the Queensland Sugar Industry to 1920', in Saunders (ed.).

Saunders, K. (ed.), 1984, *Indentured Labour in the British Empire 1834–1920*, London: Croom Helm.

Scott, J.C., 1985, *Weapons of the Weak*, New Haven, CT: Yale University Press.

Siddique, Muhammed Abu B., 1990, *Evolution of Land Grants and Labour Policy of Government: The Growth of the Tea Industry in Assam 1834–1940*, New Delhi: South Asian Publishers.

Singh, K.S., 1983, *Birsa Munda and his Movement 1874–1901*, Calcutta: Oxford University Press.

Singh, M. and K.G. Iyer, 1985, 'Migrant Labourers in Rural Punjab', in Patnaik and Dingwaney (eds.).

Solow, B.L. and S.L. Engerman (eds.), 1987, *British Capitalism and Caribbean Slavery: The Legacy of Eric Williams*, Cambridge: Cambridge University Press.

Starobin, R.S., 1970, *Industrial Slavery in the Old South*, New York: Oxford University Press.

Stoler, A.L., 1985, *Capitalism and Confrontation in Sumatra's Plantation Belt, 1870–1979*, New Haven, CT: Yale University Press.

Sundaram, J.K., 1986, *A Question of Class: Capital, the State and Uneven Development in Malaya*, Singapore: Oxford University Press.

Sutch, R., 1975, 'The Treatment Received by American Slaves: A Critical Review of the Evidence Presented in *Time on the Cross*', *Explorations in Economic History*, Vol.12, No.4.

Taylor, L., 1984, 'Literature as History: Ciro Alegria's View of Rural Society in the Northern Peruvian Andes', *Ibero-Amerikanisches Archiv*, Vol.10, No.3.

Thompson, V., 1947, *Labor Problems in Southeast Asia*, New Haven, CT: Yale University Press.

Tinker, H., 1974, *A New System of Slavery: The Export of Indian Labour Overseas 1830–1920*, London: Oxford University Press.

Tinker, H., 1984, 'Into Servitude: Indian Labour in the Sugar Industry, 1833–1970', in Marks and Richardson (eds.).

Tise, L.E., 1987, *Proslavery: A History of the Defense of Slavery in America, 1701–1840*, Athens, GA: University of Georgia Press.

Trotsky, L.D., 1962, *The Permanent Revolution and Results and Prospects*, London: New Park Publications.

US Congress (Subcommittee on Migratory Labor), 1970, *Migrant and Seasonal Farmworker Powerlessness*, (Parts 1–6A), Washington: US Government Printing Office.

van den Boogaart, E. and P.C. Emmer, 1986, 'Colonialism and Migration: An Overview', in P.C. Emmer (ed.).

van Onselen, C., 1976, *Chibaro: African Mine Labour in Southern Rhodesia 1900–1933*, London: Pluto Press.

Watson, J.L. (ed.), 1980, *Asian and African Systems of Slavery*, Oxford: Basil Blackwell.

West India Royal Commission, 1897, *Report of the West India Royal Commission* (Appendix C, Vol.III, Parts VI to XIII – Proceedings, Evidence and Documents relating to the Windward Islands, the Leeward Islands, and Jamaica), London: Her Majesty's Stationery Office.

Williams, E., 1964, *Capitalism and Slavery*, London: Andre Deutsch.

Zegeye, A. and S. Ishemo (eds.), 1989, *Forced Labour and Migration: Patterns of Movement within Africa*, London: Hans Zell Publishers.

'White Gold' or 'White Blood'?:[1] The Rubber Plantations of Colonial Indochina, 1910–40

MARTIN J. MURRAY

CAPITALIST PLANTATIONS: THEORETICAL DEBATES

In the main, the study of plantation systems, both at the level of individual case studies and within the wider global context, has concentrated on those large-scale agricultural estates in the Americas and the Caribbean specialising in the cultivation of particular crops for sale on the world market [*Breman*, 1989: 1–2].[2] Generally, the scholarly literature on plantation economies and systems has focused on (1) distinguishing plantations extrinsically from other types of estate agriculture (*latifundia, haciendas*) in terms of land tenure, type of labour force, and degree of commercialisation; and (2) investigating the ways in which plantation units – as historically-specific sites of agricultural production – differ intrinsically from one another according to cash crop, era of operation, region, types of management, modes of labour control, and degree of technical innovation and modernisation [*Goodyear*, 1978: 291; *Florescano*, 1975; *Wolf and Mintz*, 1957; *Hunt*, 1975].

Despite a steady growth of scholarly literature on plantation economies and systems, there is wide disagreement concerning where and when strictly capitalist plantations came into being, and how and why the transition to capitalist plantation agriculture took place – if at all [*Beckford*, 1985]. Much of the dispute stems from different understandings of capitalism, as well as from divergent vantage points for plantations. Because locating the origins of capitalist plantation agriculture is crucial to understanding the nature of agrarian social classes, the charac-

Martin J. Murray is at the Department of Sociology, State University of New York at Binghamton, Binghamton, New York 13901 USA. The author would like to thank Keith Taylor, Asian Studies, Cornell University, for assisting in interpreting the meanings of Vietnamese-language materials. He would also like to express his thanks to Valentine Daniel and Henry Bernstein for their excellent editing. David Hardiman, Michael Jimenez, Natalie Zemon Davis, and other members of the Cullom Davis Center for Historical Studies, Princeton University, also provided valuable comments an earlier draft of this essay. Finally, he would like to thank Tom Brass for very helpful remarks about the nature of 'unfree labour'. The final product is, of course, his sole responsibility.

ter of rural class struggles, and the direction of agrarian transformations, these apparently 'hair-splitting' disputes carry broader theoretical implications. The sheer scale and scope of the historiographical literature on plantation economies and systems has contributed to the blurring of differences between rival paradigms. While a significant degree of cross-fertilisation has created hybrid models which owe their existence to an ecclectic borrowing from a variety of disparate sources, it is possible to identify several influential approaches.

What has been called the 'plantation school' has made a lasting contribution to the study of the origins and development of plantation economies and systems.[3] The *raison d'être* of this approach was to locate the main sources of economic stagnation and backwardness in the Caribbean, and to provide an historical explanation for the dynamics of underdevelopment and economic dependence on a world scale. The 'plantation school' developed the 'theory of plantation economy and society' as the guiding thread of its analysis [*Benn*, 1974].[4] What distinguished the 'plantation school' from alternative perspectives was its insistence on the essential continuity of plantation economies and systems.[5]

Considerable debate has focused on both situating plantation economies and systems within the capitalist world economy and on theorising the relationship of New World plantations to the origins of capitalism on a world scale. There is widespread agreement that in investments, marketing, and corporate management, plantation agriculture was guided by the (capitalist) rational calculation of profit and loss in the determination of which cash-crops were produced for sale in the world market. Theoretical controversy within the 'plantation school' has revolved around the question of how to properly characterise the recruitment and organisation of plantation labour. Some scholars downplay differences in relations of production and focus instead on plantations as sites of commodity production for sale in the world market. In this 'circulationist' perspective, the fact that plantations engage in production for profit via the market is sufficient to characterise those as capitalist enterprises.[6] Others argue that in so far as plantations rely on 'extra-economic' coercion to secure and control labour they are indelibly inscribed with distinctive 'non-capitalist' features [*Stoler*, 1985: 11–12]. Thus, to distinguish plantation agriculture from various types of agrarian capitalist enterprise, these scholars tend to situate 'plantation economy' in an intermediate, non-classifiable position between feudal and capitalist production systems [*Mandle*, 1972; 1973; *Beckford*, 1985].

Similar debates occupy a central place in the historiographical literature on plantation agriculture in south and southeast Asia [*Stoler*, 1985].

Clifford Geertz, for example, argues that the Java sugar industry did not 'forcibly marshall' the peasantry 'onto enclave plantations and degrade it into a fully proletarianised, essentially landless labour force'. Instead, the 'Javanese cane-worker remained a peasant at the same time that he became a coolie, persisted as a community-oriented household farmer at the same time as he became an industrial wage-labourer. He had one foot in the rice terrace and the other in the mill' [*Geertz*, 1963: 89].[7] Similarly, Alec Gordon suggests that because Java's sugar industry was 'built upon non-economic coercion enforced by colonial state power' plantation agriculture was not a 'pure capitalist enterprise' and that the final phase of Dutch rule in Java generated a specifically 'colonial mode of production' (or 'hybrid capitalism') [*Gordon*, 1979; 1982]. These two approaches have been labelled 'dualist' and 'articulationist' respectively, because they suggest either that capitalism and pre-capitalism existed side-by-side as separate spheres in uneasy symbiosis, or that pre-capitalist 'modes of production' were infused, and thoroughly imbricated, into a bastardized form of 'quasi-capitalism' [*G. D. Scott*, 1976].

The theoretical weaknesses of these various approaches sketched above can be traced to their reductionist and essentialist biases. By taking a single element (for instance, the market, production relations, labour systems, etc.) as their analytic point of departure, these rival paradigms tend to rely upon a single, all-embracing, formula to explain the inner-dynamics of plantations and their relationship to the evolving capitalist world economy. Each perspective captures an essential part of the historical process of plantation development, but because they ignore others, fail to grasp the significance of the whole.

The 'plantation school' takes as its point of departure a concrete historical model of the plantation as an economic entity, and from this historical characterisation it derives a general paradigm based on identifying those structural characteristics governing the relationship between metropole and hinterland [*Benn*, 1974: 253–4]. The 'pure plantation economy' contains the essential features of all subsequent plantation systems and economies. By emphasising an essential continuity of plantation economies and systems, the 'plantation school' depends upon a set of universalistic, static concepts that are imbued with with apparent timeless validity. While this paradigm provides a useful orientation for investigating the structural linkages between metropole and hinterland, it falls short as a theoretical approach to understanding the historical specificity of capitalist plantation agriculture. The 'plantation school' is more concerned with arriving at descriptive generalisations about the political economy of underdevelopment than with explaining the origins of capitalist plantation agriculture.

The 'circulationist' perspective is theoretically flawed for different reasons.[8] By treating capitalism as synonymous with production for profit, the 'circulationists' subsume all commercialised agriculture under the generic rubric of capitalist production, ignoring vastly different ways in which dominant classes organise the labour process and appropriate surplus labour from subordinate classes. The bold yet untenable contention that plantation slavery constitutes 'the heart and essence of capitalism as a mode of production' epitomises this way of thinking [*Wallerstein*, 1980: 218]. To conflate the origins of capitalism with the rise of the world market, and, correlatively, the identification of capitalist development with the geographical spread and deepening of exchange relaitons, leads inevitably to conceptual ambiguity [*Mann*, 1987]. At a fundamental level, the 'circulationist' approach is flawed because it is unable to theoretically specify 'the particular, historically developed class structures through which the processes [of capitalist accumulation and expansion] actually worked themselves out and through which their fundamental character was actually determined' [*Brenner*, 1977: 91].

Another particularly strong current in the scholarly literature associates the advent of capitalist plantation agriculture with the eclipse of various forms of bonded servitude and enslavement and the consequent dawning of 'free labour' [*Stolcke and Hall*, 1983]. At first glance, this focus on the transition of labour relations *per se* signals a welcome advance over the 'circulationist' approach, a viewpoint which downplays class relations and different forms of labour organisation. The strength of the 'free labour' perspective is its recognition of the substantive differences between slavery, debt peonage, sharecropping, and wage-labour, along with its emphasis on the changing historical patterns of how plantations secured labour [*Fraginals et al.*, 1985]. Much of the literature on the transition to 'free labour', however, begs the question. What exactly is 'free labour'? Without a precise and unambiguous definition, the idea of a transition from one type of labour to another opens a Pandora's Box of competing interpretations. One common theme identifies the transition to 'free labour' with the elimination of non-market coercion, that is, the clearing away of impediments to the purchase and sale of labour-power in the marketplace, including the lifting of juridical-legal restrictions on labour mobility, thereby enabling labouring subjects to commodify their own labour-power.[9] The problems with this approach are two-fold: first, it is reductionist because it establishes a single formal criterion, namely, the unrestricted buying and selling of labour-power in the marketplace, as the key element to explain the origins of capitalist plantation agriculture. Second, given this single-minded focus on market relations, it conceives of labour solely in terms of the conditions of its circulation, ignoring the historically specific ways in which capital takes command of the labour process and subordinates wage-labour to the logic of surplus-value extraction.[10]

Put broadly, to associate 'unfree labour' with 'non-economic (or non-market) coercion' and, conversely, 'free labour' (and capitalism) with unrestrained market activities is untenable. The division between 'market' 'and non-market' poses an artifical dualism which hinders rather than clarifies the understanding of the complex historical processes through which capitalism originates and evolves. To suggest that intrinsically non-market coercion is neccssarily incompatible with the growth and expansion of capitalism comes dangerously close to neo-classical economic theory. Those scholars who rely upon the market/nonmarket distinction as the cornerstone for their definition of free labour and capitalism falsely dichotomize what is essentially an integral process wherc market/non-market and free/coerced relations are inextricably intwined. Not only has capitalist production proper always and everywhere been preceded as well as conditioned by non-market coercive mechanisms, but the maintenence of capitalist relations also presupposes extra-economic domination.

Under the historical circumstances of a volatile or underdeveloped labour market, 'labour-repressive' policies are often the *sine qua non* of capitalist development [*Murray and Post*, 1983a; 1983b]. The origins and development of capitalist production processes are often accompanied by the introduction of entirely novel ways of securing a workforce and harnessing these assembled workers at the point of production. Rather than being understood as the persistence of atavistic, 'feudalist', or pre-capitalist relics, some historically-specific mechanisms of non-economic coercion owe their existence to the rise of capitalism. Scholarly literature on the southern and central African mining industry, for example, demonstrates conclusively that capitalist production processes went hand-in-glove with repressive mechanisms of labour control, including closed compounds and legal restrictions on geographical mobility. The dependence of mining capital on migrant labour retarded the development of a full-fledged capitalist labour market, thereby slowing the growth of a stabilised labour force completely separated from alternative means of survival [*van Onselen*, 1976; *Phimister and van Onselen*, 1978; *Worger*, 1987; *Turrell*, 1987; *Parpart*, 1983].

In studying the origins of the Java sugar industry, G.R. Knight approaches this problem from a different angle. He suggests that it was the economic necessity of large numbers of landless villagers to sell their labour-power that 'underpinned the system of so-called "forced" labour'. The mechanisms of 'non-economic coercion' did not pry peasants away from readily available 'preferred means of subsistence' through direct compulsion, but instead diverted an already existing landless workforce away from other opportunities of employment toward the sugar industry [*Knight*, 1988: 264–5; *Knight*, 1980].

> It is plain that the worker does not have to be bonded to separate him [or her] from the means of production. Quite the contrary, for it is the fact that he [or she] has *already* been separated from the means of production and left with only his [or her] labour to sell which has rendered him [or her] vulnerable in the first place' [*Knight*, 1989: 52].[11]

WORLD RUBBER PRODUCTION, *c.*1890–1940

At the end of the nineteenth century, a wave of technical innovations, including the use of new materials and new ways of making old materials, generated the accelerated demand for rubber products.[12] Before the onset of this 'Second Industrial Revolution' [*Landes*, 1969: 235, 249, 269–70], virtually the entire world production of rubber of only a few hundred tons annually, came from wild plants located mainly in Brazil and West Africa [*Weinstein*, 1983].[13] The wild rubber boom hit the African continent in response to rising demand and buoyant world market prices by the last decade of the nineteenth century [*Henry*, 1906; *Christy*, 1911; *Harms*, 1975; 73–5; *Knorr*, 1945: 9–16]. The desire for quick profits led to the adoption of extremely crude methods of gathering rubber, including overtapping, uprooting of vines, and felling of trees, which, in turn, engendered the wholesale squandering and eventual depletion of rubber supplies. After 1910, increasing quantities of inexpensive, high-quality rubber from *hevea* plantations in Southeast Asia began to arrive in Europe, thereby undercutting alternative sources of supply, particularly in Africa [*Harms*, 1975: 74–5].

Rubber was first introduced into colonial Indochina in 1897 when *Hevea Brasiliensis* plants were imported from Malaya. Rubber production only began in earnest in Cochinchina after 1907 when the expansion of British-owned rubber plantations in Malaya proved very profitable. Many small planters pioneered rubber cultivation, and within a few years, a number of modest plantations of less than 100 hectares were established within a radius of 30 miles northwest, north, and east of Saigon [*Spire and Spire*, 1905; *Institut français du caoutchouc*, 1945].[14] The colonial administration took an active part in encouraging large-scale metropolitan investment in rubber plantations by giving away huge tracts of land and requisitioning local *corvée* labour to construct a road network extending into the highly valued *terres rouges* ['red lands'] of eastern Cochinchina. The early results of these efforts were mixed.[15] But after 1910 rubber production in colonial Indochina experienced a dramatic take-off. The combination of unprecedented military production along with the rapid development of the automobile industry, particularly in the United States, created the accelerating world demand for rubber. By 1921, approximately 29,000 hectares in Cochinchina were planted with rubber trees, of which about 20,000 hectares were concentrated in the three provinces of Baria, Bien Hoa, and Thudaumot [*Montaigut*, 1929: 19].[16]

By the early 1920s, the rapid expansion of rubber production, particularly the British plantations of Malaya and the Dutch operations of the East Indies, glutted the market with too much rubber to sustain price levels [*Pommery*, 1952: 154; *Gérard*, 1924; *Robequain*, 1944: 205; *Knorr*, 1945: 90–93]. The rubber planters of Indochina were not spared the hazards of vicious price competition, and only an infusion of colonial subsidies, together with loans from the *Banque de l'Indochine*, ensured their economic survival [*Thompson*, 1968: 132]. In 1922, the world's major rubber producers reached an agreement, known as the Stevenson Plan, to jointly reduce output in order to maintain steady prices. Ironically, this artificial reduction in scale of rubber production slowed output on the plantations of Malaya and Ceylon precisely at the time when the astronomical growth of automobile production created upward pressure on demand [*Knorr*, 1945: 93–101]. The resulting sharp rebound in the world price triggered a huge expansion of rubber plantations in Indochina and contributed to record profit levels [*Pommery*, 1952: 154]. Between 1925 and 1929, the land under rubber cultivation expanded from 15,000 hectares to 90,225 hectares with a capital investment of 700 million francs. During this brief period, metropolitan companies reported fabulous profits. Montaigut reported that dividends for the most successful plantations exceeded 160 percent of the initial capital investment [*Isoart*, 1961; *Bourbon*, 1938; *Montaigut*, 1929: 26–7].[17]

The combination of the end of the Stevenson Plan output restrictions in 1928 and the 1929 world depression brought rubber production to the brink of financial disaster as world market prices failed to even approach production costs [*Pommery*, 1952: 154; 182; *Bourbon*, 1938: 95; *Bauer*, 1948].[18] In 1934, five countries controlling 98.7 per cent of world rubber exports reached an accord, known as the International Rubber Regulation Agreement (IRRA), which established fixed export quotas for each rubber-producing region and created an international commission to supervise enforcement and further renegotiation [*Bourbon*, 1938: 103–8; *Knorr*, 1945: 108–10; *Bauer*, 1948: 88–110].[19] Due to the particularly shrewd negotiations of representatives of the French rubber plantations in Indochina, France received a relatively large quota of 30,000 tons, which exceeded current production in Indochina by almost 100 per cent [*Bourbon*, 1938: 108–9]. In 1938, the terms of the agreement were renegotiated and the IRRA authorised Indochina to export 60,000 tons without penalty or restriction [*Knorr*, 1945: 132].[20]

By 1937, a total of 1,007 separate plantations had brought approximately 127,000 hectares under rubber cultivation, of which 98,000 were located in Cochinchina, 27,000 in Cambodia (where there had been only 18 hectares in 1921), and only 1,700 in Annam. On balance, enterprises devoted to rubber production can be classified into two groups: small and large. Approximately 70 per cent of the total number of rubber plantations in operation in 1937 were less than 40 hectares, with an average size of about 11 hectares, and accounted for only six per cent of the total area devoted to rubber cultivation. Before the first World War, these small-scale and medium-scale planters formed their own business association, called the *Syndicat des Planteurs de Caoutchouc* of Saigon. In contrast, large-scale metropolitan concessionary companies, including Michelin, produced almost the entire output of Indochina's latex. Twenty-seven enterprises owned between 1,000 and 5,0000 hectares, and four owned more than 5,000 hectares. In 1930, the largest firms organised themselves into the *Section Autonome de l'Union des Planteurs de Caoutchouc de l'Indochine* which in turn formed part of the Paris-based *Comité de l'Indochine*.

Yet even this highly uneven distribution of land and output does not sufficiently indicate the degree of concentration of capital characteristic of rubber production. Large-scale joint-stock companies often owned many separate plantations and were frequently financially interrelated with or dominated by still larger holding companies [*Isoart*, 1961: 294; *Robequain*, 1944: 207; *Montaigut*, 1929: 99–105].[21] By the late 1930s, Indochina ranked fifth among world rubber producers in acreage, fourth in output, and first among major producers in output per acre [*Andrus*, 1943: 25–30].[22]

THE PLANTATION LABOUR SYSTEM

While the rapid expansion of rubber production in colonial Indochina was the direct consequence of rising world market demand, the plantation system, including the peculiar fusion of class exploitation with racial hierarchy, the methods of procurement of labour, and the nature of the work regime, can only be understood by reference to local conditions. The rubber plantations were literally carved out of remote, rugged, overgrown, and scarcely populated regions to the north and east of Saigon. The first plantations brought into production before the First World War were typically of modest size and, as a general rule, were located in the *terres grises* (or 'grey lands') within a 15-kilometre radius of the principal port at Saigon. Over time, the center of gravity of rubber cultivation steadily shifted outward, pushing the rubber frontier further northward. For the most part, the large plantations which emerged after the First World War were situated in the more fertile though isolated *terres rouges* (or 'red lands') region. The *terres rouges* extended from the southeast to the northwest of Saigon in a gentle sloping arc approximately 300 kilometres long and 30 to 40 kilometres wide. Consequently, a considerable proportion of the large plantations in the rubber zone (notably, the Chup, Snoul, and Minot plantations inside Cambodia; the Budop, Quan-Loi, and Phu-Rieng plantations in northern Cochinchina) were located more than 100 kilometres from Saigon [*Camus*, 1949].[23]

Until around 1918, the rubber plantations experienced only minor difficulties in attracting sufficient supplies of labour-power. The medium-sized rubber estates in the *terres grises* of Cochinchina were close to overcrowded villages where a large resident population was in nearly continuous search for part-time wage-labour. The rapid expansion of rubber cultivation after the First World War took place in eastern Cochinchina whose population density averaged a little more than 33 persons per square kilometre. The population, however, was unevenly distributed.[24] Before the First World War, the 'force of economic circumstances' (for example, the need for cash to pay colonial taxes) was sufficiently powerful to drive landless, or virtually landless, non-Viet ethnic minorities (the Stiengs, Chomas, Muong) into seeking temporary wage-paid employment on the relatively small number of rubber plantations of eastern Cochinchina [*Montaigut*, 1929: 32]. After 1918, the rapid expansion of plantation cultivation into both the *terres grises* and *terres rouges* areas pushed the plantation frontier further north. This swift increase in plantation production led to the exhaustion of local labour supplies. By the early 1920s, the rubber plantations employed an estimated 18,000 *coolies*, of whom a little more than 3000 were hired under three-year contracts and the rest, the 'free workers', were employed as casual labourers.

These early labour hiring practices conformed to an ethnic separation of job-tasks. The local non-Viet people were hired to perform the arduous labour of clearing the virgin forests and the rough work required prior to planting. In contrast, Vietnamese labourers were hired for plantation maintenance and the semi-skilled, delicate labour of tapping rubber trees. Each rubber tapper was required to treat 400 trees (or two planted hectares) per day, a figure far above that of Java or Malaya [*Montaigut*, 1929: 34].[25] By the mid-1920s, the plantations faced severe labour shortages. On the one hand, poor working conditions, the high incidence of malaria, and low wages contributed to high labour turnover, and, on the other hand, they discouraged local labour from seeking employment on the rubber plantations. At the same time, the colonial administration increased its forced labour drafts (the *corvée*) amongst resident ethnic minorities in northern and eastern Cochinchina in order to complete the network of roadways from Saigon to the plantation region, thereby further depleting available sources of labour-power and creating more consistent upward pressure on local wage rates. Most importantly, work-seekers in the region were attracted to the expanding employment opportunities in the growing agro-industrial enterprises and in transport, commercial, and trading sectors in the Saigon/Cholon area [*Montaigut*, 1929: 35].

The main obstacle to capitalist expansion and accumulation was the limited supplies of labour. Faced with shrinking sources of low-cost labour, the rubber plantations hoped to entice Javanese *coolies* because of their experience with rubber-tapping. This scheme failed because of the reluctance of Dutch officials and the high cost of transport. Consequently, the rubber plantations turned their attention to the overcrowded villages of the Red River delta in Tonkin and the coastal lowlands of Annam. Almost overnight, the number of workers recruited from the north under three-year contracts skyrocketed: 3,242 in 1922; 5,500 in 1923; 9,200 in 1924; 14,011 in 1925; 29,168 in 1926; 36,201 in 1927; and 41,750 in 1928 [*Montaigut*, 1929: 34–6].

Even though the high population densities in the rice-producing areas of the Red River Delta and the relatively higher wages in Cochinchina were conditioning factors,[26] they alone cannot adequately explain why work-seekers left the relative security of their natal villages in search of wage-paid employment on the rubber plantations. The sheer scope of labour migration indicated that the far-reaching socio-economic transformation of agrarian relations in the northern countryside was well underway. The geographical circulation of labour on a large and expanding scale was not possible until capitalist penetration had breached the seemingly impenetrable walls of village autonomy.

Put in theoretical terms, the social foundations of the migrant labour system did not entail an 'articulation of [discrete] modes of production' as some analysts have suggested. Rather the siphoning of labour out of the northern countryside was the direct result of the insertion of new forms of capital into the socio-economic fabric of village life. Seen from one angle, the emergence of embryonic forms of agrarian capitalism encouraged class differentiation in the northern villages, of which the most visible sign was the growing polarisation of land ownership. On the one hand, the propertied classes ensured their reproduction and survival by accumulating and hoarding cultivatable land. On the other, the loss of land led inexorably to growing immiseration and marginalisation where the propertyless, or virtually propertyless, villagers were compelled to sell their labour in order to subsist.

Seen from another angle, the expansion of money-dealing capital acted as a principal lubricant facilitating exchange relations, with the inevitable result that moneylending at usurous rates accelerated the growing polarisation between creditors and debtors. Increasing numbers of villagers became enmeshed in a vicious cycle of rural indebtedness from which they were never able to escape. Even after a normal harvest, small landowners could live for only two or three months, after which they were compelled to borrow to keep themselves alive until the next harvest. Village small-holders were unable to save any appreciable amount of working capital; hence they were compelled to borrow money with which to prepare for the following year's harvest, pay colonial taxes, and to repay previous debts. Wealthier landlords extended credit in both kind and cash. The terms for borrowing rice were generally six months, and occasionally only three or four months, with interest rates of 50 to 100 per cent. Moneylending at usurious rates led inevitably to the expropriation of small-scale owner-occupiers, and the debt-bondage of tenants, sharecroppers, and wage-labourers [*Gourou*, 1955: 359, 375, 382, 401–3; *Murray*, 1980: 398–402, 471; *Hickey*, 1958: 54–8, 83; *Truong Chinh and Vo Nguyen Giap*, 1974: 36–7]. As a consequence, those villagers caught in the trap of rural indebtedness were often cast adrift from the relative security of landownership, and were pushed into the surplus population. Indebted landless villagers were virtually powerless to resist the enticements of itinerant labour recruiters [*Murray*, 1980: 374–472; *Brocheux*, 1983].[27]

Even before they set foot on the rubber plantations, contract workers were subjected to indignities and rigid social regulations characteristic of the efforts of metropolitan capital to create and maintain a docile labour force. Recruited *coolies* were transported under armed escort from the provincial recruiting depots in rural Tonkin and Annam to various ports of embarkation along the coast for the voyage south. They disembarked at a special immigration camp at Xom-Chieu, near Saigon, where all arrivals underwent a comprehensive medical examination designed to leave only the healthiest workers for the plantations [*Goudal*, 1938: 55–65]. These recruits were housed in military-style barracks surrounded by barbed wire, under constant surveilliance by special police to prevent early desertions. Beginning in 1929–30, the *Sûreté* became involved in this screening process in order to weed out workers deemed politically undesirable and potential trouble-makers.[28]

Contract workers and 'free' labour existed side-by-side on the plantations. Whether bonded or 'free', wage-labour is always and everywhere defined by its formal separation from the principal means of production such that (1) it is dependent upon the sphere of circulation to sell its labour-power and to purchase its necessary means of subsistence, and (2) it is subordinated to capital at the point of production [*Marx*, 1973: 325–6; 419–20; 464–9; 471–514; 547–8]. Non-contract workers were hired to meet temporary, seasonal demands for casual labour. Payment was in both money-wages and kind. Contract workers pledged to work for three years in exchange for periodic cash payments, daily rations, shelter, and miscellaneous services. The contract system not only ensured the plantations of a steady supply of labourers under specific conditions, but also legally bound wage-labourers to the point of production for periods that almost always exceeded the initial three-year agreement.

The contract subjected recruited northern workers to a spider's web of obligations which were nothing less than indentured servitude.[29] It included binding clauses stipulating the nature, duration, and conditions of work on the plantations. Infringement of the agreed terms amounted to a breach of contract and, in many instances, constituted a criminal offense. As a way of ensuring supervision, contract workers were required by colonial law to have in their possession 'work-books' with pertinent information (name, age, place of employment, date of expiration of contract, arrest record, etc.). Under the terms of the contract, wage-labourers were prohibited from engaging in collective bargaining and participating in work stoppages. In order to maintain a smooth functioning migrant labour system, colonial statutes authorised state officials to pursue and arrest contract workers who dared to break the terms regulating their employment on the plantations. Those who left the plantation grounds without official permission – escapees – were labelled deserters and, hence, were subjected to criminal proceedings which often resulted in long prison terms and severe physical beatings [*Tran Tu Binh*, 1985: 15–31].[30] *Coolies* borrowed money to buy supplementary rations, to pay fines, satsify the extortionist demands of their overseers, and for a host of other reasons. Thus, workers almost always found themselves hopelessly mired in debt at the expiry of their contracts, hence were compelled to 're-engage' in order to repay loans to moneylenders [*Murray*, 1980: 287, 292].

In order to operate efficiently and without necessary delays in production, the rubber plantations required a large, steady, and dependable supply of unskilled and semiskilled wage-labourers. In general, plantations were unable to depend upon the influx of voluntary immigrants to satisfy thier labour demands. Work-seekers explored other options, including trying their luck in the burgeoning casual labour market of the urban areas. For the enterprising and adventurous, the newly-opened frontier rice-growing regions of western Cochinchina offered the possibility of acquiring freehold rights to land for subsistence cultivation. The rubber plantations thus came to rely upon the contract system, along with indebtedness, *as powerful levers to guarantee sufficient quantities of labour-power in the absence of a well-defined and well-functioning labour market* [*Montaigut*, 1929: 36; *Murray*, 1980: 272–3].

PLANTATION COMPOUNDS

The plantation compounds themselves resembled gigantic prison-like enclosures where labourers were separated to a considerable degree from social contact with the outside world. The typical plantation complex stretched for miles in all directions from its central core of work-sheds, shortage warehouses, living quarters, and administrative headquarters. Most plantations were arranged in a carefully selected geographical pattern. The outer reaches were set aside for the planted lots of mature rubber trees, for the nursery gardens where the young saplings grew to maturity, and for the vast stretches of unused forest lands, which, at least theoretically, were to be gradually cleared, planted, and brought into production [*Murray*, 1980: 275].

The inner core of the plantation 'was founded as a "micro-society" in the image of the global colonial society. It was built according to a schema of a spacio-professional and spacio-racial kind' [*Brocheux*, 1975: 68]. This central core – literally miles from the plantation's main entrances – was divided between the working area (*bâtiments du travail*) and the living area (*l'habitât du personnel*). The first area included the various plantation infrastructural facilities (first-stage processing equipment, work-sheds, storage warehouses, garages, repair shops, laboratories, experimental stations, central electrical works) and the administrative facilities (managerial offices, police headquarters, hospital). The second area was sub-divided into two separate zones: one reserved for the residences and recreational facilities of the European professional and managerial staff, the other for the resident Indochinese workers. In the 'European area', plantation salaried and professional staff were provided with individual bungalows of stonework construction. Whenever possible, these were built on the tops of hills in comfortable surroundings and at some distance from the administration and other centers of plantation activity. Even within the European zone, a rigid hierarchy prevailed: high-ranking plantation officials enjoyed the most comfortable and aesthetically pleasing lodgings, while the 'petit-blancs' and 'half-breeds' (or *métis*) had the lower-quality residences.

In the 'native area', the living conditions were entirely different. For the most part, plantation workers were housed in replicas of rural villages with between 300 and 500 *coolies* each. Accommodations consisted of long buildings, called *trai*, in which dozens of workers (in the early years, young men recruited in the prime of life) slept side by side in small stall-like enclosures. These buildings were generally made of brick or wood, with roofs of tiles or corrugated iron, and were grouped together to simulate a rural village [*Goudal*, 1938: 75].[31] In the pioneer days of the rubber industry, plantation managers found the barracks-like accommodations with their long tiers of bunks cost effective because both materials and skilled labour for their construction were difficult and expensive to obtain [*Carton*, 1924a: 450–56].

Beginning in the late 1920s, as the increased recruitment of women workers contributed to the establishment of a semi-permanent resident plantation labour force, plantation managers replaced the barracks-like central compounds with individual dwellings constructed of mud and thatch, each separated from the next by a yard and a small garden for vegetables [*Robequain*, 1944: 217; *Le Fevre*, 1927; *Monet*, 1930]. These smaller huts became the most popular type of housing for plantation workers because they offered a relatively private form of living arrangement. These dwellings were generally occupied by 'three or four unmarried *coolies*, between 20 and 30 years old, and a woman who does the cooking, and who many or may not be the wife of one of them' [*Robequain*, 1944: 217].

The arduous nature of plantation work, the excessively long hours of toil, the poor quality of the workers' diet, and the unhealthy living conditions resulted in abnormally high death and disease rates on the plantations, particularly during the 1920s, when the area of virging land brought under rubber cultivation expanded most rapidly. From the late 1920s onward, plantation hospitals and clinic staff were taxed to their limits and health care required relatively large plantation budgetary outlays. Unhealthy living and working conditions on the rubber plantations sapped the workers of their strength, resulting in lower labour productivity and contributing to rising desertion rates [*Murray*, 1980: 276–9]. On the whole, the rubber plantations experienced relatively high labour turnover and recurrent labour shortages. During the 1930s, the state administration intervened in order to protect the long-term interests of large-scale capital. By establishing minimum standards for housing, in addition to food, medical attention, and overall working conditions, the colonial officials sought to regulate the labour market and thereby stabilise employment on the rubber plantations [*Murray*, 1980: 274–8].

CONTESTING CAPITALIST POWER

Capitalist Coercion and Plantation Labour

From the point of view of capital, workers were assembled at the point of production solely because of their capacity to labour. Plantation overseers recognised that time and work-discipline were integrally connected. Accelerating the pace of work and extending the length of the working-day meant greater output and, hence, larger profits. One plain fact of plantation life took precedence over all others: the daily routine on the rubber plantations revolved around the ceaseless efforts of overseers to compel reluctant *coolies* to work. At all levels, plantation management accumulated an arsenal of weapons designed to ensure docility on the plantation grounds and discipline in the labour process. At the bottom of the administrative hierarchy, Vietnamese *cais* functioned as plantation management's first line of defense in the maintanence of social control over the captive labour force. These *cais* occupied an intermediate place in the technical division of labour on the rubber plantations. They made use of intimidation, harassment, and physical violence in order to squeeze labour-time out of direct producers. More often than not, they took advantage of their intimate knowledge of workers under their direct supervision to extort money and favours. At the next level in the administrative hierarchy, European overseers resorted to an extensive system of fines to punish *coolies* for infractions and alleged violations of plantation rules. These low-level functionaries used the flimsiest excuses to make deductions from the money-wages of workers for a vast array of 'offenses', including insubordination or impertinence, loafing, feigned illness, self-inflicted injuries, deliberate wastefulness, gross carelessness in handling tools, and abuse of plantation equipment [*Murray*, 1980: 278–9; *Monet*, 1930: 27–8; *Durtain*, 1930: 199–208; *Goudal*, 1938: 79; *Tran Tu Binh*, 1985: 21, 25–7, 38, 44; *Ngo Vinh Long*, 1974: 43].

European supervisors also subjected *coolies* to beatings, floggings with special leather whips, and other sorts of physical abuse [*Tran Tu Binh*, 1985: 30, 33–45, 62, 65].[32] Nearly all large plantations built well-constructed private prisons that were used to incarcerate *coolies* captured after attempting to escape, accused of refusing or neglecting work, or charged with wilful destruction of plantation property. Private plantation punishments also included the use of handcuffs, iron collars and chains, the aim of which was to discipline recalcitrant *coolies* or to prevent them from attempting to desert [*Tran Tu Binh*, 1985: 30–31].[33]

Individual and Collective Responses of Workers

In order to understand how workers responded to the conditions they encountered on the rubber plantations, it is necessary to reconstruct the socio-economic conditions they came from. Contract workers emigrated from closed corporate villages in Tonkin and northern Annam which, despite internal tensions and developing class antagonisms, were characterised by a certain degree of communal solidarity. Those who left the relative security of village life (at least until the mid-1930s) were generally young able-bodied males in the prime of their working lives who were usually unable to gain access to land in the village. They nearly always left behind their wives, children, and parents. The overwhelming majority of these *coolies* longed to return to their natal villages at the end of their contracts, or as soon as they had acquired sufficient cash savings to meet their yearly colonial tax obligations and to service village debts. These so-called 'target workers' had little or no intention of remaining in the position of full-time proletarians.

Once they reached the plantations, workers tended to group themselves into various sociocultural associations, mutual aid societies, and sport and dance clubs. These somewhat informal and voluntary organisations were ostensibly formed to at least partially compensate for the absence of the twin pillars of customary social life: the village commune and the extended family. These associations had great appeal to young workers recently uprooted from the fortified and insular social matrix of customary family and village life. For the most part, the primary function of most of these informal workers' associations was collective protection against the real threat of mistreatment, including extortion, bullying, and intimidation, often engaged in by malevolent *cais* and supervisors. In particular, various mutual aid societies offered at least rudimentary social insurance against destitution and indebtedness for those unfortunate enough to suffer work-related injuries or unable to work due to prolonged illness.

The plantation management raised few objections to these expressions of workers' self-organisation, particularly when they provided the type of social-security protection which plantation managers themselves were unwilling to offer. Nevertheless, the effect of these autonomous workers' associations was to greatly enlarge the horizons of the *coolies*, in part because they transcended the customary village and provincial affinities and in part because they were virtual 'learning centers' for co-operation and mutual assistance. Perhaps most significantly, plantation workers who experimented with new organisational forms and patterns of leadership gained practical experience which proved invaluable in the clandestine planning for the major work stoppages that took place on a number of plantations during the late 1920s and 1930s [*Tranh Tu Binh*, 1985: 55–85].

Workers did not submit to the autocratic and arbitrary rule of metropolitan capital without a struggle. In their fight against those who harassed, threatened, and exploited them, rubber plantation workers generally avoided direct and open confrontations. Instead, they relied upon what some [*Scott*, 1985; *White*, 1986] have called 'everyday forms of resistance'. The nature and character of these largely spontaneous and clandestine actions varied considerably in time and place, but their common thread was that they were largely defensive acts, requiring little or no co-operation, and undertaken not to directly challenge capitalist power and authority *per se*, but to alter the 'rules of the game' in workers' favour, to ensure their individual survival, and to exact revenge. Confronted with the seemingly all-encompassing dictatorial rule of the plantation, these often desperate acts were truly 'weapons of the weak' [*Scott*, 1985].

The most rudimentary forms of workers' resistance were individualised acts of sabotage of company property, self-multilation (including suicide), and desertion. Sabotage and self-mutilation are mentioned only occasionally in official sources. Yet there is no doubt that they were frequent enough for the colonial administration to specifically list them as criminal offenses punishable by heavy fines, hard labour, and imprisonment. The clearest and most unambiguous indication of the overall troubled state of mind of plantation workers was the sheer scale of desertions. Desertion became the most widespread and successful class-grounded response to the oppressive realities of plantation life [*Murray*, 1980: 296–8]. For example, in 1925 and 1926, when more than one-third of all *coolies* employed were disabled due to sickness or death, about one in every ten plantation workers deserted [*Lasker*, 1972: 258–9]. As an index of social discontent, these figures for the mid-1920s do not seem exaggerated: in 1929, according to Goudal, approximately 11.2 per cent of the work force deserted; and, in 1930, an estimated 8.3 per cent did the same [*Goudal*, 1938: 84–5].[34] This steady flow of desertions forced plantation managers to spent an inordinate amount of time and energy on recruiting replacements. In 1930, for example, the combination of desertion, debilitating disease, and crippling injuries meant that plantations had to recruit 75,000 *coolies* in order to maintain a work-force of 22,000 contract labourers [*Brocheux*, 1975: 63].

The official explanation for high desertion rates was that most deserters were enticed away by cunning rival recruiting agents and were hired as noncontract workers on plantations which did not generally employ contract labour.[35] To the extent that this explanation is appropriate, it appears to indicate that *coolies* pragmatically sought to reach a market where the sale of their labour-power could fetch a higher price. This worker strategy was in direct opposition to the combined efforts of European planters and colonial officials, to use coercive legislation to greatly reduce the market-place bargaining power of workers [*Murray*, 1980: 237–50]. European planters were particularly disturbed by the fact that a large percentage of desertions took place amongst 'new arrivals' on the plantations. Having been lured under false pretences into signing contracts, many workers were appalled by the reality of their working and living conditions, and seized the earliest opportunity to desert. There was another significant reason for the quick exit of workers from the plantations. Recruiting agents almost always offered prospective *coolies* an 'engagement bonus' of ten piastres (the equivalent of one month's money-wages) at the time of the signing of the three-year contract. Plantation workers were rarely if ever forewarned that deductions were automatically made from their cash-wages during the first few months of employment to repay this original advance. Numerous workers just left the plantation once they realised that these deductions forced them into indebtedness from the very onset of the contract period [*Murray*, 1980: 297–8].

To counteract high desertion rates, the colonial administration introduced specific legislative measures to curtail what planters considered the fraudulent retention of advances. For example, the colonial decree of June, 1932, authorised the arrest of deserters who had fled to their homes in Tonkin for breach of contract and their return to the southern plantations. The decree also stipulated that contract workers who failed to repay their 'engagement bonuses' were liable to imprisonment for not less that two months and not more than two years and a fine of 25 to 3,000 francs, or to either of these penalties singly [*Goudal*, 1938: 82–3].[36]

The Strike: Offensive Weapon of Plantation Workers

Plantation workers forged sophisticated responses to the dictatorship of metropolitan capital. *Coolies* seized upon such diverse activities as sabotage, petty theft, 'loafing', and desertion in order to demonstrate their discontent with the appalling working conditions on the plantations. Workers accumulated countless grievances, with the petty despotism of *cai* supervisors high on their list of complaints. Workers sometimes reacted to the brutal and inhuman treatment which they suffered at the hands of their *cai* supervisors by assaulting and even killing them [*Tran*

Tu Binh, 1985: 38–9, 55–85]. As a general rule, however, the fact that such responses were uncoordinated, individualised, and *sub rosa* actions reveals their limits as effective methods of challenging the rule of capital.

The workers' strike was an economic weapon of a different magnitude. The collective decision to stop work – whether spontaneously or premeditated – involved a much greater calculated risk for the participants. Work stoppages represented a frontal assault on the prerogatives of capital. While pilfering, 'loafing', and wilful destruction of plantation property had the effect of keeping the plantation management off balance, workers' strikes brought the machinery of exploitation to a temporary halt, and simultaneously demonstrated in a peculiarly dramatic fashion the collective power of the working class. Between 1930 and 1937, accounts of at least eight large-scale strikes on the rubber plantations can be found in official colonial records. While these actions achieved only modest material gains for plantation workers, they contributed signficantly to the growth of the Vietnam workers' movement in the 1930s. The regularity of these work stoppages, coupled with the inventiveness of those *coolies* who planned and participated in these strikes (for example, collecting crude weapons, storing food, planning escape routes, etc.) offer unambiguous testimony to the existence of a distinctive working class consciousness which came to understand the exploitative nature of capitalism [*Tran Tu Binh*, 1985; *Hemery*, 1975: 347–8; *Murray*, 1980: 299–311].

CONCLUSION: THE CLASS POLITICS OF PLANTATION PRODUCTION

These agro-industrial estates were European operations *par excellence*: capital investments, technical knowledge, and organisational skills were imported from outside; local unskilled labour was mobilised, put to work in an alien environment producing commodities for export; and profits were repatriated. These agro-industrial estates brought the rigors of industrialised production into an isolated, rural setting. The rubber plantations were mainly located in the foothills of the sparsely-populated northern and eastern Cochinchina and away from the settled, lowland, wet-rice cultivation areas of the Mekong delta to the south. This specific location necessitated the use of various bonded forms of hire in order to ensure a steady supply of low-cost labour. The use of extra-economic coercion to recruit workers in the sphere of circulation was logically distinct from the subordination of wage-labour to the rule of capital at the point of production. Once these recruited workers were assembled at the point of production, they were put to work in an alien environment with the aim of producing commodities for sale in the world market. The rubber plantations were capitalist enterprises: the propertied classes took direct command over the labour process, subjecting direct producers to a work regime under their control where surplus-labour was appropriated as surplus value through wage labour. The rubber plantations of colonial Indochina quickly became one of the main engines of economic growth and a principal source of capitalist profits for European entrepreneurs. Workers, both singly and collectively, found ways to resist.

For the European capitalists who sought easy profits through investments in colonial rubber cultivation, the latex which oozed from the rubber trees became known as 'white gold'. Plantation owners and the colonial state administration worked hand-in-glove to ensure that latex flowed without interruption. Whenever the collective power of the working class threatened to disrupt production, the state administration co-operated fully with plantation managements to counter workers' actions by dispatching civil guards, establishing local militia posts, and supplying police spies.[37] The frequent use of state power to break strikes and settle industrial disputes became an integral part of crushing workers' resistance in Indochina [*Brocheux*, 1975: 84–5].

Following the first wave of labour unrest which erupted in the late 1920s, a number of large-scale rubber-producing enterprises hastily introduced improvements in living and working conditions on their plantations. Yet because these rival enterprises operated within the constraints of cost-efficiency and market-prices, plantation managers were under continuous pressure to maximise rubber output and to minimise production costs. If European capitalists called the precious latex 'white gold', plantation workers found a different name to describe it: 'white blood' [*Woodside*, 1976: 210–11]. This expression symbolised in microcosm how workers experienced the rule of capital at the point of production. Plantations – just like the mines and industrial enterprises – served as schools for the development and evolution of workers' consciousness in colonial Indochina.[38] The experimentation with different strategies and tactics at the point of production added, incrementally, to the overall experience of the workers' movement. The class resistance of plantation *coolies* was an important building-block in forging the linkage between the anti-colonial and anti-capitalist struggles in Indochina [*Murray*, 1980: 310–11].

NOTES

1. The Vietnamese scholar, Diep Lien Anh, juxtaposes *mau trang* (literally, 'white blood', and figuratively, 'white sap' from the rubber trees) with *mau dao* (literally, 'peach blood', but figuratively, 'human blood'). Diep Lien Anh suggests that the 'white

blood' of the rubber trees comes at the human cost of the pink/red blood of workers [*Woodside*, 1976: 323]. The point of departure for this paper is Murray [1980], but I have revised many of my earlier ideas. The following archival abbreviations are used in the text:

Archives Nationales, Section d'Outre-Mer (AOM) [Paris, France];
Fonds de L'Agence de la France d'Outre-Mer (Agence FOM);
Indochine Nouveau Fonds (Indochine NF);
Fonds de la direction des Affairs Economiques, Indochine (Affairs Economiques);
Fonds de la Commission d'Enquête Parlementaire dans les Territoires d'Outre-Mer (Fonds Guernut).

2. I am indebted to Cesar Ayala for bringing many references to my attention. For an excellent review of this scholarly literature, see Ayala [1990].
3. The 'plantation school' is primarily associated with the Caribbean New World Group, particularly Lloyd Best [1968] and George Beckford [1985]. The idea of plantation 'systems' refers to not only the ensemble of economic relations constituting the production, circulation, and distribution of commodities but also the evolving institutional arrangements holding these in place.
4. In the most original formulation of the New World Group, Lloyd Best situated the 'pure plantation economy' as a sub-type of the general category of 'export-propelled economies', and, structurally, as an integral part of an 'overseas economy' linking metropole and hinterland in terms of a definite pattern of economic relationships.
5. The concept of 'plantation economy' was flexible enough to accommodate both merchantilist-era slave plantations in the Caribbean and the Americas as well as twentieth-century vertically-integrated corporate enterprises engaged in export-oriented agriculture [*Beckford*, 1985: 9, 37, 46, 61–4].
6. For a wider discussion of this issue, see Mintz [1978: 81–98], Brenner [1977], and Wallerstein [1980].
7. Elsewhere, Geertz contends that the Java sugar industry represented a mixture of 'Western capitalist enterprise with Eastern precapitalist land and labour' [*Geertz*, 1965: 46].
8. Whereas the 'plantation school' treats capitalist plantations as a distinctive sub-type of the 'pure plantation economy' because the former display the essential characteristics of the latter, the 'circulationist' approach regards all agricultural estates engaged in commodity production for profit via the market as capitalist enterprises *tout court*.
9. The scholarly literature on this topic is indeed complex. See, *inter alia*, Miles [1987] and Brass [1986; 1988; 1990]. The problem with those who define unfree labour as the loss of freedom both to sell labour-power and to withdraw from either the labour market or the labour process is that it is too narrow and restrictive. By this definition, those wage-earners, for example, who agree to 'no-strike' pledges or to long-term contracts as a condition of employment would be regarded as 'unfree' labour. On the face of it, this definition does not appear to be helpful.
10. Strictly speaking, capitalist production and accumulation presupposes, *inter alia*, capital and wage-labour. Wage-labourers are 'free' in a double sense: first, they are 'cut loose' from various bonds of servitude (whether enslaved as the personal property of another, or enserfed as an appendage of the soil), and once 'set free' from these encumbrances, they gain control over their own labouring capacity to utilise as they see fit; and second, they are separated from the means of production and of subsistence, and hence are 'free to choose' alternative means of livelihood in order to ensure their reproduction and survival. What should be obvious is that this putative 'freedom' is a two-edged sword: while direct producers (that is, workers), abstractly at least, set at least minimal conditions on the exchange of their labouring capacity in the market-place, they are compelled by the force of economic circumstances to sell their labour-power in order to obtain their means of subsistence.
11. In his investigation of plantation agriculture on Sumatra's East Coast, Breman takes a similar position. 'Labour relations', he argues, 'continued to be imbued with elements

of force, legalised or not, even when they were structured along capitalist lines' [*Breman*, 1989: xvi].

12. For the expanded industrial uses of rubber, see Harms [1975], Woodruff [1958], Coats [1987], and Knorr [1945].
13. Because it is an organic product that flourished only under specific climatic and ecological conditions, rubber in its natural state had to be sought where it could be found.
14. AOM, Agence FOM, carton 190, dossier 106. Henri Krieg, 'L'Organization du Travail dans les Exploitations Agricoles Coloniales', *Bulletin de la Société d'encourage pour l'industrie Nationale* (juin 1930).
15. By 1910, the colonial administration had generously granted an estimated 61,268 hectares in Cochinchina for rubber cultivation, but less than 5,000 hectares were planted with 1,700,000 trees. The six-year hiatus between the planting of rubber lots and the initial yields meant that entrepreneurs had to endure a relatively long gestation period before initial outlays of capital and equipment became profitable investments. They preferred instead to seek immediate returns on invested capital. While rubber production struggled to get off the ground in Cochinchina, French capitalists invested an estimated 100 million francs in existing Malayan rubber plantations [*Robequain*, 1944: 201–5; *Thompson*, 1968: 130–33; *Montaigut*, 1929: 1–20].
16. See Carle [1912], Carton [1924b], and AOM, Agence FOM, carton 190, dossier 106, Congrès d'Agriculture Colonial, 'Le Caoutchouc de Plantations en Indochine [1917]'.
17. Between 1910 and 1927, the colonial administration provided nearly 180,000 hectares in free land concessions to metropolitan companies for agricultural development, mainly for rubber cultivation. By the end of 1927, the colonial administration had recorded applications for an additional 220,000 hectares, of which metropolitan concessionary companies had applied for 180,000, and the giant rubber company, *Michelin*, had requested 60,000 hectares alone. In 1928, the colonial administration issued a decree cancelling previous restrictions on the size of tracts that could be given away in free concessions. By a stroke of the pen, the colonial authorities endorsed large-scale colonisation of eastern Cochinchina and the virtually unrestrained growth of plantation agriculture, thereby unleashing one of the most concentrated movements of capital investment in French colonial history [*Schweitzer*, 1971: 509].
18. In Indochina, only the uninterrupted flow of generous subsidies from the French Treasury enabled large-scale plantations to weather the world economic crisis [*Thery*, 1931: 140; *Bourbon*, 1938: 97; *Thompson*, 1968: 132].
19. The IRRA was a treaty between five signatory governments (Great Britain, the Netherlands, France, India, and Thailand), covering British Malaya, Ceylon, British India, Burma, North Borneo, Srawak, the Netherlands Indies, Thailand, and French Indochina [*Knorr*, 1945: 107–9].
20. By the late 1930s, rubber had become Indochina's second most valuable export behind rice. In terms of value, it constituted 18 per cent of all colonial exports [*Bourbon*, 1938: 115–17; *Reny*, 1938: 179–81; *Schweitzer*, 1971: 528].
21. AOM, Affairs Economiques Indochine, carton 59, dossier 7. Les Cahiers Coloniaux, février, 1944, Institut Colonial de Marseille, *Le Caoutchouc*, pp.272–91. Some of these metropolitan concessionary companies operated other types of plantations, including coffee and tea estates, and also nonagricultural enterprises in Indochina and elsewhere [*Robequain*, 1944: 207–8]. These financial groups with considerable capital investments in rubber plantations were also closely linked to the most powerful French banking houses [*Montaigut*, 1929: 99].
22. AOM, Affairs Economique Indochine, carton 59, dossier 7, for production figures and the state of rubber output in 1940.
23. AOM, Agence FOM, carton 190, dossier 106. Rapport de Inspecteur des Colonies Rheinhart; sur la Cochinchine, 2 décembre 1920: Caoutchouc.
24. For example, the province of Giadinh near Saigon had a population density of 160 persons per square kilometer, whereas the population density was less than 22 persons per square kilometer in the province of Thudamot and 15 per square kilometer in Bien

Hoa, two major outlying rubber-producing areas [*Robequain*, 1944: 213; *Vasil'eva*, 1947: 104].

25. For labour competition, see AOM, Agence FOM, carton 167 dossier 19. Communique No. 62 de la presse Indochinoise, 31 août 1923.
26. Some provinces in Tonkin registered over 400 inhabitants per square kilometer. Wages in Cochinchina were 40 to 60 cents *per diem* compared with 25 to 35 cents in Tonkin.
27. The rubber plantations worked hand-in-glove with private recruiting agencies which, in turn, dispersed their Vietnamese recruiting agents (called *cais*) throughout the principal source areas. These *cais*, who received a capitation fee of 16 piastres per *coolie*, often bribed village notables to identify debt-ridden villagers vulnerable to recruitment. During the early years of the rubber boom, recruiting agents almost always used lying, deceit, false promises, and even intimidation and coercion to persuade reluctant villagers to agree to work on the southern plantations [*Monet*, 1930: 40–41; *Woodside*, 1976: 211; *Tran Tu Binh*, 1985; *Murray*, 1980: 253].
28. See AOM, Agence FOM, carton 190, dossier 206, for a description of recruitment practices.
29. AOM, Agence FOM, carton 168, dossier 22. 'La Régime et l'Organization du Travail des Indigènes en Indochinc'. Rapport par H. Sambuc, juin 1929.
30. AOM, Agence FOM, carton 167, dossier 19. Immigration: Arrêté concernant l'émigration de la main d'oeuvre Tonkinoise, 25 octobre 1927.
31. AOM, Affaires Economiques, carton 96 dossier 3. *Rapport sur une visite des plantations Michelin*, L'Inspecteur Général du Travail, Dalamarre, Saigon, 17 juillet 1928, p.10.
32. 'Report of Mr. Delamarre' [cited in Gouldal, *Labour Conditions*, pp.307–8]; AOM, Affairs Economiques Indochine, carton 96, dossier 3. *Rapport sur une visite des plantations Michelin*. L'Inspecteur Général du Travail, Delamarre. Saigon. 17 juillet 1928, p.14.
33. 'Report of Mr. Delamarre' [cited in Goudal, 1937: 302–309]; AOM, Indochine NF, 2404. Télégramme Officiel, No. 1215, 2 novembre 1936. Women workers were also subjected to this brutal regime of terror and violence. They suffered the same indignities, maltreatment, and corporal punishments as their male counterparts. In addition, it was not uncommon for resident *cais* and their European superiors to sexually assault women workers [*Tran Tu Binh*, 1985: 24–5; *Ngo Vinh Long*, 1974]. Brocheux [1975: 68] uses polite phraseology to describe these asaults on women workers: 'at times [women] were used to satisfy the needs and wishes of the overseers'. Tran Tu Binh [1985: 61–3] reports that French managers also sexually abused male *coolies* as well.
34. See Murray [1980: 297 (Table 6–4)] for even higher figures.
35. AOM, Agence FOM, carton 168, dossier 22. *Le probléme de la main-d'oeuvre en Indochinc. Les monographes de la Dépêche Coloniale*.
36. AOM, Fonds Guernut Bd. *Xtrait Revue Colonial Médecine*, 15 mai 1937. Dr. Cambandon, *médecin des plantations*. This report provides figures on desertion rates at the *Minot* and *Kantroy* plantations, 1927–31.
37. See, for example, AOM, NF Indochine 2404. *Rapport par M. Erard, Le Chef du 1er Bureau de la Direction des Affairs Economique et Administratives*, 28 août 1936.
38. This observation is, of course, not to suggest that *coolies* did not acquire elements of all sorts of different kinds of consciousness as a result of working on the rubber plantations. It also does not imply that the material conditions of capitalist exploitation necessarily engender socialist consciousness, or that workers' consciousness necessarily leads to socialist consciousness.

REFERENCES

Andrus, J. Russell, 1943, *Preliminary Survey of the Economy of French Indochina*, Washington, DC: United States Department of State, Division of Economic Studies, Far Eastern Unit, Bureau of Foreign and Domestic Commerce.

Ayala, Cesar, 1990, 'Industrial Oligopoly and Vertical Integration: The Origins of the American Sugar Kingdom in the Caribbean, 1881–1921', Unpublished Ph.D. dissertation, State University of New York at Binghamton.

Bauer, P.T., 1948, *The Rubber Industry: A Study in Competition and Monopoly*, Cambridge, MA: Harvard University Press.

Beckford, George, 1972, *Persistent Poverty: Underdevelopment in Plantation Economics of the Third World*, London: Oxford University Press.

Beckford, George, 1985, 'Caribbean Peasantry in the Confines of the Plantation Mode of Production', *International Social Science Journal*, Vol.37, No.3, pp.401–14.

Benn, Denis, 1974, 'The Theory of Plantation Economy and Society: A Methodological Critique', *Journal of Commonwealth and Compative Politics*, Vol.12, No.3, pp.249–51.

Best, Lloyd, 1968, 'The Mechanisms of Plantation-Type Economies: Outlines of a Model of Pure Plantation Economy', *Social and Economic Studies*, Vol.17, No.3, pp.283–326.

Bourbon, André, 1938, *Le Redressement Economique de l'Indochine, 1934–1937*, Lyon: Bose Frers M. and I. Rious.

Brass, Tom, 1986, 'Unfree Labour and Capitalist Restructuring in the Agrarian Sector: Peru and India', *Journal of Peasant Studies*, Vol.14, No.1.

Brass, Tom, 1988, 'Slavery Now: Unfree labour and Modern Capitalism', *Slavery & Abolition*, Vol.9, No.2.

Brass, Tom, 1990, 'The Latin American *Enganche* System: Some Revisionist Reinterpretations Revisited', *Slavery & Abolition*, Vol.11, No.1.

Breman, Jan, 1989, *Taming the Coolie Beast: Plantation Society and the Colonial Order in Southeast Asia*, Delhi: Oxford University Press.

Brenner, Robert, 1977, 'The Origins of Capitalist Development: A Critique of Neo-Smithian Marxism', *New Left Review*, No.104, 25–93.

Brocheux, Pierre, 1975, 'Le Prolétariat des plantations d'hévéas au Vietnam méridional: aspects sociaux et politiques (1927–1937)', *Le Mouvement Social* No.90 (janvier–Mars), pp.55–86.

Brocheux, Pierre, 1983, 'Moral Economy or Political Economy? The Peasants are always Rational', *Journal of Asian Studies* Vol.42, No.4, pp.791–803.

Camus, Jean-Jacques, 1949, *Les Plantations de Caoutchouc en Indochine: l'oeuvre humaine et sociale*, Paris: Bibliotèque d'Afrique et L'Outre-Mer.

Carle, Emile, 1912, *L'hévéa brasiliensis en Indochine*, Saigon: Portail.

Carton, Pierre, 1924a, 'Le Caoutchouc en Indochine', *Bulletin Economique de l'Indochine*, pp.450–56.

Carton, Pierre, 1924b, *Le caoutchouc en Indochine*, Hanoi: Imprimerie d'Extrême-Orient.

Christy, Cuthbert, 1911, *The African Rubber Industry and Funtumia Elastica*, London: J. Bale, Sons & Danielson.

Coates, Austin, 1987, *The Commerce in Rubber: The First 250 Years*, Singapore: Oxford University Press.

Durtain, Luc, 1930, *Dieux Blancs, Hommens Jaunes*, Paris: Ernest Flammarion.

Florescano, Enrique (ed.), 1975, *Haciendas, Latifundios y Plantaciones en America Latina*, Mexico: Siglo Veintiuno Editores.

Fraginals, Manuel Moreno, Frank Moya Pons, and Stanley Engerman (eds.), 1985, *Between Slavery and Free Labor: The Spanish-Speaking Caribbean in the Nineteenth Century*, Baltimore, MD and London: Johns Hopkins University Press.

Geertz, Clifford, 1963, *Agricultural Involution: The Process of Ecological Change in Indonesia*, Berkeley and Los Angeles, CA: University of California Press.

Geertz, Clifford, 1965, *The Social History of a Javanese Town*, Cambridge, MA: M.I.T. Press.

Gérard Emile, 1924. *Les Plantations d'Hévéas en Indochine*, Paris: Société d'Editions Maritimes et Coloniales.

Girard, E., 1924, *Académie des Sciences Coloniales, Extrait des Comptes rendues du Tome II. Les Plantations d'Hévéas en Cochinchine*, Paris: Augustin Challamel.

Goodyear, James, 1978, 'Recent Trends in Plantation Historiography', *The New Scholar*, Vol.7, Nos.1–2, pp.291–96.

Gordon, Alec, 1979, 'The Collapse of Java's Colonial Sugar System and the Breakdown of Independent Indonesia's Economy', in F. van Anrooij, *et al.*, *Between People and Statistics: Essays on Modern Indonesia's History*, The Hague: Martinus Nijhoff.

Gordon, Alec, 1982, 'Indonesia, Plantations and the "Post-Colonial" Mode of Production', *Journal of Contemporary Asia*, Vol.12, No.2, pp.168–75.

Goudal, Jean, 1938, *Labour Conditions in Indo-China*, Geneva: League of Nations, International Labour Office, Studies and Reports, Series B, No.26.

Gourou, Pierre, 1955, *Peasants of the Tonkin Delta, Volume I.* [Translated by Richard Miller]. New Haven, CT: Human Relations Area Files.

Harms, Robert, 1975, 'The End of Red Rubber: A Reassessment', *Journal of African History*, Vol.16, No.1, pp.73–88.

Hemery, Daniel, 1975, *Révolutionnaires Vietnamines et pouvoir colonial en Indochine: Communistes, Trotskytes, nationalistes a Saigon de 1932 à 1937*, Paris: François Maspero.

Henry, Yves, 1906, *Le caoutchouc dans l'Afrique occidentale française*, Paris: A. Challamel.

Hickey, Gerald, 1958, 'Social Systems of Northern Vietnam: A Study of System in Contact', Unpublished Ph.D. dissertation, University of Chicago.

Hunt, Shane, 1975, 'La Economia de Las Haciendas y Plantaciones en America Latina', *Historia y Cultura* [Peru], Vol.9, pp.7–66.

Isoart, Paul, 1961, *Le Phénomène National Vietnamien: De l'Indépendance unitair à l'Indépendance Fractionée*, Paris: Librairie Générale de Droit et de Jurisprudence.

Institut français du caoutchouc (Paris), 1945, *Le Caoutchouc d'hévéa: Initiation aux méthodes d'exploitation en Indochine*, Paris: l'Institut.

Knight, G.R., 1980, 'From Plantations to Padi-Field: The Origins of the Nineteenth Century Transformation of Java's Sugar Industry', *Modern Asian Studies*, Vol.14, No.2, pp.177–204.

Knight, G.R. 1988, 'Peasant Labour and Capitalist Production in Late Colonial Indonesia: the "Campaign" at a North Java Sugar Factory, 1840–1870', *Journal of Southeast Asian Studies*, Vol.19, No.2, pp.245–65.

Knight, G.R., 1989, 'Sugar, Peasants, and Proletarians: Colonial Southeastern Asia, 1830–1940', *Critique of Anthropology*, Vol.9, No.2, pp.39–63.

Knorr, K.E., 1945, *World Rubber and its Regulation*, Stanford University, California, CA: Stanford University Press.

Landes, David, 1969, *The Unbound Prometheus: Technological Change and Industrial Development in Western Europe from 1750 to the Present*, Cambridge: Cambridge University Press.

Lasker, Bruno, 1972, *Human Bondage in Southeast Asia*, Wesport, CT: Greenwood Press.

Le Fèvre, George, 1927, *Démolisseurs et Batisseurs*, Paris: André Délpeuch.

Mandle, Jay, 1972, 'The Plantation Economy: An Essay in Definition', *Science & Society*, Vol.34, No.1, pp.57–60.

Mandle, Jay, 1973, *The Plantation Economy: Population and Economic Change in Guyana, 1838–1960*, Philadelphia, PA: Temple University Press.

Mandle, Jay, 1982, *Patterns of Caribbean Development: An Interpretive Essay on Economic Change*, New York: Gordon and Breach Science Publishers.

Mann, Susan, 1987, 'The Rise of Wage Labour in the Cotton South: A Global Analysis', *Journal of Peasant Studies*, Vol.14, No.2, pp.276–242.

Marx, Karl, 1973, *Grundrisse: Foundations of the Critique of Political Economy*, (trans.) Martin Nicolaus; Harmondsworth, Middlesex, England: Penguin Books.

Miles, Robert, 1987, *Capitalism and Unfree Labour: Anomaly or Necessity?*, London and

New York: Tavistock.
Mintz, Sidney, 1978, 'Was the Plantation Slave a Proletarian?', *Review*, Vol.2, pp.81–98.
Monet, Paul, 1930, *Les Jauniers: Histoire Vraie*, Paris: Libraire Gallimard.
Montaigut, Fernand de, 1929, *La Colonisation française dans l'est de la Cochinchine*, Limoges: Imprimerie Commerciale Pérrette.
Murray, Martin J., 1980, *The Development of Capitalism in Colonial Indochina, 1870–1940*, Berkeley, Los Angeles, CA, and London: University of California Press.
Murray, Martin J., and Charles Post, 1983a, 'The "Agrarian Question", Class Struggle and the Capitalist State in the United States and South Africa', *Insurgent Sociologist* Vol.11, No.4, pp.37–56.
Murray, Martin, and Charles Post, 1983b, 'Reply to Greenberg', *Insurgent Sociologist*, Vol.11, No.4, pp.61–2.
Ngo Vinh Long, 1973, *Before the Revolution: The Vietnamese Peasants under the French*, Cambridge, MA: MIT Press.
Ngo Vinh Long, 1974, *Vietnamese Women in Society and Revolution*, Cambridge, Massacusetts: Vietnam Resource Center.
Parpart, Jane, 1983, *Labor and Capital on the African Copperbelt*, Philadelphia, PA: Temple University Press.
Phimister, Ian and Charles van Onselen, 1978, *Studies in the History of African Mine Labour in Colonial Zimbabwe*, Gwelo, Zimbabwe: Mambo Press.
Pommery, Louis, 1952, *Aperçu d'Histoire Economique Contemporaine 1890–1939: Tome 1er*, Paris: Librairie de Medicis.
Rény, Paul, 1938, *Le Problème des Relations Commerciales entre l'Indochine et France*, Nancy: Imprimerie Georges Thomas.
Robequain, Charles, 1944, *The Economic Development of French Indo-China*, trans. Isabel Ward; London and New York: Oxford University Press.
Schweitzer, Thomas, 1971, 'The Economic Foundations of Imperialism' The French Colonialist Lobby in the 1930s:, Unpublished Ph.D dissertation, University of Wisconsin at Madison.
Scott, C.D., 1976, 'Peasants, Proletarianisation, and the Articulation of Modes of Production: The Case of Sugar-Cane Cutters in Northern Peru, 1940–1969', *Journal of Peasant Studies*, Vol.3, No.3.
Scott, James, 1985, *Weapons of the Weak: Everyday Forms of Peasant Resistance*, New Haven, CT and London: Yale University Press.
Spire, R. and A. Spire, 1905, *Le Caoutchouc en Indochine: Etude botanique, industrielle, et commerciale* (Paris: Challamel.
Stolcke, Verena and Michael Hall, 1983, 'The Introduction of Free Labour on Sao Paulo Coffee Plantations', *Journal of Peasant Studies*, Vol.10, Nos.2–3, pp.170–200.
Stoler, Ann, 1985, *Capitalism and Confrontation in Sumatra's Plantation Belt, 1870–1979*, New Haven, CT and London: Yale University Press.
Thery, René, 1931, *L'Indochine Française*, Paris: Les Editions Pittoresques.
Thompson, Virginia, 1968, *French Indo-China*, New York: Octagon Books.
Tran Tu Binh, 1985, *The Red Earth: A Vietnamese Memoir of Life on a Colonial Rubber Plantation*, Athens, Ohio: Ohio University Center for International Studies Monographs in International Studies, Center for Southeast Asian Studies, No.66.
Truong Chinh, and Vo Nguyen Giap, 1974, *The Peasant Question, 1937–1938*. [Translated by Christine Pelzer White]. Data Paper Number 94, Southeast Asia Program, Department of Asian Studies, Cornell University, Ithaca, New York.
Turrell, Robert Vicat, 1987, *Capital and Labour on the Kimberley Diamond Fields, 1871–1890*, Cambridge: Cambridge University Press.
van Onselen, Charles, 1976, *Chibaro: African Mine Labour in Southern Rhodesia, 1900–1933*, London: Pluto Press.
Vasil'eva, V. la, 1947, *Indochina*, Moscow and Leningrad: Academy of Sciences of the USSR, Institute of the World Economy and of World Politics.
Wallerstein, Immanuel, 1980, *The Capitalist World Economy*, London: Cambridge University Press.

Weinstein, Barbara, 1983, *The Amazon Rubber Boom, 1850–1920*, Palo Alto, CA: Stanford University Press.

White, Christine Pelzer, 1986, 'Everyday Resistance, Socialist Revolution and Rural Development: The Vietnamese Case', *Journal of Peasant Studies*, Vol.13, No.2, pp.49–63.

Wolf, Eric and Sidney Mintz, 1957, 'Haciendas and Plantations in Middle America and the Antilles', *Social and Economic Studies*, Vol.6, No.3, pp.380–412.

Woodruff, William, 1958, *The Rise of the British Rubber Industry during the Nineteenth Century*, Liverpool: Liverpool University Press.

Woodside, Alexander, 1976, *Community and Revolution in Modern Vietnam*, Boston, MA: Houghton Mifflin.

Worger, William, 1987, *South Africa's City of Diamonds: Mine Workers and Monopoly Capitalism in Kimberley, 1867–1895*, New Haven, CT and London: Yale University Press.

The Java Sugar Industry as a Capitalist Plantation: A Reappraisal

G.R. KNIGHT

INTRODUCTION: THE JAVA SUGAR INDUSTRY 1830–1930

The 'modern' Java sugar industry was one of the earliest-founded of the large agri-businesses run by westerners in colonial Asia. Sugar had been produced on a commercial scale in Java since the seventeenth century or earlier. By the 1820s, however, the island's industry was largely moribund. It was only in the following decades that it was revitalised under the direction of the Netherlands Indies Government, which injected capital into the industry and took steps to assure the colony's sugar manufacturers of adequate supplies of both raw material and labour. During the period of the so-called Cultivation System (1830–*c.*1880) the foundations were thus laid for the emergence of the Java industry at the forefront of the world sugar economy. At the zenith of its colonial-era development, in the 1920s, it was second only to Cuba as a supplier of cane sugar to world markets, and outstripped virtually all its rivals in terms of productivity. Moreover, although the onset of the Great Depression at the end of the decade resulted in a big fall-off in production and the closing down of many factories, it did not destroy the industry. It remained, at outbreak of the Second World War, one of the world's major producer-exporters of cane sugar, with an annual output of around one million tons.

Immediately prior to its partial collapse in the 1930s, the industry had boasted some 180 large, highly mechanised and heavily capitalised sugar factories. It engrossed more land and employed more labour than any other of Java's plantation industries. Fields planted to cane each year nominally amounted to around 200,000 hectares. In fact, since cane took between 15 and 18 months to mature, and was replanted annually, the area under sugar at any one time (or being held in preparation for cane planting) was often greatly in excess of this figure. The industry's labour

G.R. Knight, Department of History, University of Adelaide, Box 498, GPO, Adelaide, South Australia 5001, Australia. The article has benefited from comments and suggestions by the editors.

force was by far the largest employed in any plantation industry in colonial Java. It counted nearly 90,000 permanent workers and approached one million at the height of the Campaign – the five to six-month harvesting and manufacturing season which began in May each year. The sugar industry's capital – nearly all of it by the close of the nineteenth century in the hands of Dutch companies based in metropolitan Holland – amounted by the 1920s to some 325 million guilders. This represented around 25 per cent of the total overseas capital invested at that time in plantation enterprises in the Netherlands Indies, and yielded dividends surpassed in the colony as a whole only by the Sumatra tobacco companies [*Boomgaard*, 1988: 158; *Gordon*, 1979:252; *Van der Mandere*, 1928: 74; *Allen and Donnithorne*, 1957: 83].

The present study focuses on a particular sector of the Java sugar industry, that which developed in the adjoining (and after 1900 combined) north-coast Residencies of Pekalongan and Tegal during the nineteenth and early twentieth centuries. Sugar had been produced in the area for some time prior to 1800. It was only in the 1830s, however, among the rice-fields and villages of the narrow, mountain-backed coastal plain which formed the agrarian heartlands of both Residencies, that the industry began to play more than a very minor role in the local economy.

Prior to the inauguration of the Cultivation System by the Netherlands Indies Government in 1830, production of sugar in Pekalongan-Tegal had been organised by a few *peranakan* (locally-settled) Chinese entrepreneurs working in various kinds of relationship with Javanese state officials and the local Dutch Resident. Output had been insignificant and the units of production both small and primitive. After 1830, things were very different. Rapidly in Tegal, at a slower pace in Pekalongan, sugar developed into the region's major export-oriented industry. By the mid-century there were already nearly a dozen moderately well-equipped, large-scale sugar factories, owned exclusively by westerners, operating in the Residencies. Subsequent growth was considerable, so that well before the peak years of production in the 1920s – by which date the Pekalogan-Tegal sugar industry accounted for about ten percent of the colony's output – sugar factories and their attendant cane-fields had become an ubiquitous feature of Residency's lowlands.

The number of sugar factories had itself increased, so that by 1920 there were 18 of them. More significant, however, was the expansion of the area under cane which took place in the Residency between the establishment of the industry in the 1830s and the crisis which threatened to overwhelm it almost exactly a hundred years later. Whereas in the mid-nineteenth century approximately 3,600 hectares of land was planted to cane in Pekalongan-Tegal, by the eve of the Great Depression the figure had risen to nearly 18,000. Exploitation of land, moreover, had been greatly intensified. Productivity in the Pekalongan-Tegal industry – measured in terms of the amount of sugar produced from each hectare of cane – increased almost sevenfold between 1850 and 1929.

THE JAVA SUGAR INDUSTRY AND PLANTATION DEVELOPMENT IN COLONIAL ASIA

The Java sugar industry has often been ascribed a unique place in the discourse generated by plantation development in colonial Asia. Indeed, it has sometimes been denied that attribution altogether [*Courtenay*, 1980: 73]. The single most critical issue relates to the industry's ostensibly 'uncharacteristic' relations to the production factors of land and labour. For the most part, Java's colonial sugar manufacturers, from the 1830s onward, did not conduct their operations on land either owned or leased longterm by the 'plantation' companies. Neither did they, as did many of their counterparts elsewhere in more sparsely populated regions of Asia, exploit a workforce which was predominantly composed of (indentured) migrants. Having been re-established after 1830 in relatively densely settled and agriculturally well-developed parts of the island, the Java sugar industry was able to draw – ostensibly at least – on the local population for its workforce and for its raw material on peasant farmland (*sawah*), on which cane was grown on a rotational basis with rice and other 'peasant' crops.

In consequence, the prevailing image of the Java industry is one of 'sugar-with-peasants'.[1] That is to say, the industry has been conceptualised, first and foremost, in terms of the exploitation but not expropriation of the Javanese peasantry with which, in the years after 1830, it came into increasingly close and sustained contact. Integral to this delineation of the industry, moreover, has been a parallel emphasis on the role played by various forms of 'extra-economic' pressure, mediated through the agency of the colonial state, in assuring the industry of access to the resources of rural Java.

On this analysis, the very persistence of peasantry, though it guaranteed 'cheap' labour by obviating the need to meet the full reproduction costs of the industry's workforce, meant that the hands which the industry required were unlikely to be available in sufficient numbers or on a reliably regular basis without recourse to some form of 'compulsion'. It might well have been the case that increasing numbers of peasants no longer had land of their own or else had holdings totally insufficient to support themselves and their households. None the less, such people could not freely dispose of their labour, and were directed toward the sugar industry through informal 'bonding' and through the intervention of village headmen and other rural officials.

A similar analysis is also brought to bear on the question of the industry's access to land. 'Extra-economic' coercion is viewed as paramount in the process by which the sugar factories secured the rental of peasant *sawah* in competition for land-use from other crops which were potentially more profitable to the peasantry. The state, directly (as well as through its fiscal policies) brought pressure to bear on peasant landholders to rent their land to the sugar companies. At the same time, village headmen, suborned by factory managers, played a crucial role in securing the compliance of the mass of petty landholders in the industry's rental of their land at 'non-economic' rates. [*Schaik*, 1986: 82–6; 107–9; *Gordon*, 1982: 169–79; *Breman*, 1983: 5–36; *Soetrisno*, 1980: 84–109; *Mubyarto*, 1969: 37–59; *Selosoemardjan*, 1962: 271–84].

Perhaps somewhat paradoxically, therefore, any attempt at a re-appraisal of the Java sugar industry as a capitalist plantation has to contend (a) with an image of the industry as *sui generis* and largely outside the mainstream of capitalist plantation development in colonial Asia and (b) with an insistence that it shared with the rest of Asia a specific, hybrid and 'colonial capitalism', sharply differentiated from the fully-developed capitalism of the west by virtue of the key role played in the production process by 'extra-economic' coercion.[2]

RE-APPRAISAL (1): THE JAVA SUGAR INDUSTRY AND THE 'PERSISTENCE' OF PEASANTRY

At the core of the argument about Java's colonial sugar industry outlined above is a reading of its history in terms of a consolidated peasant proprietorship of the soil uniquely harnessed to the requirements of expanded production for the world market.

Closer examination reveals the problems of such an interpretation. One obvious pitfall is that the exploitation by the sugar industry of a landholding peasantry rather than its (formal) expropriation was hardly the singular development which the more insular contributors to the debate over plantation development in Java seem to assume. Indigo production in nineteenth century Bengal, for example, exhibited similar characteristics [*Chowdhury*, 1964: 134]. More importantly, the concept of a 'persisting' peasantry, and the broadly 'articulationist' perspectives which it embodies, serves to obscure rather than illuminate the relationship between colonial plantation industries and rural land and labour [e.g. *Knight*, 1989: 39–63; *Aguilar*, 1989: 41–62].

Central to the problem is the implication of the 'sugar-and-peasants' thesis that the labour process in the Java sugar industry was both fundamentally monolithic and remained substantially unaltered from the inauguration of the Cultivation System through to the eve of the Great Depression some hundred yeas later. In reality, the contrary was the case: crucial factors were the complexity of the labour process and the relation of distinctive labour processes to specific periods in the history of the industry's development.

Peasant Landholders, Petty Commodity Production and the Cultivation of Cane

In some parts of Asia, notably Taiwan, the expansion of sugar production for world markets in the late colonial era was – and remained – heavily dependent on cane grown, on their own holdings, by a myriad of peasant cultivators [*Ka*, 1987: 128–84]. In the case of the Java industry, on the other hand, any such nexus between petty commodity production and the sugar industry was always more apparent than real. Indeed, by the later stages of the industry's expansion it had become merely vestigial.

Whatever linkage existed in Java between the production of sugar for world markets and a 'class' of peasant landholding-cultivators was at its most apparent during the era of the Cultivation System. This state-organised system of compulsory planting by the peasantry of sugar-cane, indigo, coffee and other export-crops (the so-called Government Cultivations) was inaugurated in 1830 and persisted, in the case of sugar, until the 1880s. Under its aegis, Java's landholding peasants were required to grow cane for the sugar factories owned by (predominantly and eventually exclusively) European contractor-manufacturers in the vicinity of their villages. Cultivation took place on farmland, rotationally requisitioned for that purpose from the peasantry, under the direction and supervision of Javanese and Dutch state officials.

An important consequence of this was that the nexus between peasantry and landholding was consolidated rather than undermined in mid-nineteenth century Java, as state officials in charge of organising cane production sought to reinforce existing ties between cultivators and the soil. The administrative requirements of a fiscal system based on landholding (the so-called Landrent originally introduced in 1813) pointed in this direction as did, increasingly, attempts to assign both responsibility and 'rewards' for work in the Government Cultivations. In the Pekalongan-Tegal Residency itself, for example, *c.*1850, what Dutch colonial officials defined as the land-holding 'cultivator-household' was formalised as the basic unit of cane production after the failure of earlier attempts at apparently indiscriminate forced recruitment *en masse* of rural labour for work in the sugar fields.[3]

Post-1870, of course, there were major organisational changes in the sugar industry, as compulsory cultivation, at the direct behest of the state, was gradually phased out. Private sugar companies, operating on land which was rented rather than requisitioned, took over the management of cane production from the hands of government officials. At least initially, however, these changes coexisted with the continuing exploitation of the landed peasant household as the basic unit of cultivation. It was not until the early twentieth century (see below) that significant new developments in the sugar industry's labour processes became clearly apparent.

The ostensibly close and enduring nexus between the peasant 'family farm' and the cultivation of sugar cane was none the less illusory in a number of fundamental respects. Although the *labour* of the landholding peasant household may well have formed the basic unit of cane production during both the Cultivation System and its immediate aftermath, the peasant 'farm' as such rarely if ever did so. At the very least, the landholdings of individual households were amalgamated into village units (and their physical profile subject to radical alteration) for purposes of cane cultivation. At the same time, essential cultivation decisions were taken out of the peasantry's hands and responsibility for them assumed by state officials.

Developments of this kind were very apparent in the Pekalongan-Tegal industry at an early stage in its post-1830 history. Fields requisitioned for cane cultivation were merged into very large, consolidated plantations. In the 1850s, for instance, at SF Wonopringo, the 500 hectares of land planted to cane in the factory's proximity was divided up into no more than nine huge fields, the largest covering more than 85 hectares.[4] By way of comparison, the average peasant holding of arable land in Java during the mid-century decades was a little over one hectare [*Boomgaard*, 1989: 75, 85]. Huge cane 'fields' continued to characterise at least some sectors of the Pekalongan-Tegal sugar industry in the opening decades of the twentieth century. In the Residency's eastern districts, in particular, the sugar factories rented-in the entire *sawah* of whole villages 'en bloc', amalgamated it with that of its neighbours and planted it to cane.[5] The peasant farm was obliterated in the sugar industry's production process.

By the early twentieth century, moreover, the real as opposed to the ostensible nexus between the sugar industry and the farming activities of peasant landholders had been further eroded by the extensive use of 'outside' labour in all aspects of the cultivation process. In some parts of the colony at least – Pekalongan-Tegal was a notable example – land which was rented-in from peasant landholders was increasingly worked by labourers from further (and sometimes quite far) afield. This change in the nature of the labour process is further elaborated below.

The Expropriation Question

As we have seen, a crucial point in the argument for the uniqueness – and at the same time for the 'hybrid' capitalist character – of the Java sugar industry is that the landholding peasantry remained largely unexpropriated in the process of the expansion of capitalist production. The broad contention in this respect, is that in consequence it was necessary for the sugar industry to be sustained against competing 'peasant' claims to land use (for other, more profitable crops or small-holder cultivation of cane itself) by the active intervention of the colonial state [e.g. *Breman*, 1983: 31; *Gordon*, 1979: 251–3].

There is, obviously, much that is plausible about this and similar arguments. It could be maintained, for example, that the Java sugar industry was rescued from extinction in the mid-nineteenth century by the colonial government's sustained effort, over several decades, to compel the peasantry to devote large areas of good farmland to cane, regardless of the opportunity costs. Likewise, it is scarcely irrelevant to the industry's later history that it inherited the Cultivation System's mantle of profound involvement of the state in the provision of its raw material.

None the less, the contention that the Java industry's access to land for cane cultivation was determined by 'extra-economic' coercion rather than through the working of 'free market forces' oversimplifies the complex and contentious issue of 'opportunity-costs' in relation to the sugar company's land-rental. It also seriously underplays the way in which increasing differentiation in rural Javanese society was utilised by the industry in its quest for land.

Indebtedness, 'Substantial' Peasants and the Sugar Industry's Access to Land

With the phasing-out of state-organised production of cane in the 1880s, relations between peasantry and industry entered a new stage in which the sugar companies themselves entered into direct agreements with peasant landholders for the periodic rental of fields. Despite the notionally 'voluntary' nature of these agreements, contemporary critics of the industry frequently asserted that the companies paid rentals which were too low to be 'economic' [e.g. *Schaik*, 1986: 85–6]. Rentals were presumed, therefore, to be the result of some form of 'compulsion' exerted on the cultivators by state officials and village headmen. It was common knowledge at the time, for instance, that headmen received a premium from the factories for each hectare of land rented-out from 'their' village.

Yet too great an insistence on the factor of 'compulsion' overestimates the power of village heads and state officials generally – a particularly important consideration in notoriously 'unruly' areas like the eastern districts of Pekalongan-Tegal. It also underplays the role of other factors, notably the way in which high levels of rural indebtedness (consequent in part on the tax burden imposed by the state) greatly facilitated the sugar industry's ability to secure and sustain rental agreements with the peasantry. In such circumstances, it is not sufficient simply to demonstrate that cultivators would have received more profit from crops other than cane and that, in consequence, they must have been 'compelled' to rent out land by pressure exerted through the state or its agents. On the contrary, land-rental payments from the sugar companies might well fulfil a 'cash-raising and debt- servicing' function – analogous to that existing among the peasant cane growers of northern India – in which simple calculations of profit and loss became largely irrelevant [*Amin*, 1984: 70–72]. As Elson remarks of the indebted peasant landholders of turn-of-the-century east Java,

> they were heavily reliant not just on the income they received from the industry, but also on the fact that it came in a lump sum which could be used to meet their large-scale budgetary commitments. This helps explain why, even with diminished returns, peasants steadfastly and regularly continued to yield up their land to sugar manufacturers [*Elson*, 1979: 352].

The implications of the indebtedness – and probably growing indebtedness – of the mass of Java's petty landholding peasants in the opening decades of the twentieth century, went a good deal further, moreover, than predisposing them to take whatever ready cash the sugar companies offered. It also underlay an increasing process of socio-economic differentiation in a rural society which was, in any event, historically far from homogeneous.

The presence in the countryside of early twentieth century Java of 'substantial' peasants with landholdings well in excess of those of the mass of the landed cultivators has been widely recognised (although there is considerable disagreement as to whether they should be categorised as 'mere' rentiers or as 'proto-capitalists' in what remains, as yet, a fairly rudimentary debate). Likewise, there are some indications of the process – in part at least through absorption of the lands of marginal peasants – through which substantial peasants were consolidating their existence in the opening decades of the present century [*Anon*, 1905: 1079; *Geertz*, 1965: 40–41; *Elson*, 1979: 345–6, 431–5; *Elson*, 1984: 218–19; *Husken*, 1989: 305; *Breman*, 1983: 81–2]. Substantial peasants such as these, whom the colonial Dutch styled *grootgrondbezitters* [lit. large landowners], were as likely to be renters of land from impoverished and debt-ridden fellow villagers as to be landowners in the formal sense. They came to constitute a crucial factor in the sugar industry's access to land for rental.

In Pekalongan-Tegal in the early decades of the twentieth century, they were particularly prominent in the Residency's western districts. At SF Soemberhardjo in 1918, for example, the factory's arrangements for land rental were greatly complicated (so the Administrator explained to his employers) by the fact that many of the *areaal's* large peasant landholders were absentees, resulting in delays in the signing of contracts. At SF Bandjaratma, a couple of years later, similar landholders were refusing to rent their land to the factory and thereby threatening to disrupt the smooth organisation of cane production.[6]

The exact ways in which their relationship with the sugar factories, which was certainly one of considerable tension, profited peasants of this kind needs a great deal more investigation. One fundamental point, however, appears to relate to the differential between the price they paid to rent-in land from petty and marginal cultivators and the price which they extracted from the sugar companies for renting it to them [*Suiker-Enquete*, 1921: 171–3; *OMW Pekalongan*, 1908: 80]. Since they might also often be in receipt of direct loans from the factories, their transactions with the industry provided them, on several possible counts, with 'cheap' money which bolstered their position as major creditors within the village.[7]

Several conclusions can be tentatively stated. Contrary to the thrust of 'articulationist' arguments, there took place in late colonial Java a marked degree of *de facto* expropriation of peasant land. This was of great importance for the history of the sugar industry, because the expropriators were mostly substantial peasants through whose agency the industry was able to rent a significant amount of the land which it needed for cane. Insistence on the role of state-supported 'compulsion' in determining the industry's access to land has to be seen in this context: as one factor in a complex situation.

RE-APPRAISAL (2): THE SUGAR INDUSTRY AND ITS WORKFORCE

Deeply embedded in the prevailing characterisation of the Java industry is the concept of the worker who remained a peasant and of an industrial workforce which retained significant linkages with peasant farming and was hence not 'fully proletarianised'. In reality, none of these propositions is particularly useful in illuminating the labour history of the sugar industry.[8]

Work Processes and the Changing Composition of the Sugar Industry's Workforce

The work processes in Java sugar – elaborated but in most respects fundamentally the same from the mid-nineteenth century until the onset of the Great Depression – centred on the Campaign on the one hand and in preparatory field-work and subsequent cane planting on the other. The Campaign, involving the harvesting of the ripened cane by hacking it out of the ground with hand tools, binding it into bundles, loading on to carts or trucks and subsequent haulage to the factory (rails from the 1860s and steam locomotives at some factories by the early twentieth century), began in May each year and generally ended in September or October. Planting, and the elaborate and arduous preparation of the ground which preceeded it, centred on the deep trenching of the fields by manual labour, began somewhat earlier in the year, between January and April, depending on the time at which the land had been cleared of its previous planting of 'peasant' crops. If all went to plan, planting was over by the end of August, though this did not signal the end of work in the plantations. Weeding, irrigating, banking-up and fertilising the young cane employed significant quantities of labour until near the year's end.

To carry out the work of field preparation, planting, harvesting and haulage, the post-1830 Java sugar industry had never relied exclusively for labour on the landholding strata of the peasantry. On the contrary, it had always drawn on the rural landless for a substantial part of its workforce. For much of the nineteenth century landless households in Java accounted on average for around 15 per cent of the total [*Boomgaard*, 1989: 59], while in some sugar districts this percentage was very distinctly higher.[9] It was they, together with unattached vagrants, who provided the mainstay of the industry's Campaign workforce in Pekalongan-Tegal [*Knight*, 1988]. By the opening decades of the twentieth century, moreover, landless workers also formed an important element among the labourers employed in fieldwork, where they either supplemented or, more probably, largely replaced the labour of landholders themselves in the work of opening-up land for cane and in planting the crop.

Many of these early twentieth-century sugar workers, moreover, notably those involved in the heavy field-work required before the planting of the cane itself, were people from outside the immediate locality (the *areaal*) of the factory which employed them. They came and went as required, and generally found temporary accomodation in the nearby villages. Some of them came from as relatively far afield as the neighbouring Residencies of Cirebon, Banyumas and Semarang: others came from within the Residency itself, but from distances which could range up to 40 or 50 kilometres away. The exploitation by the Java sugar manufacturers of various kinds of migrant worker was in itself nothing new [*Knight*, 1980: 187–92; *Elson*, 1979: 280–81, 306]. It appears to have become much more widespread throughout the industry, however, during the early twentieth century. Some of the Residency's eighteen sugar factories did indeed continue to draw their workers predominently from the villages of their own *areaal*: the majority, however, appear to have been dependent on a workforce composed of varying proportions of local labourers and 'outsiders'. At SF Wonopringo, in the eastern part of the Residency, for example, it was reported in the late 1920s that (in contrast to the situation a decade earlier) almost no workers from within the *areaal* were employed in field-work.[10] Moreover, though this restructuring of the labour force may have been accelerated by the sugar industry's attempts to reduce costs in response to falling world prices in the lead-up to the Great Depression, it was clearly something which had begun well before then. It was a phenomena observable at several factories in different parts of the Residency as early as 1907 and 1913.[11]

In fact, there appears to be no single explanation for the sugar industry's move away from reliance on labour predominantly drawn from within the *areaal* toward a situation in which its workforce was a good deal more heterogenous in origin. One possibility is that the sugar industry's rapid expansion in the area in the early twentieth century outstripped the readily available supply of local labour. Between 1903 and 1929 the area under sugar in Pekalongan-Tegal nearly doubled (from 9,600 hectares to almost 18,000),[12] as three new factories were built and the existing ones expanded their plantations. This was a situation which was all the more serious from the industry's viewpoint because improved irrigation facilities in the Residency were creating more employment in rice cultivation than had hitherto been the case.[13] Although some Residency Officials were sceptical of the industry's general claim to suffer from a shortage of locally available labour, arguing, for instance, that that 'even taking into consideration only the population of the cane-planting districts, then there is more than enough ["overvloed"] labour-power, without counting in the sawah-holders',[14] the theme of competition between factories for labour which was in short supply is a recurring one in industry records.[15]

A second point is that the exploitation of 'outside' labour did indeed act to hold or bring down wages, particularly in the aftermath of the industrial unrest of 1920–21 which affected the sugar industry in Pekalongan-Tegal as elsewhere in Java.[16] It is notable that at SF Wonopringo, for instance, the labour component of plantation costs rose appreciably around 1921 but had been reduced by the later 1920s to the levels prevailing a decade earlier.[17] It is by no means clear, however, that outside labour was necessarily cheaper than that which was to be had locally. Indeed, at SF Soemberhardjo in 1928 it was remarked that quite the contrary was the case – operational costs were higher because of recourse to '*importvolk*'.[18]

Quite possibly, in fact, the exploitation of temporary migrant labour helped create a more malleable rather than a cheaper workforce. At SF Wonopringo, for instance, management cited the 'laziness' of the locally available workforce as the chief reason for bringing in outsiders, since 'our own people work at most from 7 to 10 in the morning and for an hour or so in the afternoon, whereas imported people do twice that'.[19]

Whatever the precise reason – and this may be assumed to have varied somewhat from one time and place to another – it is fairly clear that the Residency's sugar industry had come to rely, by the period 1900–30, on a workforce in which temporary migrants played an important part and which was also highly mobile. At some factories, for example, it was reported that preparation of the ground and subsequent planting of cane was carried out by workers who had arrived in the *areaal* after completing similar tasks elsewhere.[20]

As was the case in the colony generally, moreover, the workforce employed by the the sugar industry of Pekalongan-Tegal in the early decades of the twentieth century included a large number of women and children in addition to adult male workers. Women and children had been recorded as employed in some sectors of sugar production – most notably the drying and packing of sugar – at least as early as the 1850s. What is remarkable, on the other hand, about the early twentieth century situation is the extent of their participation in field-work. At SF Ketanggoengan-West c.1910, for example, they accounted for some 40 per cent of the field workforce, and once the cane was actually in the ground constituted the preponderant element.[21]

The full extent to which, by the decades immediately prior to the Great Depression, the Java sugar industry had become dependent on the large-scale, temporary movement of labour (as well as the substantial involvement in the workforce of women and children) is something which is relatively neglected in the literature [cf. *Tichelman*, 1987: 172–7]. In so far as it serves to point up the similarities between the Java industry and other plantation industries in colonial Asia, the omission is a serious one.

Peasantry and the Reproduction of Labour

It might yet be contended, of course, that there remained a crucial nexus between Java's 'persisting peasantry' and sugar production in the sense that the industry's workforce continued to be 'reproduced within the peasant sector', even though any direct connection between petty commodity production and the cultivation of cane had long been severed. Hence the origin of Geertz's celebrated concept of a Javanese rural worker who kept 'one foot in the rice terrace and the other in the mill' [1963: 89]. Notable here also, because of the comparative perspective which it seeks to explore, is Stoler's argument for a significant contrast between the 'plantation companies of Java, whose expansion thrived on a work force drawn from surrounding villages (where a critical portion of that labour power was maintained and renewed)' and the contemporary plantations of Sumatra's East Coast where 'workers were imported by the hundreds of thousands ... and bound by indentured status' [*Stoler*, 1985: 2].

An arguably more plausible approach, however, to the dynamics of labour in late colonial Java – and one already touched upon, for instance, by Ingleson [1986: 155] – would be to investigate the growth and/or consolidation of 'classes' of workers who were predominantly dependent for their subsistance on wages earned wherever and whenever the opportunity existed.

In early twentieth-century Pekalongan-Tegal, in particular, there existed a considerable potential for wage-work in the service of peasant landholders, sugar companies, public utilities or urban employers. Specifically, the industry found itself competing for labour by the 1910s and 1920s with rivals as various as the track-laying activities of the Cheribon-Samarang Tramway Company, the burgeoning *batik* industry in and around the city of Pekalongan (which employed some 10,000 women and men at the height of its expansion in the 1920s [*Kat Angelino*, 1931: 214–6]) and the Government Irrigation Service – in addition to the perennial attraction of the rice harvest and the planting of 'dry-season' crops in the villages. In addition, there was considerable movement of labour between the sugar factories themselves.[22]

The *impression* derived from sources within the industry, moreover, is that, in their recruitment of labour during this period, the sugar factories were dealing predominantly with people for whom the industry's wage was potentially one of several by which they could hope to subsist for most (if not all) of the year, rather than with men and women otherwise engaged in 'own-farm' labour.[23]

To pursue the argument as to whether such people as these were 'incompletely proletarianised' on the grounds that they retained some toe-hold 'within the village' is to risk missing the essential point. What matters is not whether they were exclusively dependent on wages for their subsistance/reproduction – few 'proletarians' are, as a moment's reflection on the importance of female and child labour *within* the 'classic' proletarian household of nineteenth century England, for example, will confirm. Rather, the crux of the matter, to adapt Bernstein's expression of it [1988: 261–2], is whether '... individuals are unable to exist or reproduce themselves outside the circuits of commodity economy and divisions of labour generated by the capital/wage labour relationship'. Judged by these criteria, the Pekalongan-Tegal Residency had developed an extensive 'proletarian' workforce by the immediate pre-Depression decades.

Recruitment of Labour: Intermediaries and Bonding

The historical picture of the means by which labour was recruited for work in the sugar industry is very far from complete. In the early days of the production of sugar under the aegis of the Cultivation System, both plantation and Campaign labour throughout Java was recruited ostensibly through appeal to the 'traditional' authority of Javanese state officials and village headmen (although such a statement leaves many questions unresolved concerning the basis of that authority). From the 1860s onward, however, increasing landlessness among a growing population, together with the marginalisation of landed peasants and reduced opportunities for work on 'peasant' crops as more and more land was devoted to sugar, caused increasing numbers of rural workers to seek wage labour in one form or another. The sugar industry undoubtedly benefited from these developments, although 'economic' pressures did not exclude the possibility that village heads and state officials may have continued to be instrumental in the industry's recruitment and disciplining of labour [*Elson*, 1989; *Knight*, 1988; *Elson*, 1986].

In some areas of Pekalongan-Tegal, however, the role of these local power-holders in assuring the industry of an adequate workforce appears to have become fairly peripheral at a quite early date. During the 1867 Campaign at SF Wonopringo, at that date one of the biggest factories in the Residency, the owner's records contain a graphic description of 'a crowd of men assembled on the roadway in front of the factory' waiting for a chance to be taken on – but they also indicate that their administrator at Wonopringo paid a premium to *mandoors* (foremen/gangers) for each labourer they brought in [*Knight*, 1988: 257].

Recruitment-agents were likewise prominent in securing the field (as well as Campaign) workforce in the Residency early in the twentieth century. At SF Ketanggoengang-West in 1921, for example, the management congratulated itself on having greatly facilitated the recruitment of field labour by securing the services of several good '*koeli mandoors*' who had previously worked for other factories in the area.[24]

At some factories, at least, arrangements for recruiting Campaign labour were quite separate from those in force in the other sectors of production. At SF Wonopringo, from *c*. 1900 until 1930, the supply of Javanese labour for cane cutting was in the hands of a local Chinese subcontractor. On the very eve of the Great Depression, however, the factory dispensed with his services, apparently because he could no longer ensure a steady and/or adequate supply of labour, a responsibility the factory management took over directly.[25]

None of this as yet fragmentary evidence is particularly conclusive. It suggests, however, certain possible parallels with the Indian experience of the ubiquity of the intermediary in the labour market, while perhaps implying that in Java such intermediaries were considerably less powerful and indispensable than their counterparts on the sub-continent. There seems little sense, nonetheless, in concluding that workers recruited (often, presumably, on the basis of some kind of bonding) through such intermediaries were 'sub-capitalist' workers and, as such, further evidence of the hybrid nature of a specifically 'colonial' capitalism. This confuses specific mechanisms for directing and controlling labour with the underlying reasons (landlessness, lack of sufficient land and so on) which impelled individuals to seek wages and which hence made them vulnerable to bonding. Far more persuasive in this context, for instance, is Brass's argument that 'against the view that characterises [bonded labour] ... as a pre-capitalist relationship, unfree labour is [best] regarded ... as a crucial aspect of class struggle between capital and labour in particular agrarian contexts' [*Brass*, 1986: 51].

CONCLUSION

The idea that the large-scale production of sugar for export in nineteenth and early twentieth century Java differed fundamentally from other plantation industries in colonial Asia because of a close and continuing nexus between sugar and the petty commodity production of a persisting landed peasantry is a deep rooted one.

The argument outlined here, however, is that for all its apparently distinguishing features, the Java sugar industry belonged to a mainstream of plantation development in colonial Asia which was itself fundamentally (rather than hybrid or 'colonial') capitalist in character. A number of points are crucial to that argument, none more so, however, than that which stresses that the ostensible nexus between sugar and 'peasants' in late colonial Java was little more than a facade. Behind it can be discerned complex and changing labour processes which did not differ significantly from others found, in one or other period or place, throughout the history of the development of capitalism.

Abbreviations:

ANJ Arsip Nasional Republic Indonesia, Jakarta.
ARA Algemeen Rijksarchief, The Hague.
JV Jaarverslag (Annual Report).
KOLONIALE BANK Archief Koloniale Bank/Cultuurbank NV (NV Cultuurmaatschappij 'Bandjaratma').
KOLONIEN Archief Ministerie van Kolonien.
NHM Archief Nederlands Handelmaatschappij.
SF Suikerfabriek (Sugar Factory).
TJOMAL Archief Maatschappij tot Exploitatie Suikeronderneming Tjomal.

NOTES

NB: For reasons of clarity, the colonial-era spelling of the names of sugar factories has been retained throughout.

1. The *locus classicus* of these ideas is, of course, Geertz [1963: *passim*]. See in particular his contention [1963:87] that the industry 'was an odd, centauric social unit which was neither a proper plantation nor yet a peasant community ... The centaur's head was the mill ... [the] body was the peasantry upon whom the mill drew not just for land but for the casual, seasonally extremely variable labour force it needed ...'. Less eloquently expressed, however, similar ideas had earlier found expression in Allen and Donnithorne's survey of *Western Enterprise in Indonesia and Malaya*: 'The organisation of the [sugar] industry was peculiar and differed from that of the [other] main plantation crops ... [in that] the bond between sugar cultivation for export and the subsistence farming of the village remained intimate ... Whereas the ordinary Western plantation remained an enclave in a traditional agricultural environment, sugar growing continued to be an integral part of the village economy' [1957: 78, 81].
2. For a brief discussion see Gordon [1979: 251]. There is, however, no parallel, in the case of colonial Indonesia, with the rich debate on this score which has been carried on among social scientists working on India. See, for example, Chua [1986], Das Gupta [1986] and Bagchi [1988].
3. Besluiten GG 9.11.1841/16.ARA KOLONIEN 1625; Politiek Verslag Pekalongan

1852 and 1856 [unpaginated]. ANJ. Archief Cultures 1624; Umbgrove 'Monographie' Fabriek Wonopringo B/II. ARA KOLONIEN 1181.
4. NHM Batavia to Amsterdam 26.1.1858/619. ARA NHM 4889.
5. Resident of Pekalongan to GG, 12.2.1913, no.1605/18, in Verbaal MK 6.5.1918/14. ARA KOLONIEN 1824.
6. JV SF Soemberhardjo 1918: 13, ARA NHM 9227; SF Bandjaratma Verslag no.26 (18–20/6/1920): 4, ARA KOLONIALE BANK.
7. For a brief discussion of moneylending among the peasantry, see Burger [1930: 396].
8. I have discussed some of the issues arising in the following section in greater detail in 'Wage Labour in Colonial Indonesia: Sugar Industry Workers in North Java in the Early Twentieth Century', Paper presented to the First International Conference on the Economic History of Indonesia, Jakarta 1991.
9. According to the Umbgrove Commission's mid-1850s report on the sugar industry, in the countryside surrounding some of the Cirebon sugar factories, for instance, there were more landless than landed households. See the 'Monographie' for the Tersana, Karang Semboeng and Sindanglaut factories, II/C and III/B, ARA KOLONIEN 1181.
10. JV SF Wonopringo 1928:44, ARA NHM 9234.
11. JV SF Kalimati 1907:10 ARA NHM 9206; JV SF Ketanggoengan-West 1913:10 ARA NHM 9210.
12. *Koloniale Verslag* 1904 and 1930 Bijl PP & T.
13. See, for example, the situation, as outlined in industry sources, at SF Kemantran c. 1908 and at SF's Tjomal, Bandjardawa and Petaroekan *c.*1921, in JV SF Kemantran 1908:4. ARA NHM 9208 and JV SF Tjomal 1921:22. ARA TJOMAL 21.
14. Resident Stibbe to Assistants-Resident of Pekalongan, 31.3.1917/3574/16, in Exh. 6.5.1918/14 ARA KOLONIEN 1824.
15. JV SF Soemberhardjo 1914:7, ARA NHM 9227; JV SF Doekoewringin 1926:10, ARA NHM 7211.
16. There were a number of strikes and very widepread unrest – usually focussing on a demand for higher wages – among almost all sections of the industry's workforce in Pekalongan-Tegal in 1920–21. See JV SF Tjomal 1920:7, 39 and 1921:9, ARA TJOMAL 21; JV SF Kemantran 1920:3, ARA NHM 9209; Administrateur SF Bandjaratma to Koloniale Bank, Sourabaya 25.7.1920 and 2.8.1920, ARA KOLONIALE BANK 1017. On the strikes in general, see Boomgaard [1988: 164]; Elson [1984: 200–202]; Ingleson [1986: 155–209; Levert [1934: 207–212].
17. JV SF Wonopringo 1908–32, ARA NHM 9234–6.
18. JV SF Soemberharjo 1928:25, ARA NHM 9229.
19. JV SF Wonopringo 1930:7, ARA 9234; Local officials of the colonial government put a rather different gloss on the situation by arguing that local labourers found the work involved in ground preparation for cane too heavy and (presumably) found alternative employment elsewhere. Memorie v. Overgave ... Popelier, 31.12.1932: 31–2. ARA KOLONIEN AA227.
20. See, for example. JV SF Kalimatie 1906 (unpaginated) ARA NHM 9206; JV SF Tirto 1906:1 and 1919:8, ARA NHM 9233; JV SF Ketanggoengan West 1906 (unpaginated), ARA NHM 9210 – where specific mention is made of the 'zeer mobile [werk]volk' on whom the SF relied for much of its plantation labour.
21. JV SF Ketanggoengan West 1909: 3–4 and 1911:11, ARA NHM 9210. This is the only factory in Pekalongan-Tegal where I have, as yet, found records (and then only for a few years) which provide this sort of breakdown of the workforce.
20. Based on information from the following sugar factory annual reports: JV SF Kemantran 1924:9, ARA NHM 9209; JV SF Tirto 1916:7 and 1918:6, ARA NHM 9233; JV SF Wonopringo 1928:44, NHM 9234; JV SF Doekoewringen 1926:10, ARA NHM 7211; JV SF Kalimati 1907:10 and 1923:9, ARA NHM 9206–7; JV SF Katanggoengan-West 1908:8 and 1920:11, ARA NHM 9210; JV SF Soemberhardjo 1914:6–7, ARA NHM 9227.
23. See JV SF Tirto 1919:8, ARA NHM 9233 and JV SF Katanggoengan–West 1921:4, ARA NHM 9211. Important in this context are the findings of the Indies Government's

'Coolie Budget Commission' (dating from the mid-1930s) that over two-thirds of the plantation workers it surveyed held (formally, at least) land from which they could notionally rely for part of their subsistance in addition to wages. Their landholdings, however, were on average miniscule even by the standards of late colonial Java and totally inadequate to sustain and reproduce a household [*Van Niel*, 1956: 38–49].

24. JV SF Ketanggoengan-West 1921:17. NHM ARA 9211.
25. JV SF Wonopringo 1930:69; 1931:57. ARA NHM 9234.

REFERENCES

Aguilar F.V., Jr., 1989, 'The Philippine Peasant as Capitalist: Beyond Categories of Ideal-Type Capitalism', *Journal of Peasant Studies* Vol.17, No.1.

Allen G.C. and A.G. Donnithorne, 1957, *Western Enterprise in Indonesia and Malaya*, London: George Allen & Unwin.

Amin S., 1984, *Sugarcane and Sugar in Gorakhpur*, Dehli: Oxford University Press.

Anon, 1905, 'Mandelijkshe Revue', *Indische Gids*, Vol.27, No.2.

Bagchi, A.K., 1988, 'Colonialism and the Nature of the "Capitalist" Enterprise in India', *Economic and Political Weekly*, Vol. 23, No. 31.

Bernstein, H., 1986, 'Capitalism and Petty Bourgeois Production: Class Relations and Divisions of Labour', *Journal of Peasant Studies*, Vol.15, No.2.

Boomgaard, P., 1988, 'Treacherous Cane: The Java Sugar Industry', in B. Albert and A. Graves, *The World Sugar Economy in War and Depression*, London: Routledge.

Boomgaard, P., 1989, *Children of the Colonial State* (CASA Monographs No.1), Amsterdam: Free University Press.

Brass, T., 1986, 'Unfree Labour and Capitalist Restructuring in the Agrarian Sector: Peru and India', *Journal of Peasant Studies*, Vol.14, No.1.

Breman, J., 1978, 'Seasonal Migration and Co-operative Capitalism', *Journal of Peasant Studies*, Vol.6, Nos.1 and 2.

Breman, J., 1983, *Control of Land and Labour in Colonial Java*, Doordrecht, Verhandelingen van het KITLV No.101, Floris Publications.

Burger, D.H., 1930, 'Het Niet-Officiele Credit in het Regentschap Pati in 1927', *Koloniale Studien* 14 (2).

Chowdhury, B., 1964, *The Growth of Commercial Agriculture in Bengal*, Calcutta: Maitra.

Chua, C., 1986, 'Development of Capitalism in Indian Agriculture: Gujarat 1850–1900', *Economic and Political Weekly*, Vol. 21, No. 48.

Courtenay, P.P., 1980, *Plantation Agriculture*, second revised edition, London: Bell.

Das Gupta, R., 1986, 'From Peasants and Tribesmen to Plantation Workers: Colonial Capitalism, Reproduction of Labour Power and Proletarianisation in North East India, 1850's to 1947', *Economic and Political Weekly*, Vol. 21, No. 4.

Elson, R.E., 1979, 'Sugar and Peasants: The Social Impact of the Western Sugar Industry on the Peasantry of the Pasuruan Area, East Java, from the Cultivation System to the Great Depression', Ph. D. Thesis, Monash University.

Elson, R.E., 1984, *Javanese Peasants and the Colonial Sugar Industry*, Singapore: Oxford University Press.

Elson, R.E., 1986, 'Sugar Factory Workers and the Emergence of "Free Labour" in Nineteenth Century Java', *Modern Asian Studies*, Vol.20, No.1.

Elson, R.E., 1989, 'The Mobilisation and Control of Peasant Labour in Java in the Early Cultivation System', in R. J. May and W.J. O'Malley (eds.), *Observing Change in Asia: Essays in Honour of J.A.C. Mackie*, Bathurst: Crawford House.

Geertz, C., 1963 (eds.), *Agricultural Involution*, Berkeley and Los Angeles, CA: University of California Press.

Gordon, A., 1979, 'The Collapse of Java's Colonial Sugar System and the Breakdown of Independent Indonesia's Economy' in Anrooij F. (ed.), *Between People and Statistics*, The Hague: Nijhoff.

Gordon, A., 1982, 'Indonesia, Plantations and "Post-Colonial" Modes of Production', *Journal of Contemporary Asia*, Vol.12, No.2.

Husken, F., 1989, 'Cycles of Commercialisation and Accumulation in a Central Javanese Village', in G. Hart, A. Turton and B. White (eds.), *Agrarian Transformations*, Berkeley and Los Angeles, CA: California University Press.

Ingleson, J., 1986, *In Search of Justice*, Singapore: Oxford University Press.

Ka, C.-H., 1987, 'Land Tenure, Development and Dependency in Colonial Taiwan, 1895–1945', Ph.D. thesis, State University of New York at Binghamton.

Kat Angelino, P. de, 1931, *Batikrapport: (II) Midden-Java*, Batavia: Landsdrukkerij.

Knight, G.R., 1980, 'From Plantation to Padi-Field: the Origins of the Nineteenth Century Transformation of Java's Sugar Industry', *Modern Asian Studies*, Vol.14, No.2.

Knight, G.R., 1982, 'Capitalism and Commodity Production in Java', in H. Alavi *et al.*, *Capitalism and Colonial Production*, London and Canberra: Croom Helm.

Knight, G.R., 1988, 'Peasant Labour and Capitalist Production in Late Colonial Indonesia', *Journal of Southeast Asian Studies*, Vol.19, No.2.

Knight, G.R., 1989, 'Sugar, Peasants and Proletarians: Colonial Southeastern Asia 1830–1940', *Critique of Anthropology*, Vol.9, No.2.

Mubyarto, 1969, 'The Sugar Industry', *Bulletin of Indonesian Economic Studies*, Vol.5, No.2.

Levert, P., 1934, *Inheemsche Arbeid in de Java Suikerindustrie*, Wageningen.

OMW Pekalongan, 1908, *Onderzoek naar de Mindere Welvaart der Inlandsche Bevolking op Java en Madura. Samentrekking ... Economie van der Desa in de Residentie Pekalongan*, Weltevreden: Smits.

Schaik, A. van, 1986, 'Colonial Control and Peasant Resources in Java: Agricultural Involution Reconsidered', Proefschrift, Universiteit van Amsterdam.

Selosoemardjan, 1962, *Social Changes in Jogjakarta*, Ithaca, NY: Cornell University Press.

Soetrisno, L., 1980, 'The Sugar Industry and Rural Development: The Impact of Cane Cultivation for Export on Rural Java: 1830–1934', Ph. D. thesis, Cornell University.

Stoler, A.L., 1985, *Capitalism and Confrontation in Sumatra's Plantation Belt*, New Haven, CT and London: Yale University Press.

Suiker-Enquete, 1921, *Verslag van der Suiker-Enquete Commissie 1921*, Soerabaja: Fuhri.

Sutter, J., 1959, *Indonesianisasi: Politics in a Changing Economy 1940–1955*, Ithaca, NY, Cornell University, Department of Far Eastern Studies, Data Paper 36.

Tichelman, F., 1987, 'Problems of Javanese Labour. Continuity and Change in the Nineteenth Century', in M. Hasan *et al.*, *India and Indonesia from the 1830's to 1914: The Heyday of Colonial Rule,* Leiden: Brill.

Van der Mandere, J., 1928, *De Java-Suikerindustrie in Heden en Verleden*, Amsterdam.

Van Niel, R. (translator), 1956, *Living Conditions of Plantation Workers and Peasants in Java in 1939–40*, Ithaca, NY, Cornell University, Department of Far Eastern Studies, Modern Indonesia Program, Translation Series.

Labour Control and Labour Resistance in the Plantations of Colonial Malaya

P. RAMASAMY

INTRODUCTION

In this essay I examine labour control and labour resistance in plantations in pre-war colonial Malaya. I argue that the nature and scope of resistance among Indian and Chinese labour was to a large extent determined by the system of control in question. Individual forms of resistance of a passive nature were dominant under the indentured system of labour control. These forms of resistance ranged from simple acts of foot-dragging to the more conspicous act of desertion [*Scott*, 1985]. Under the (Indian) *kangani* and the (Chinese) contract systems of labour control, collective forms of resistance of both passive and confrontational types began to emerge [*Adas*, 1986]. These instances of collective resistance were forcibly put down by the colonial state acting in the interests of the plantation capitalists in the belief that these were inspired by agitators and communists.

COLONIAL PLANTATION IN MALAYA: ITS HISTORY, STRUCTURE AND OWNERSHIP

By the beginning of the twentieth century, the British succeeded in extending political control over the Malay peninsula both by crushing peasant rebellions and by placating the Malay rulers. Having done this, Britain turned next to the question of development; given that Malaya was richly endowed with forest, a plantation agricultural strategy was pursued. The pro-plantation policy was also predicated on the need to reduce Malaya's over-dependence on tin and most importantly to recoup the losses of European capitalists' failures in mining [*Lim Teck Ghee*, 1977: 15; *Chai Hon Chan*, 1964].

Even before the establishment of political control, Europeans were involved in the planting of sugar in places like Singapore, Penang, Province Wellesley and Perak, using Chinese and Indian indentured

---

P. Ramasamy, Department of Political Science, National University of Malaysia, Bangi, Selangor. The author thanks the editors for their substantial comments on earlier drafts.

labourers in the second half of the nineteenth century. By the end of the century, the industry declined due to the difficulty of procuring an adequate labour supply, increasing competition from beet sugar production in Europe and from large scale production of sugar in Cuba and Java. The decline of the industry was also precipitated by the official policy in the 1890s that sought to reserve more areas for rice planting; areas once reserved for sugar growing were turned over for rice cultivation [*Jackson*, 1968: 170].

As the sugar industry declined, many European planters experimented with coffee as a plantation crop. Initially, coffee growing was impeded by planters' ignorance of local conditions, the unsatisfactory land terms offered by the British administration, and again inadequate labour supply. However, by 1890, coffee growing picked up due to the availability of cheap land, the government's attempt to provide cheap labour from India, and the replacement of Arabian coffee by the Liberian variety that was better suited to local conditions [*Jackson*, 1968: 183]. For about ten years, coffee enjoyed a pre-eminent position in Malayan plantation agriculture before it was replaced by rubber. Coffee production declined, among other reasons, due to dropping prices as a result of increased production in Brazil, the inability of the Liberian coffee variety to compete with other good varieties grown elsewhere, the devastations wrought by pests, and the planters' lack of initiative in improving the processing and marketing aspects of the industry [*Jackson*, 1968: 200].

During its pioneer stage, rubber was interplanted with crops like coffee and coconut. The commercial value of rubber or Hevea was only realised following a series of experiments at both the local level and international level. In Singapore, Henry Ridley, the Director of Singapore Botanical Gardens, through a number of experiments showed that rubber could be cultivated on a commercial scale. At the international level, the invention of the pneumatic tyre using rubber by Dunlop in 1888 was the important discovery that propelled the growth of the natural rubber industry. In 1990, rubber exported to the United States and Europe came mostly from Brazil [*Drabble*, 1973; *Voon*, 1976]. Until 1913, rubber exports from Malaya were below those of South America, but in 1914 Malaya exceeded South American exports by nearly 6,000 tons [*Drabble*, 1973: 109]. In 1918, exports of rubber from Malaya amounted to about 50 per cent of the world's total rubber consumption [*Drabble*, 1973: 130].

The big demand for rubber pushed the expansion of the plantation industry. In 1877 rubber planting amounted to only about 350 acres but went up to about 20,000 acres by 1904 [*Jackson*, 1968: 226], about 43,000 acres by 1910 [*Drabble*, 1973: 76–133], and by 1940 reached over two million acres [*Radkrishnan*, 1975: 32]. The rubber industry that took shape in Malaya was composed of predominantly European owned large estates, non-European, mainly Chinese owned, medium and small estates, and smallholdings held by the peasantry. In 1921, one-fifth of total estate acreage in Malaya was in Chinese owned estates with an average size of about 400 to 500 acres [*Jackson*, 1968: 266]. Capital in these estates was mainly derived from family business ventures in tin mining and other commercial pursuits. In 1940, smallholdings constituted about 39 per cent (1,350,000 acres) of the total rubber acreage of 3,500,000 [*Radkrishnan*, 1975: 32].

Two other factors were important in contributing to the growth of the rubber industry. One was the role of the colonial state which not only provided an appropriate political climate for the development of the industry [*Lim Teck Ghee*, 1977], but also facilitated the flow of capital by giving special land concessions for plantation development [*Jackson*, 1968: 235; *Lim Teck Ghee*, 1976: 84–9], establishing a loan fund [*Drabble*, 1973: 39], and laying good lines of communications [*Jackson*, 1968: 236]. During the slump periods, it protected the rubber industry by introducing restriction schemes (the Stevenson Restriction Scheme in 1922 and the International Rubber Regulation Agreement in 1934; see Murray in this volume). These schemes restricted rubber output so as to raise prices and maintain the competitiveness of rubber in the world market. As argued by Bauer [1948] and Lim Teck Ghee [1977], restrictions on rubber production were mainly felt by smallholders who vented their frustrations by engaging in strikes and demonstrations.

The second factor was the change in capital ownership from proprietary to corporate ownership in the early twentieth century, which meant, among other things, monopolistic control of the rubber industry by a select group of people in Britain through a system of interlocking directorships with certain key personnel in the agency houses. According to Puthucheary [1974: 27–8], in 1953, about 75 per cent of the European estate acreage was dominated by agency houses, with only about 25 per cent not owned by public limited companies. It was also estimated that some 20 agency firms controlled the rubber industry, of which the largest were Harrisons & Crossfield, Boustead-Buttery, Guthries, R.E.A-Cumerbatch and Sime Darby. As further pointed out by Voon [1976] 77 directors (seven per cent of the total) controlled 30 per cent of total directorships, each holding five to 22 directorships.

FORMATION OF THE PLANTATION PROLETARIAT

The successful development of rubber plantations in late nineteenth-century Malaya relied on the availability of cheap labour. Having experienced difficulty in recruiting local native Malay labour whom the British dismissed as 'lazy' and 'unsuitable', attempts were made to bring in overseas labour. While it was true that Malays generally were not interested in plantation work, this did not mean that they had an aversion to wage labour. They objected to plantation work because of its oppressive nature – monotonous work, low wages, high mortality rate, regimented life, and other features. De Silva [1982] in his study of the political economy of Sri Lanka also concludes that Sinhalese peasants did not join the plantation workforce because of the nature of the plantation system, rather than any like or dislike for wage labour.

Initial attempts to recruit Javanese labour failed because of restrictions imposed by the Dutch authorities [*Jackson*, 1961: 131]. Subsequently, a Labour Commission, set up in 1910, recommended that efforts should be made to recruit Chinese labour. However, the presence in Malaya of large numbers of Chinese workers previously recruited by private agents, and their purported involvement in secret societies in places like Singapore and Penang, displeased London, even though the colonial authorities in Malaya were in favour of extending official recruitment to the Chinese. London's veto prevailed despite the protests of the Planters Association of Malaya [*Ramasamy*, 1990: 64].

Although official recruitment was never extended to the Chinese, there were many Chinese immigrants in Malaya by the turn of the century. Due to the heavy demand for labour, Chinese contractors working in league with brokers in places like Singapore and Penang recruited migrant labourers for Malaya. The costs of bringing labourers to Malaya were then recovered from those employers who 'purchased' them, and labourers were contractually tied to their employers until such debts were discharged. According to Blythe, in 1911 25 per cent of the estate population was Chinese, and in 1931, this increased to 35 per cent in the Federated Malay States or FMS [*Blythe*, 1947: 68].

In European planting circles, South Indian labour was considered ideal for plantation work. South Indian labour had all the 'right qualities' such as docility, submissiveness, malleability and little self-reliance [*Sandhu*, 1969: 57]. As early as 1887 Sir Frederick Weld, a senior colonial official, recommended migration from India 'because Indians are a peaceable and easily governed race'. 'It is advisable' said Weld, 'that, in a country like this, the preponderance of any one Eastern nationality should not be excessive' [*Stenson*, 1980: 16]. European planters' experience with South Indian labour in Ceylon also led to this preference. Furthermore, economic considerations notwithstanding, the British policy of 'divide and rule' encouraged the importation of Indians to counterbalance the growing Chinese population.

South Indian labourers were brought to Malaya first under indenture and later under the *kangani* system. The abolition of the former was not based as much on humanitarian grounds as on economic ones. The costs and problems of ensuring an adequate supply of labour made it necessary for the colonial state to recognise the requirements of plantation capital. The answer was the introduction of the *kangani* system which evolved slowly from the late 1860s. The vast majority of Indian plantation workers were brought in under this system. According to Sandhu, between 1898 and 1938, about 1,153,770 *kangani* recruits landed in Malaya at an average of 28,000 per year [*Sandhu*, 1969: 96]. The *kangani* system in turn was abolished in 1938 due to a dispute between the Malayan government and the Government of India over wages.

As stated by Breman [1989: 131] plantation enterprise depended on the bondage of labourers. Subservience, servility and malleability were characteristics that European planters looked for in their labour force. To these Europeans, this was not too much to expect because they had invested considerable capital in developing plantations and securing labour. Labourers were bonded as soon as they were recruited and shipped to Malaya. Under indenture, various tactics were used by agents and brokers to snare unsuspecting migrants [e.g. *Tinker*, 1974; *Parmer*, 1960].

CHINESE WORKERS: CONTROL AND RESISTANCE

Chinese workers unable to pay their passage became indebted to brokers or organisations involved in their transportation to Malaya. Many unsuspecting Chinese were kidnapped and shipped to Malaya by brokers bent on making a quick profit. Once in Malaya the immigrants were placed in lodging houses until they were purchased by employers. Those purchased in this manner were taken to their work places and required to toil until they had discharged their debts. Labourers recruited under this system were known as '*Chu Tsai*' or 'piglets' and the persons involved in the transportation of immigrants were known as '*Chue Tsai Thau*' or 'heads of piglets' [*Blythe*, 1947: 69].

Hugh Tinker aptly characterises indenture as a new form of slavery [1974]. Under this system of labour control that lasted until the first decade of the twentieth century, there were few differences in the way Chinese and Indian workers were treated. Unlike Indian labourers who were employed directly by planters, Chinese labourers were employed through a headman (*kepala*). Employment contracts were signed between employers and Chinese labourers and lodged in the Chinese Protectorate Offices. Wages were not paid directly to labourers by European employers but to Chinese headmen who then paid their workers after deducting expenses and debts incurred by workers [*Blythe*, 1947: 80], giving headmen an interest in retaining labourers for as long as possible to make the maximum profit. Labourers were often retained even after the period of indenture had expired by debts they had acquired. The Labour Commission of 1910 reported that:

> It appears also that they are frequently detained long after the year has expired for which they contracted, and after they have done more than 360 days' fair work. This is done by keeping them in debt by advances of cash or provisions, so that the cooly at the end of his year owes more than he did at the beginning [*FMS, Labour Commission*, 1910: 147].

Apart from contractual obligations imposed on Chinese indentured labourers, other mechanisms served to control them. The operation of secret societies based on clan and kinship ties was one important source of control. The Chinese headmen's important leadership role in secret societies, by virtue of their economic power, served as an effective instrument of labour control. While British officials might not have condoned all the practices of the secret societies, they none the less shared the opinion of European planters that these organisations were effective in tying labourers to the plantation. This perhaps tell us why there was little official effort to crack down on the activities of the secret societies until the 1920s. Secret societies served the dual purpose of protecting the Chinese employers and effectively disciplining labour [*Jackson*, 1961: Ch.IV].

Opium-smoking, despite its harmful effect on the health of labourers, was another notorious practice encouraged by Chinese headmen and condoned by European officials to effect control. In the early 1920s, the British Malaya Opium Committee was quite concerned over widespread opium-smoking among Chinese labourers, but did not categorically oppose opium-smoking because not all its members, particularly those representing propertied interests, were convinced that opium-smoking was harmful. After having spoken to a number of Chinese headmen, the Committee even decided that opium-smoking was beneficial to labourers: 'The extraordinary number of healed lungs which are found in opium smokers lends colour to the belief that there may also be some curative agent in the smoke' [*FMS, British Malaya Opium Committee*, 1924: 224].

There is no evidence to suggest that the British Malaya Opium Committe did anything concrete to eradicate opium smoking. Opium-smoking, as argued by Jackson, served the capitalists in the plantations in two ways. First, extreme opium addiction inevitably forced labourers to become indebted to their employers. Such indebtedness, as noted, allowed employers to re-indenture labourers for a longer period. Second, opium-smoking, again in the opinion of Chinese headmen, made it possible for labourers to work long hours [*Jackson*, 1961: Ch. IV]. A Chinese headman told the British Malaya Opium Committee that 'when we have an extra dose of wet weather or anything like that, opium smokers will see their work through and get it finished ...'. Similarly another Chinese headman remarked to the Committee that labourers who smoked opium were better workers than those who did not [*FMS British Malaya Opium Committee*, 1924: 225]. In short it was not surprising that plantations employers encouraged this habit [*Jackson*, 1961: 154].

Another form of labour control among the Chinese was the operation of *kongsi* houses or barracks where labourers were lodged by the headmen. According to the testimony of a Chinese contractor in 1910: 'The coolies are locked up after 6 p.m. They are let out to work at 6 a.m. The working hours are from 6 a.m. to 3 p.m. – 9 hours a day' [*Jackson*, 1961: 154). The *kongsis* were fenced and guarded to prevent the workers from escaping. According to the Assistant Protector of Chinese, Negeri Sembilan, a worker tried to escape from the *kongsi* 'because by doing so he can earn 60 cents a day on the next estate as a *laukeh* (free labourer), instead of 5 cents and food as a *sinkeh* (unfree labourer)' [*Jackson*, 1961: 154].

Colonial records like those cited above tend to argue higher wages as a reason for Chinese labourers to escape, but that the inhuman treatment of labourers was as much an incentive to flee cannot be discounted. In 1910, the acting Protector of Chinese, Perak, revealed to a Labour Commission how *sinkehs* were beaten and forced to eat human excrement by a headman for being too ill to take a bath. On another visit to an estate in Krian district he found that the labourers were terrorised by the headmen, who was subsequently charged and prosecuted for ill-treatment of workers [*FMS, Labour Commission*, 1910: 151].

One drawback of the Labour Commission of 1910 is that it was biased towards European-owned estates and singled out Chinese-owned estates for ill-treating labourers: "There appears to be no ill-treatment of Chinese indentured labour on European-managed estates ... There are, however, some matters which require attention' [*FMS, Labour Commission*, 1910: 153]. Although the Commission targeted the Chinese-owned estates for criticism, the persons prosecuted for labour offences were not Chinese managers but Chinese headmen [*FMS, Labour Commission*, 1910: 156–7].

The indenture system of labour control and rentention was abolished for Chinese in 1914, which meant that employers could no longer use written contracts to retain their employees. The contract system, which first evolved in the tin mines, was extended to the plantations to replace indenture as the means of labour recruitment. Plantation employers now commissioned agents or private recruiters to recruit labour from China or from lodging houses in the Straits Settlement terrorities. This system allowed the recruitment of families, and those contracted were under a moral obligation to repay their employers for advances (written contracts being no longer allowed). The contract system was a slight improvement over indenture, but '... if the labourers were not sufficiently docile there was close surveillance by Chinese Kepalas and Sikh watchmen and the locking of the Kongsi doors at night' [*Blythe*, 1947: 99].

The contract system, which became an important feature of the plantations from the 1920s onwards, contained numerous restraints on the labour force. According to a report submitted by David Gray of the Chinese Protectorate Office, Negeri Sembilan, in 1937, the contract system worked to the disadvantage of labourers in the following ways: (a) the contractor delayed payment to the workers on the pretext that the estate manager had not paid him; (b) in the allocation of tasks to workers, the contractor tended to increase the tasks for workers (without the approval of the manager); (c) the contractor underpaid workers (for example, the manager would pay the contractor five cents per pound for rubber brought in by tappers, but the contractor would only pay four cents); (d) the contractor got the workers indebted by forcing them to buy goods from him. [*NSSF 1349/1937, Report of Inspection of Estates, Mines and Factories Employing Chinese Labour (or Gray's Report), 1937*, Enclosure (Encl) 4].

Gray's report is a useful and detailed document that detailed the contract system but again spared European estate management criticism. This was on the premise that European estate managers could not be blamed for the unrest in 1937, because the contract system functioned independently of them. However, the contract system, despite the indirect nature of its recruitment of labour, was part and parcel of the European dominated plantation system. It can be argued the contract system, with its premium on labour subservience and docility, worked to the advantage of European-controlled plantations. While Europeans occasionally blamed labour unrest on the practices of the contractors, there is no doubt that without the sanction and blessing of Europeans the contract system would not have taken root in the plantations in the first place [*Ramasamy*, 1990: 226].

Given the near absolute nature of management control in plantations, overt violence was rare and occured only under extreme conditions of oppression. Passive resistance in the form of foot-dragging, not reporting to work on the pretext of being sick, running away from the work place, and the like, was very common on plantations. Chinese indentured labourers' favoured form of resistance was simply to run away from the *kongsis*, both to escape the harshness and brutality of *kongsi* confinement and to become free labourers and earn better wages. Because of the frequency of desertion among Chinese indentured workers, *kongsis* had to be strictly guarded by fences and watchmen.

What accounts for the absence of collective resistance among Chinese labourers before 1937 (see below)? It can be argued that various social mechanisms and practices within the Chinese community helped to reduce overt forms of class antagonisms. The operation of Chinese guilds based on linguistic ties, the important role of secret societies based on tribal affiliation, and the prevalent practice of smoking opium, cushioned the Chinese community from intense forms of class conflict. The secret societies headed by Chinese capitalists served both to control and protect labourers. In the plantation *kongsis* the headmen were invariably the heads of secret societies which enhanced their control of labour: 'the social and economic "closed shop" which they attempted to enforce by internal discipline in fact amounted to serious oppression' [*Jackson*, 1961: 19]. The strong internal discipline exercised by the heads of secret societies subdued overt resistance, with resistance otherwise displaced in excesses of opium-smoking and gambling [*Jackson*, 1961: 52].

There is no evidence of collective forms of resistance among the Chinese for the first three decades of the twentieth century. The first recorded instance of a major collective action by Chinese plantation workers against European capital was in 1937, when hundreds of Chinese rubber tappers in a number of estates in the Ulu Langat district in the state of Selangor went on strike for better wages and living conditions. The strike later spread to other areas in Selangor, Negeri Sembilan, Malacca, Johore and Pahang. By late March 1937, an estimated 30,000 Chinese labourers were on strike in Selangor and Negeri Sembilan, of whom 25,000 were employed in rubber estates [*Selangor Secretariat File* (SSF) 194/1937, *Strikes on Rubber Estates in Kajang Area*, 1937, Encl. 2]. Others were factory and mine workers, organised in the Malayan General Labour Union (MGLU, a labour affiliate of the Malayan Communist Party) who struck in sympathy with the Ulu Langat plantation labourers. In some of the sympathy strikes, Indian and Javanese workers also participated, although in small numbers [*FMS, Labour Department Report*, 1937: 54].

The speed with which the strikes spread from district to district, state to state, and estate to factory, suggests that the MCP (Malayan Communist Party) had considerable influence in directing and coordinating them. Yeo Kim Wah [1976: 50] argues that the strikes were directed by the MCP, and that the celebration of International Women's Day, the raising of the Red Flag in a number of estates, and the formation of the Rubber Workers' Union (RWU), were all organised by communists. Workers demands included reduction of workloads, (daily tapping of 250 trees instead of 400), extra compensation for cleaning tapping equipment, benefits for pregnant women, improvement in living quarters, and work-free public holidays [*SSF* 194/1937, Encl. 3].

A significant target of the labour unrest was the contract system, and in particular estate *kepalas* (contractors) and Chinese *towkays* (capitalists), who were the traditional leaders of their communities. As pointed out by Yeo Kim Wah [1976: 55] these leaders lost much of their considerable power in the community as they were virtually displaced by the communist leadership, which effectively organised workers to challenge authority in the plantations. When some Chinese associations attempted to play a conciliatory role, they were rebuffed by workers as representing the capitalists. European managers tried to use the *kepalas* to intervene and persuade labourers to return to work, but the labourers refused [*SSF* 194/1937, Encl. 4].

The strikes in the Ulu Langat district lasted for more than a month. The colonial authorities in league with estate managers saw the strikes as manifestations of communist influence and, fearful of losing their political control, crushed the labour uprising. Police raids were carried out throughout the country, and important working class leaders were detained and eventually banished. After the raids, Selangor's Chief Police Officer remarked: 'The raid had an excellent effect – and destroyed completely the morale of the labour force' [*SSF* 194/1937, Encl. 6]. Following police repression, labourers had no alternative but to negotiate with the employers as arranged by the Chinese Protectorate Office.

As is evident from the colonial documents, the strikes and militancy exhibited by Chinese plantation workers caught the authorities by surprise [*SSF* 194/1937, Encl. 7, Report by the Chief Police Officer]. It never occurred to them that workers in Malaya, known for their docility, could go so far as to pose a danger to the colonial order. The High Commissioner felt that any public inquiry would embarass not only the government but also the employers. In fact, he directed his anger against the officials of the Chinese Protectorate Office, who were at times at odds with the employers over their treatment of labour. The High Commissioner's view was that officials had not understood the Chinese community well; if they had, the March strikes could have been averted [*Yeo Kim Wah*, 1976: 70].

The High Commissioner directed the various government departments to monitor developments in the Chinese community. When some departments proposed introducing social legislation to prevent disturbances in the plantations, he rejected them saying 'inspection is in my opinion the prime need' [*NSSF* 1349/1937, Gray's Report, Encl. 4]. He advised the British Residents in the affected states to adopt a similar approach to that of the British Resident in Pahang, who directed all his district officers to give top priority to inspections [*NSSF* 1349/1937, Gray's Report, Encl. 4]. In fact, David Gray's report on the conditions of Chinese labour in the plantations, was undertaken in the context of the appeal for more inspection reports on labour conditions.

INDIAN WORKERS: CONTROL AND RESISTANCE

Indians, unlike the Chinese, were recruited and shipped to Malaya with the blessings of the colonial governments in India and Malaya. However, the much used rhetoric of 'official' in no way prevented gross violations of labourers' fundamental rights. The Indian Immigration Ordinance, passed in the late nineteenth century to 'protect' newly arrived immigrants in Malaya, served the interests of the capitalist-planters by punishing labourers who broke their contract. An Indian Immigration Agent in Penang who was supposed to enforce the provisions of the Ordinance was ineffectual in looking after the welfare of immigrants [*Jackson*, 1961: 69].

Unlike the Chinese, Indians were directly employed by European estates where labour control was much more systematic and severe. Before the 1890s, a 'joint and several' type of recruitment contract was imposed on labourers, under which each member of a gang or a group of workers was liable for the default of others. A Labour Commission of 1890 described this system as 'capable of very inhuman application, even to the making one man in a hundred work out the defaults of ninety-nine' [*FMS*, *Labour Commission*, 1890: 38].

In the 1890s the 'joint and several' type of contract was replaced by the Indian Immigration Ordinance, which stipulated that immigrants should only sign indentures on their arrival. Contracts established under the indenture system were weighted in favour of the employers; labourers found that their breach of contract was severely punished while employers' offences against labour went unpunished. Colonial officials were no doubt convinced that the planters were doing a public service in bringing workers into Malaya. Lamb, an outspoken member of the Province Wellesley Planters' Club, remarked:

> The Planters are the chief importers of labour into this country, and the immigration is in no way subsidised by Government as in most other Colonies importing Indian or Chinese labour. The Planter has thus not only all the risk and trouble of importation, but also that of passing the men through the critical climatising period, whilst ultimately only a portion of the imported labour remains on estates, the rest being absorbed into the general labour supply to the benefit of the public at large [*Jackson*, 1961: 65].

The Protector of Immigrants of the Straits Settlements who once remarked that Indian coolies were well treated in Malaya, later observed that coolies received very low wages in Malaya and were 'far less liberally treated' in comparison to other colonies. He asked 'Are then desertions from the estates to be wondered at?' [*Jackson*, 1961: 65].

The indenture contract facilitated control over the labour force, especially since indentured labourers found it impossible to regain their status as free labourers. European employers, despite everything working in their favour, were the most vociferous critics of the government, wanting the authorities to enact even more stringent laws to punish labourers who breached contracts. By the late 1880s, the only way for labourers to escape the harshness of the plantation system was to desert and move elsewhere. Rather than examining labour conditions, the authorities concluded that higher wages in nearby plantations were the main cause for desertions. This explanation, typical of colonial officials, absolved European employers of any fault.

The indenture system not only helped overcome the problem of labour shortages but, more importantly, helped keep a tight lid on wages. Because of miserably low wages, workers were perpetually in debt which in turn reinforced low wage levels. The Principal Medical Officer of the Straits Settlements noted '... it is more likely the new cooly has nothing at all left from his pay at the end of the month and is probably in debt, with no surplus to buy anything with (sic)'. He also noted that employers cut labourers' wages for very minor offences [*Jackson*, 1961: 64].

In enforcing their control, employers frequently resorted to violence. A commission appointed to investigate labour conditions in Malaya in 1890 pointed out that labourers were often beaten in estates. However, the commission blamed native supervisors and not European managers or their assistants [*FMS, Labour Commission*, 1890: 51]. A subsequent commission, however, documented cases of assault by European managers against labourers [*FMS, Labour Commission*, 1910: 4]. It is clear that violence was very much part of the plantation system. As Breman [1989: 164–7] has pointed out, violence served to maintain the social distance between coolies and their bosses; moreover, violence was considered by planters as a necessary evil to 'educate' rebellious coolies.

The *kangani* system replaced indenture in 1910, and helped check some of the grosser abuses of labour; coercion nevertheless remained an important feature of labour control. As shown by Jain [1970: Ch.4], the *kangani* used his kinship and caste ties to bind plantation labourers to him. The *kangani*'s money lending activities constituted another important aspect of his paternal role. Money given on credit was recovered from labourers either by deduction from their wages or enrolling them in a savings scheme. Selvaratnam [1983: 27], in a study of two estates, shows that in the early twentieth century, the management paid the wages of the labourers to the *kangani* so that he could deduct a proportion of the debt owed him. Labourers always repaid their debts to the *kangani* due to their social dependence on him [*Jain*, 1990: 21].

There was an uglier side to the *kangani*'s role. The *kangani* was given the authority by management to mete out severe and drastic punishments, and in the first two decades of the century was, even given the authority in some cases to kill estate workers who engaged in premarital sexual intercourse. On occasion, the *kangani* inflicted such severe beatings that victims had to be hospitalised or died [*Jain*, 1970: 280].

Assaults against workers were common in plantations. Estate employers gave the impression to the colonial authorities that they were as interested as the latter in eliminating physical assaults on labourers. European employers blamed the *kanganis* who, they argued, got carried away in their zeal as estate foremen [*FMS, Labour Commission*, 1910: 6]. Physical punishments meted out by the *kanganis* were in accordance with the dictates of the employers. The *kangani* was not a mediator of conflicts between labour and capital; he was very much the agent of capital and his primary role was to subject labour to the rigorous discipline required by the plantation production system.

The colonial authorities became concerned about the strong hold of *kanganis* over their labour force, and warned employers to check the power of *kanganis*:

> The affluence and power acquired by many of the kanganies is a matter which requires the careful attention of all employers of labour; kanganies can make themselves very useful in obtaining

> labour and in every matter affecting its employment on the estates; but it is dangerous to allow them to get too much control into their hands at the expense of the personal influence of the manager of the estate. There are indications of such a state of affairs existing at present at some places; it will be a serious menace to stability of labour forces throughout the country if this tendency becomes much accentuated [*Jackson*, 1961: 118].

Indeed, there were instances where the *kangani* used his power to move workers out of estates in defiance of the management. There were also occasions when *kanganis* organised labour protests and wildcat strikes. Despite such instances of defiance against management, this was not the norm. Colonial records on *kangani* disputes with employers show that they occurred not to defend labour but in pursuit of the personal interests of *kanganis* [*Ramasamy*, 1990: 190–93].

The plantation class structure was one in which Europeans occupied managerial roles, educated Asians filled clerical and supervisory roles, while the South Indian Tamils constituted the labouring masses. This class structure was reinforced in racial and ethnic terms as managers were whites and Asian supervisory staff were mainly from Kerala and Ceylon. The estate management ensured labour acquiescence by the provision of estate shops, places of worship, toddy shops and creches, which among other things created a psychology of dependence. As noted by Jain [1970: 285], the degree of management 'generosity' fluctuated with its need for labour; in times of acute need for a stable labour force, as in the 1920s, European managers showed more interest in the welfare of labour.

As with Chinese labour under indenture, lack of evidence prevents us from providing a detailed picture of individual forms of resistance among Indian labour in the plantations. Both European officialdom and planters in Malaya had good reasons to paint a picture of worker well-being and prosperity in Malaya. However, there are bits and pieces in the colonial records which confirm that individual and collective forms of resistance were present.

One such piece of evidence is to be found in Table 1. The figures provide an indication of the extent of the 'problem' of desertion among Indian labourers. According to the report, desertions were more common among labourers who were stronger and healthier. Since planters wished to make it clear that such attempts were not worth repeating, those caught escaping were punished severely and forced to work out the period of their contract.

Workers wanted to desert their work-places because of appalling conditions. For example, according to the report of the Labour Commis-

TABLE 1
DESERTION AND RECAPTURE UNDER THE INDENTURE SYSTEM

Year	*Average Population of Indentured Labourers*	*Desertions*	*Recaptures*
1902	4,008	387	91
1904	1,184	244	61
1906	3,515	444	63
1908	4,983	788	189
1910	3,041	351	43

Source: FMS, Labour Commission 1910, p. 56 (National Archives, Kuala Lumpur).

were 351 deaths and 6,569 admissions to government and estate hospitals. In 1908, out of an average population of 5,000 there were 423 deaths and 6,058 admissions to hospitals [*FMS*, *Labour Commission*, 1910, 45]. Given the pro-plantation bias of the Commission it can be argued that the figures obtained were, if anything, conservative. Many who did not have the will or energy to desert, literally worked themselves to death as reported by the Labour Commission of 1910: 'these conditions engender a state of mind which renders many indentured Indian labourers careless as to whether they live or die and lessens their power of resisting disease' [*FMS*, *Labour Commission*, 1910, 45].

In regard to Indian labour, there are few recorded instances of collective resistance of the avoidance-type under the indentured system. In December 1912 about 1500 Telegu labourers of the Rantau Panjang-Sungei Tinggi stopped work and marched in the direction of Klang [*SSF* 5679/1912, Strike at Rantau Panjang Estate, Encl. 1]. A few months later, a similar incident took place in the Escot estate, Tanjung Malim, Perak. On 15 April 1913, about 280 labourers, mostly Telegus, stopped work and marched in the direction of Tapah, taking with them their women and children and possessions [SSF 214/1913, Coolies Strike at Escot Estate, Tanjung Malim, Encl. 1]. In 1915, over 100 workers of the Sipantas estate, Sabak Bernam, left the estate [*SSF* 4841/1915, Labour Trouble on an Estate at Sabak Bernam, Encl. 1]. In the 1920s, with the *kangani* system in force, mass desertions also reflected conflicts between *kanganis* and employers, for example, on the Sogmana estate [*Ramasamy*, 1990: 191].

What were the causes of these desertions? In the case of the Rantau Panjang–Sungei Tinggi estate labourers, the employers' non-payment of wages for six months and their harsh treatment precipitated the walk-out. In the case of the Escot estate, the labourers quit because they were unable to tolerate the high mortality rate, low wages, management's broken promises, and the frequent assaults against them. The labourers of the Sipantas estate left because they were unable to stand the unhealthy and harsh working and living conditions [*SSF* 5679/1912, SSF 241/1913, *SSF* 4841/1915].

In these three cases of labour resistance, employers felt that labourers deserted because they were instigated by certain ringleaders. The manager of Rantau Panjang–Sungei Tinggi estate convinced the Assistant Commissioner of Labour for Klang that Telegu labourers were the source of trouble – Telegus were judged 'an exceptionally bad class of labourers' [SSF 5679/1912, Encl. 3]. The Assistant Commissioner, like the manager, felt that certain ringleaders were responsible for the unrest and advised the District Officer of Kuala Selangor to take appropriate action against them. The District Officer in turn ordered the police to arrest and detain the labourers, especially those suspected of being leaders. Over a hundred workers were charged and jailed for breaching their contracts under section 229/30 of the Labour Code [SSF 5679/1912, Encl. 3]. The Deputy Controller of Labour was not pleased with this action because the alleged offences were committed before the Labour Code of 1912 came into force, and he urged the Resident Selangor to intervene to seek the release of the jailed labourers. The Resident did not. Later, just before their release, when approached by the Assistant Commissioner of Labour for Klang, they refused to return to their former estate, saying that they preferred to stay in jail or 'even walk into the sea and be drowned' [SSF 5679/1912, Encl. 2].

The manager of the Escot estate was also of the opinion that the unrest in his estate was the work of certain ringleaders. He sought the permission of the Controller of Labour to arrest and detain five men whom he suspected. However, this was denied and the Controller further warned him that if the alleged ringleaders were arrested he would not hesitate to institute legal proceedings against the manager [*SSF* 241/1913]. In the Sepintas estate, the manager also sought the permission of the Labour Department to arrest and detain labourers who fled but was turned down [*SSF* 4841/1915, Encl. 1].

As with Chinese plantation labour, it is possible to speak of collective resistance among Indian labour as more confrontational in nature in the 1930s and onwards. Factors such as the severe effects of the depression on the Indian labour force, its more domiciled nature, and restrictions on movement between estates, contributed to a situation where Indian workers developed more confrontational types of collective resistance, which invariably took the form of strikes. According to the Labour Department Annual Reports, Malaya witnessed ten strikes in 1934, three strikes in 1936, and quite a number of strikes in 1937 and 1938 [*Reports of the Labour Department, 1934, 1936, 1937, 1938*].

While the vast majority of the strikes were non-violent in nature there were a few cases of violent protests. In 1936, work stoppage in the Nova Scotia estate, Teluk Anson, Perak, turned into a riot as a result of the behaviour of a supervisor. Later, the labourers involved were dismissed. In 1937, a major strike in the Sungei Palu estate, Cameron Highlands, Pahang, turned violent and labourers' housing was burned. In the same year, about 100 labourers were involved in an assault against the Assistant Manager of the Belmont estate, Selangor. Apparently, the assault occurred after the labourers were punished for being late for muster [*Labour Department Reports, 1936, 1937*].

The most famous confrontational resistance engaged in by Indian labourers, and one that received international attention, was the Klang district strikes in 1941. Involving thousands of labourers, they began on 8 April and lasted until 17 May 1941 affecting an area of 1,500 square miles, stretching from Klang to Sepang in the state of Selangor. In Wilson's words: '... had the workers been better organised, with effective channels of communication to concentrations of estate labourers, the strikes would have been pan-Malayan in scope' [*Wilson*, 1981: 5].

Without negating the role of Indian nationalism through the work of the Central Indian Association of Malaya (CIAM) and its affiliate in Klang, the Klang Indian District Union (KIDU), the major impetus behind the strikes was opposition to new restrictions introduced to boost wartime production. Company representatives in London made it known that wartime conditions precluded wage increases. Labour's bargaining position became more difficult when it became known that wage fixing powers had been transfered from estate managers to company representatives in London. Further, the classification of rubber as an 'essential good' was a warning to labour not to disrupt production in the plantations. The inequity became most pronounced when companies making huge profits due to heavy demand for rubber refused to increase wages. In order to pre-empt a strike the employers offered a five cents increase (five per cent) which was rejected outright by the KIDU [*Stenson*, 1980: 60].

The ensuing strikes were crushed by the authorities using Indian troops, and resulted in five deaths and numerous injuries. A number of labourers were deported to India and a number restricted to other districts in Malaya. The state bought wholesale the planters' argument that the strikes were the work of Indian nationalist agitators. The High Commissioner, as ever sympathetic to the planters, delayed and eventually refused to order a public inquiry into the strikes as demanded by the Government of India through London, and in a letter to the Colonial Office in London remarked that there were 'no alleged grievances worth mentioning'. As Wilson remarked, 'the advantages of a "cover-up" clearly commended themselves to His Excellency' [*Wilson*, 1981: 26]

CONCLUSION

Control of and resistance by Chinese and Indian plantation labourers in colonial Malaya exhibit important similarities. Under indenture resistance seems to have been individualistic rather than collective. Labourers unable to stand the brutalities of plantation life simply deserted to try to become free labourers and earn better wages. Those caught escaping were severely punished. Collective resistance of both the avoidance and confrontational types emerged under the (Chinese) contract and (Indian) *kangani* systems of labour control. While there is no evidence of collective resistance among the Chinese before the 1930s, there is some evidence that Indians engaged in collective resistance of the avoidance type in the 1910s and 1920s, in the form of mass desertions.

The most important collective resistance of the confrontational type took place in the late 1930s among the Chinese and in the early 1940s among the Indians. In the case of the Chinese, the strikes in the Ulu Langat district were organised by the Malayan Communist Party, mobilising workers' grievances. In 1941, thousands of Indian labourers organised by the radical Indian organisation – Central Indian Association of Malaya – staged strikes in the Klang district to press for higher wages and better living conditions. In both cases, the strikes were forcibly suppressed by the colonial authorities on the grounds that they threatened the political and economic stability of Malaya.

REFERENCES

Adas, Michael, 1986, 'From Footdragging to Flight: The Evasive History of Peasant Avoidance Protests in South and South East Asia', *Journal of Peasant Studies*, Vol.13, No.2.

Bauer, P.T., 1948, *The Rubber Industry: A Study in Competition and Monopoly*, Cambridge, MA: Harvard University Press.

Blythe, W.L., 1947, 'Historical Sketch of Chinese Labour in Malaya', *Journal of Malayan Branch Royal Asiatic Society*, Vol.XX, Part I.

Breman, Jan. 1989. *Taming the Coolie Beast: Plantation Society and the Colonial Order in Southeast Asia*, Delhi: Oxford University Press.

Chai Hon Chan, 1964, *The Development of British Malaya 1896–1909*, Kuala Lumpur: Oxford University Press.

De Silva, S.B.D., 1982, *The Political Economy of Underdevelopment*, London: Routledge & Kegan Paul.

Drabble, J. 1973. *Rubber in Malaya 1876–1922*, Kuala Lumpur: Oxford University Press.
Jackson, R.N., 1961, *Immigrant Labour and the Development of Malaya*, Kuala Lumpur: The Government Press.
Jackson, J.C., 1968, *Planters and Speculators: Chinese and European Agricultural Enterprise in Malaya 1786–1921*, Kuala Lumpur: University of Malaya Press.
Jain, R.K., 1970, *South Indians on the Plantation Frontier in Malaya*, New Haven and London: Yale University Press.
Jain, R.K., 1990, 'Colonial Ambience and Immigrant Tamil Labour: A View from Malayan Plantations', unpublished manuscript.
Lim Teck Ghee, 1976, *Origins of a Colonial Economy: Land and Agriculture in Perak 1874–1897*, Penang: Universiti Sains Malaysia.
Lim Teck Ghee, 1977, *Peasants and Their Agricultural Economy in Colonial Malaya 1874–1941*, Kuala Lumpur: Oxford University Press.
Parmer, Norman, 1960, *Colonial Labour Policy and Administration: A History of Labour in the Rubber Plantation Industry in Malaya, c. 1910–1941*, New York: J.J. Augustin.
Puthucheary, J.J., 1974, *Ownership and Control in the Malayan Economy*, Kuala Lumpur: University of Malaya Bookstore Ltd.
Radkrishnan, P., 1975, 'The Role of Rubber in the West Malaysian Economy', Ph.D. thesis, Stanford University.
Ramasamy, P. 1990. 'The State, Unions and Plantation Labour in Malaysia', Ph.D. thesis, University of Malaya, Kuala Lumpur.
Sandhu, Kernial Singh. 1969. *Indians in Malaya: Some Aspects of Their Immigration and Settlement*, Cambridge: Cambridge University Press.
Selvaratnam, V. 1983. 'The Study of the Plantation System in Peninsular Malaysia: A Total Institution', unpublished manuscript.
Scott, James C., 1985, *Weapons of the Weak: The Everyday Forms of Peasant Resistance*, New Haven, CT: Yale University Press.
Stenson, Michael. 1980. *Class, Race and Colonialism in West Malaysia*, St. Lucia: University of Queensland Press.
Tinker, H., 1974, *A New System of Slavery: The Export of Indian Labour Overseas 1830–1920*, London: Oxford University Press.
Voon, P.K., 1976, *Western Rubber Planting Enterprises in Southeast Asia 1876–1921*, Kuala Lumpur: University of Malaya Press.
Wilson, H.E., 1981, *The Klang Strikes of 1941: Labour and Capital in Colonial Malaya*, Singapore: ISEAS.
Yeo Kim Wah, 1976, 'The Communist Challenge in the Malayan Labour Scene, September 1936–March 1937', *Journal of Malayan Branch of Royal Asiatic Society*, Vol.XLIX, Part 2.

Government Documents

British Resident Office, Negeri Sembilan State Secretariat File (NSSF) No.1349/1937, *Report on Inspection of Estates, Mines and Factories Employing Chinese Labour* (Gray's Report).
FMS, *Report of the Commission of Enquiry into the State of Labour in the Straits Settlements and Protected Native States 1890* (Labour Commission of 1890).
FMS, *Report of the Commission Appointed to Enquire into the Conditions of Indentured Labour in the FMS 1910* (Labour Commission of 1910).
FMS, *Annual Report of the Labour Department for the years 1934, 1936, 1937 and 1938*.
FMS, *British Malaya Opium Committee 1924*.
Selangor Secretariat File (SSF) 194/1937, *Strikes on Rubber Estates in Kajang Area*.
SSF 5679/1912, *Strike at Rantau Panjang Estate*.
SSF 241/1913, *Strike at Escot Estate, Tanjung Malim*.
SSF 4841/1915, *Labour Trouble on Estate at Sabak Bernam*.

Sugar Barons: Formation of a Native Planter Class in the Colonial Philippines

ALFRED W. McCOY

INTRODUCTION

During the depths of the Depression in July 1930, the Philippine Secretary of Agriculture, himself a wealthy sugar miller, gave a remarkable address to the sugar planters of Negros Occidental Province, the colony's leading plantation district:[1]

> Not as friends, but as brothers, I want to speak to you with my hand on my heart. You are leading an extravagant and ridiculous life. And until now you have not taken proper steps to change that mode of living which, if you keep [it] up, in five or six years may rob you of all your property and render you foreigners on your own land, as the Hawaiians are on theirs. You still have time to save your property. Leave the town and live modestly on your farms as our fathers and grandfathers did ten and twenty years ago. Give up your automobiles and stay on your *haciendas*. I led the same life when I was with you. But I came to learn that the life I was leading was extravagant and ridiculous after I had visited all the provinces of the archipelago.

For a senior government official to urge financial restraint upon a colony's planters at a time of grave economic crisis was, of course, unexceptional. The remarkable aspect of this speech was not its content, but the speaker and his audience. While a similar gathering of prominent planters in Sumatra, Cochin China, or Assam would have been European, this audience, like Agriculture Secretary Rafael Alunan himself, was Filipino.

As Secretary Alunan hinted with his references to their 'fathers and grandfathers', the Philippines, unique among the lands that gird the

---

Alfred W. McCoy, Department of History, University of Wisconsin – Madison, 455 North Park Street, Madison, WI 53706, USA. The author is grateful to the editors for their comments.

globe's tropical latitudes, had produced an indigenous planter class under colonial rule. While there were small rubber producers among the Malay peasantry and an ephemeral group of native planters in nineteenth century India, nowhere else in colonial Asia did an indigenous elite develop and then dominate a modern plantation industry. Even sugar milling factories, the bastion of colonial control in Java and Fiji [*Geertz,* 1963: 83–123; *Elson,* 1984: 36–8, 80–82, 134–5; *Moynagh,* 1981: 12–68], were largely Filipino-owned by independence in 1946. Beyond their contribution to the colonial economy, Philippine sugar plantations fostered resilient regional elites who would survive independence to play a central role in their country's post-war politics.

While historians of Ceylon's colonial plantations can still debate whether or not they were 'enclaves' insulated from surrounding Sri Lankan society (Meyer, this volume) this question simply does not arise in the study of Philippine plantations. Indigenous ownership in the Philippines gave its plantations a depth of social penetration and breadth of political ramification not found elsewhere in tropical Asia. In India or Indonesia plantations have left a problematic, even pernicious social legacy, but they are, in most cases, a part of the colonial past. In the Philippines, by contrast, Filipino sugar planters continue to exercise extraordinary political influence, as they have over the past century and a half. In studying Philippine plantations, the issue is not their marginality, but their centrality in the emergence of the modern Filipino nation state.

The formation of a Filipino planter class has important implications for any comparative analysis of the impact of plantations upon the colonised societies of Asia. Concerned with the long-term processes of social change within subject societies, contemporary historians of Asia have largely focused upon the native labourers and tended to ignore the expatriate planters, whether Dutch, French, or English. Often transients on limited-term contracts, these faceless Europeans, like the forgotten colonial companies that hired them, have disappeared from the current historical record. In the Philippines, however, we cannot ignore the planters. Not only did they they create a social institution with a lasting impact upon Filipino society, the planters shaped the processes of social change for workers and tenants. Instead of ameliorating the pressure of the plantation regime upon the workers, Filipino ownership excluded the *haciendas* from close scrutiny by either colonial officials or nationalist leaders. For nationalists engaged in an anti-colonial struggle, the simple fact of Filipino ownership exempted plantations from critical examination. Since the planters served as surrogate district officers for the colonial state, its officials did not interfere with their *de facto* autonomy. Judging from the record of violent uprisings in the sugar districts of

Negros Island, indigenous ownership intensified the harshness of plantation life.

Relations with the state, colonial and national, have been an essential factor in the evolution of Philippine plantations. Since their emergence in the nineteenth century, Philippine planters have survived three empires and two republics, manupulating each to protect their fragile plantation economy. While building a colonial industry, Filipino planters were well served by the colonial state and prospered in the last decades of Spain's rule. In particular, their alliance with the Spanish state won them land titles and troops for labour control. After the American conquest of 1898, Filipino planters struck a similar bargain with the new colonial state, winning, over the next 20 years, military support for suppression of a workers' revolt, duty-free access to the US sugar market, modern sugar milling technology, and an infusion of capital. As the Philippine state developed under American tutelage after 1907, the Negros planters acquired sufficient power to secure the requisites of their peculiar production system: subsidised national capital, access to the US market, and local labour control. Already entrenched in the Philippines' two sugar regions for a full half century, the planters of Negros and Luzon used their exceptional wealth and provincial power to dominate the new National Assembly after the 1907 elections, shaping the legislature in ways that made it a redoubt for the defence of their industry. As the country's first major capitalists, the sugar planters thus had the power, through their local control and national wealth, to overcome any threats to their industry's survival, whether nationalist revolution, militant unions, global war, national independence, or communist revolt.

Unlike other Filipino elites who produced coconut and rice, the sugar planters have been unified in their pursuit of common interests. Their role in shaping the conservative character of the modern Philippine state should not be underestimated. Since the 1930s, sugar planters have acted as a reactionary force in national politics, advocating a close alliance with the United States to protect their main market and blocking passage of domestic labour reforms to minimise their costs. Evolving over the past 150 years, this mix of provincial political power and national economic influence has thus given the sugar planters, particularly those from Negros Occidental Province, a preeminent position among the country's ruling elites.

Within their home region, Negros Occidental's planters are a clearly defined class who control most of the province's land, capital, and labour. The process of land aggregation that formed the large plantations in the late nineteenth century also created the basis for a closed upper class which has passed its property onward through the generations for over a century, fostering a familial style of power so strong that planter political control within Negros Occidental is near absolute. Outside their home region, the Negros planters act as an influential regional-cum-product elite who compete with rival elites for national support in the form of tariffs, capital, tax subsidies, and labour legislation. By concentrating its resources on a single crop, Negros Occidental has, since the late nineteenth century, produced over half of the Philippines' sugar, until recently the country's leading export. Thus, by the 1920s the Negros planters had become both a regional upper class and national economic elite.

Although their economic interests are similar at the national level, there are nonetheless differences in the character of the planters and the structure of their plantations in the country's two main sugar regions, Negros Occidental and Central Luzon. Instead of being created by foreign investment, Philippine plantations were a product of indigenous capital formed in disparate provincial settings. While Central Luzon's planters descended from Manila merchants who had acquired lands with their profits from mixed petty commerce, Filipino entrepreneurs in the Visayan Islands shifted more substantial assets from a major handicraft textile industry into the acquisition of larger sugar plantations on Negros Island. Unlike Central Luzon, where a lower capital base forced local planters to work their lands indirectly with tenants, the Negros elite quickly moved from tenancy to a directly-administered *hacienda* system.

Formed between the 1850s and 1880s, the Negros *hacienda* became a social institution without parallel in the Philippine archipelago. Instead of the usual Philippine pattern of dividing large land holdings into tenancies of two to four hectares, the Negros planters formed integrated, centrally administered farms. Usually sited along one of the island's rivers with a loading dock for small sailcraft, these fields surrounded a central compound with its own great house, workers' barracks, steam-driven mill, barns, chapel, and security force. Although fragmentation from bilateral inheritance eventually threatened the survival of the large Negros *haciendas*, the formation of family corporations with paid managers after the Second World War maintained their integrity and encouraged a slow shift to a less paternalistic style of labour management.

As capitalists in a colonial production system, the Negros planters acquired a Janus-faced role as both subjects and beneficiaries of imperial rule. Under Spanish law, the planters remained natives, *indios* or *mestizos*, who were still in theory subject to arbitrary administrative or legal action. Within the export economy, however, the planters were key brokers between the world market's demand for sugar and local labour supply, and thus became beneficiaries of the colonial order. From this privileged position, the Negros planters came to identify with the imperial system, thereby developing a life style, language, and set of social relations with their workers that was essentially colonial. Long after the colonial plantation disappeared from Sumatra or Cochin China, the Negros *hacienda* continues as a legacy of this problematic past.

Unlike British plantation managers in India or their Dutch counterparts in Indonesia, the Negros planters were thus political brokers, not mere corporate employees. Consequently, the Negros plantation served a political purpose that was almost as significant as its economic function. Even under Spain, the Negros planters served the state by generating revenues and maintaining local order, in effect becoming the equivalent of colonial district officers. As their influence expanded under both US rule and the Philippine Republic, the planters used their local control over land and labour to win political power at the national level. By the 1920s, the Negros elite's roles as local planters and national politicians were mutually reinforcing. Local control in Negros provided the planters with the electoral and financial base to become a dominant force in the national legislature. Simultaneously, legislative power gave the planters leverage to extract state resources that reinforced their local control. Thus, for the Negros planters their sugar *haciendas* came to assume a rationality that was not exclusively economic. On the labour-deficit Negros frontier of the late nineteenth century, the *hacienda* served as a bastion to provide the planters with both the physical security and stable labour force needed to produce a demanding colonial crop under chaotic conditions. After 1916, when elections became the basis for access to national power, the planters began to deploy their plantations as vote banks to form a perpetual, powerful sugar bloc in the national legislature. Since the end of colonial rule in 1946, it has been this political rationality that has assured the survival of the Negros *hacienda* despite its growing economic inefficiency.

Until independence, the economic and political rationalities of this colonial production system were generally consonant. By the 1950s, however, it was clear that the Negros plantations, with their antiquated labour-intensive methods, now required privileged access to a protected US market as the condition of their economic survival. Despite the inefficiency of operating a quasi-colonial plantation system within an independent state, the Negros planters had sufficient political power to preserve their *haciendas*. Political protection insulated the Negros plantation from changes in the world sugar economy, preserving an antiquated colonial production system and elaborating its economic irrationality for nearly half a century after independence.

Thus, we cannot understand the social history of the Negros plantation without an awareness of its overarching political logic. Just as Kelly (this volume) explains the Fiji sugar plantation as an expression of the politics of race and empire, so we must be aware of a political rationale underlying the Negros *hacienda* which has, since the 1920s, superseded its economic rationality. Whether under colonial regime or independent Republic, the simple fact of spatial control has provided the Negros planters with the means – land, capital and labour – to act as political mediators between capital and countryside. The Negros *hacienda's* application of distinct production methods, that were both labour and capital-intensive, to extensive lands made it an exceptionally powerful political unit. Given the planters' role as political brokers, a full history of the Negros plantation must cut back and forth between local and national levels, juxtaposing the planters' political courtship in Manila with their mix of coercion and control in the province.

The character of the planters and their plantation system has played a strong role in shaping the lives of the Negros sugar workers. Sugar cultivation in Negros required, until the technological changes of the 1980s, legions of ill-paid workers. To secure a permanent workforce on a labour-deficit frontier in the 1880s, the planters used a mix of paternalism, paramilitary control, and debt bonding. While living on these isolated plantations, workers willingly accepted the planters' largesse of free housing, rice rations, and guaranteed employment. At the same time, the coercive aspects of plantation life – beatings, rigid discipline, and paramilitary controls – inspired protracted worker resistance. Living in a mono-crop province with little alternative to sugar work, Negros labourers have, since the 1890s, practiced a politics that swings sharply between an almost craven acceptance of planter paternalism and episodes of sudden, violent resistance. Despite their compliance with both the social and political demands of plantation life, Negros workers have produced some of the Philippines' largest and most protracted uprisings – from the Papa Isio messianic revolt (1896–1907), through militant general strikes (1930–31), to the current revolutionary actions by the New People's Army (since 1972). Unable to withstand the planters' mix of local political controls and national military intervention, these volatile movements have usually collapsed without winning significant concessions for social change.

ORIGINS OF A REGIONAL ELITE

Over the past two centuries, the character of Filipino elites in the Western Visayas region, comprising the islands of Panay and Negros, has been shaped by the dramatic changes in the region's export trade. As in other economically advanced areas of the archipelago, the emergence of the Chinese *mestizos*, the offspring of unions between Chinese immigrants and native Filipinas, coincided with major changes in the economy of the Western Visayas during the late eighteenth century. Although the processes of change were somewhat similar in Central Luzon and Cebu where new rice and sugar producing elites appeared, there are also significant differences.[2]

Instead of moving directly from petty commerce to agriculture, the *mestizo* merchants of the Western Visayas passed through a distinctive phase as textile traders and manufacturers. Exploiting the improved logistics of European world trade, Chinese *mestizo* merchants in the towns of Jaro and Molo, today suburbs of Iloilo City, adapted traditional hand-loom weaving for export to markets in the Philippines, Europe, Asia, and Latin America. When competition from cheaper Manchester textiles captured their local markets, the Jaro-Molo mestizos shifted their capital into plantation sugar production on the neighbouring island of Negros. By the 1880s, the *mestizos*, in conjunction with a smaller group of Spanish colonial planters, would partition Negros Occidental's broad western plains into large plantations by dispossessing the peasant pioneers who had cleared the province's tropical forests. Once sugar land became the basis of wealth in the region, upward mobility into its upper class closed. Although some families did decline through inheritance or bad management, the regional elite became a landed aristocracy that has shared wealth and power within the region for four or five generations. Despite strong sectoral and factional conflicts, the Negros planter class has historically demonstrated a remarkable consensus in the defense of its interests against lower class challenge or external threat.

The rise of the region's textile industry is one of the most significant and least understood developments in modern Philippine history. In contravention of accepted scholarship, the Western Visayas experienced an indigenous proto-capitalist transformation that created all the prerequisites for the subsequent sugar industry – local capital, entrepreneurial skills, and a labour force both disciplined and dependent upon cash income [*Rivera*, 1982: 1–13]. By drawing up to a dozen weavers into crude Jaro-Molo factories, the *mestizo* merchants made Iloilo a substantial city of 71,060 in the 1850s, comparable to Chicago's 1856 population of 84,000. British consular records from the 1850s reveal that there were an estimated 60,000 looms in Iloilo Province, equivalent to one for every two women over the age of 12.[3]

Only a decade after the opening of Iloilo to direct foreign trade in 1855, its textile industry began a precipitous decline. Faced with competition from cheap Manchester goods, textile exports from Iloilo declined from

151,826 *piezas* (approximately, 696,836 metres) in 1861 to 12,700 *piezas* in 1869. Simultaneously, Manchester's local agent, British vice-consul Nicholas Loney, encouraged the *mestizo* merchants to invest their capital in sugar production. Loney sought these exports to fill the holds of British vessels discharging textiles at Iloilo and thereby finance the region's continuing imports of finished cloth. Between 1855 and 1898 the region's sugar exports climbed from 12,000 to 2,470,400 piculs, a change made possible by an increase in Negros Island's sugar mills from 7 to 821 and Negros Occidental's population from 18,805 to 308,272 [*McCoy*, 1982b: 307–14].

The entrepreneurial character of the Negros planter class and its uniquely important role in the emergence of the Philippine nation-state is thus a product of this region's early capitalist development. The century of textile production that preceded the sugar epoch gave the Negros planters a greater capital accumulation than elites in other regions. Depending on definitions, the textile industry experienced a proto-capitalist transformation while other regions such as Central Luzon were still in a less developed phase variously described as 'semi-feudal' or 'rent capitalism'. By drawing weavers from the countryside into crude Jaro-Molo factories, the textile merchants not only created a large city but they also began to transform the household economy of the island's villages. In the 1860s when the textile industry collapsed, many peasant families in western Panay were living on small farms, usually less than one hectare,[4] that were not viable without women's income from textiles. Although many weavers had remained in their villages, the character of their households had changed to the point that the capacity for autonomous subsistence was lost and partial wage dependence had developed.

Without this initial phase of capitalist development from the textile trade, the rapid growth of the Negros sugar industry would not have been possible. Capital and urban legal skills enabled the Jaro-Molo *mestizos* to divide an entire province into extensive, fully equipped plantations in only two decades. Migrating from the impoverished weaving villages of western Panay Island, it was peasant pioneers who cleared the first farms from the virgin tropical forests of Negros Occidental in the 1850s and 1860s. Since uncleared land was initially of little value on a labour deficit frontier, the *mestizo* merchants concentrated on acquisition of small peasant farms – by purchase of contiguous plots, foreclosure on usurious mortgages, and violent expropriation ratified with perjured land claims. In central Negros, for example, a minor Jaro cloth trader named Teodoro Benedicto laid the basis for generations of family wealth by fraudulently expropriating 11,200 hectares in less than a decade.[5]

Once the land was cleared and consolidated, *mestizo* planters shifted rapidly from a less extractive share tenancy to centralised plantations worked by debt-bonded wage labourers. While notarised loan documents from the 1870s required that peasant borrowers in arrears work as share tenants on the lender's plantation for an indefinite period, similar agreements a decade later instead required wage labour at fixed daily rates.[6] A striking exception to the tenancy patterns that characterised most Philippine regions, the centralised Negros *hacienda*, a kind of factory in the field, used a system of labour control that had first appeared in the crude Jaro-Molo textile factories of the 1850s.

Although our understanding of the reasons for this shift from tenancy to plantation production remains speculative, there is strong historical evidence that most planters came from the *mestizo* communities of Jaro and Molo. Indeed, one Jaro family has published a detailed history which allows us to trace the precise genealogical linkages from Jaro-Molo merchants to Negros planters. According to the Lopez genealogy, the family's founder was Basilio, a prominent *mestizo* timber merchant who served as *cabeza de barangay* (ward leader) in Jaro from 1842 to 1862, later acquiring a 148 hectare *hacienda* in Sarabia, Negros, sometime before his death in the 1870s. While Basilio's wealth was substantial, most of his ten surviving children built far larger fortunes and held a total of 2,210 hectares in the 1897 Negros census, still only a partial survey of family lands.[7] His eldest daughter Clara engaged in moneylending from her home in Iloilo and died in 1887 with a substantial estate valued at P125,750. Similarly, his eldest son Eugenio purchased 4,000 hectares of sugar land in Iloilo and Negros Occidental between 1879 and 1891, leaving 3,000 hectares to heirs at his death in 1906 [*Lopez*, 1982: xxxv–lviii].

Over the next six generations, the Lopez family's nine lines produced 2,676 descendants and some of the region's most important entrepreneurs and political leaders. As the most dynamic of Basilio's ten children, Eugenio acquired one of the largest landholdings in the region and established the most resilient line. Fourth among Eugenio's 12 children, Benito was elected governor of Iloilo in 1906 at the same time his first cousin was elected governor of Negros Occidental. Assassinated in 1908, Benito left two children, Eugenio and Fernando, who built a modest inheritance of two *haciendas* into one of the largest private fortunes in the post-war Philippines [*Lopez*, 1982: xliv–lii, 167–71].

Spanish colonial statistics show the emergence of the Negros planters as the region's dominant class by the 1890s. Of the 80,313.96 hectares enumerated in the incomplete Negros Occidental census of 1897, 62,162 hectares, or 77.4 per cent, was held in parcels larger than 50 hectares With holdings averaging 192 hectares, only 324 proprietors among a provincial population of 308,272 held nearly 80 per cent of the province's agricultural lands – an enormous concentration in comparison with other Philippine regions.[8]

Since the 1897 census was never finished, it is difficult to use for more detailed statements about the whole of Negros society. But by taking the statistics for Hinigaran, a major sugar-producing municipality with complete data, we can reconstruct a partial profile of Negros plantation society at the century's turn. Confirming data from other sources, Hinigaran shows a marked stratification in land holdings: 284 of the town's 321 farm owners, 88.5 per cent of proprietors, had lands under 20 hectares. Of the remaining 37 proprietors, 23 held 21 to 50 hectares, 12 had 51 to 499 hectares, and two owned over 700 hectares – Victoriano Seguenza with three *haciendas* totalling 723 hectares; and Teodoro Yulo, a Molo *mestizo* who was then the province's largest landowner with 18 *haciendas* totalling 2,525 hectares in three adjacent towns.

In any Negros municapality with extensive *haciendas*, one would expect to find a considerable number of landless rural workers who eluded enumeration. The 1896–97 census for Hinigaran also surveyed owners of four legged animals – horses, water buffaloes, cattle, pigs, and goats. The animal survey, more likely to reveal landless households than any other category, showed 748 owners of animals, a number well over double the 321 farm-owning households. If we arbitrarily assume five members per household, then we find that 5 × 748 animal owners = 3,740 people, leaving some 12,000 people not accounted for among Hinigaran's total population of 15,875. Again assuming an average household of five, there is a balance of 2,427 households, 76.4 per cent of the total, without lands or animals – most likely permanent plantation residents who were dependent upon the *hacienda* for their income and survival.[9] With these data, we can construct a tentative profile of household assets in Hinigaran that suggests a marked social stratification (see Table 1). When 76.4 per cent of households (2,427) have no agricultural assets of any description, ownership of 20 hectares or more by only 1.16 per cent of the total (37 households) represents an enormous concentration of wealth.

In a complex export economy, a survey of land holdings is at best an incomplete index of social stratification. Although Negros Occidental was the region's main sugar producer, it was dependent on Iloilo City for all its secondary services. Given the province's close economic integration with the city, we cannot consider the Negros plantations in isolation under the mistaken assumption that they constituted a self-contained society. If we define our unit of social analysis as the sugar industry, instead of a purely spatial entity, the plantation, then our image of social stratification changes. Although nearly absent from the list of Negros landowners, the European sugar brokers, bankers, and merchants resident in Iloilo City were clearly the dominant force in the region's sugar economy.

TABLE 1
AGRICULTURAL ASSETS BY HOUSEHOLD, HINIGARAN, 1897

Area	*Number of Households*	*Percentage*
Land holdings over 100 ha	8	00.25
Land 51 to 99 ha	6	00.19
Land 21 to 50 ha	23	00.72
Land 0.01 to 20 ha	284	08.9
Animals but no lands	427	13.4
Without lands or animals	2,427	76.4
Total	3,175	100.0

Source: Estadisticas 1896–97, Negros Occidental, Philippine National Archives.

Let us take the case of Hinigaran's preeminent planter Teodoro Yulo. The 1891 *fincas urbanas* (urban property taxes) for Iloilo show that he owned one warehouse valued at P9,000, nearly double the average for the city's 70 warehouses. Moreover, he owned four houses on Calle Real, the main commercial street, with a total value of P24,000 that approached the cost of the city's two most expensive structures – the provincial capitol and the offices of the Spanish firm Ynchausti worth P40,000 each.[10] But property is not the most revealing index of wealth and power in an entrepot city. Yulo might have been the lord of his Negros haciendas and lived grandly in Iloilo City, but he had little influence within the wider regional economy. During the 1890s, he did not record any loans, nor did he act as broker for the export of any sugar.

By the 1890s access to capital and foreign markets were just as important as control over land and labour. Indeed, if we are to apply the concept of 'colonial plantation society' more broadly to include the industry's service sector, then its apex was foreign, not Filipino. At the summit of the region's sugar society were two British banks, the Hong Kong & Shanghai and The Chartered, that loaned funds largely to the foreign merchant houses and sugar exporters, the *casas de comercio*. The banks and allied *casas* enjoyed substantial profits, high salaries, and a vast amount of property – residential and commercial houses, warehouses, retail shops, coastal shipping, liquid capital, and sugar lands. With only one exception, Isidro de la Rama, all the major Iloilo sugar merchants were foreigners – English, Spanish, Swiss, Scots, and Chinese.

Foreign control over this sugar society is shown in Iloilo City's trade statistics for the 1890s. During this period all Negros *haciendas* shipped their sugar across the Guimaras Strait for storage and export from Iloilo's river-front warehouses. Judging from the crop loan records and export statistics, the most powerful actors in the sugar economy were Iloilo City's five or six foreign import-export houses, the *casas*, which controlled almost all sugar exports, finished-good imports, and crop-loan capital. In 1895, six *casas* – four English, one Swiss and one Spanish – handled all but 55,151 piculs of the 1.75 million piculs of sugar exported. By 1919 two Filipino brokers, Jaro *mestizo* Julio Javellana and Molo *mestizo* Esteban de la Rama, son of Isidro, had emerged as significant exporters. Between them, however, they still controlled only 148,517 piculs of the 2.2 million exported – just 6.8 per cent of the total.[11]

The region's plantation economy was bound together by a chain of credit reaching downward from the European banks to the *casas*, the major *mestizo* planters and, finally, *hacienda* workers or tenants. Moving downward on this hierarchy of capital, at each level the amount of credit contracted and interest rates escalated. Through archival documents and the account ledgers of Ynchausti y Compania, a Spanish *casa*, we can trace the lines of credit control within this sugar society. In mid-1898, Ynchausti maintained accounts with two banks, the Hong Kong & Shanghai and the Banco Espanol Filipino, with curret balances of P222,172, of which P42,310 was an overdraft at six to eight per cent interest. Using this credit and a facility from the company's head office in Manila, Ynchausti's Iloilo branch, in turn, offered working capital to a half dozen major Negros planters at ten per cent interest payable in sugar deliveries at market price. With credit from Ynchausti ranging from P99,843 to P227,866 between 1895 and 1900, the Molo *mestizo* Manuel Locsin, owner of 320 hectares in central Negros, made loans totalling P49,578 in 1894–95 to smaller planters at 12 to 20 per cent interest which he secured by mortgages on their lands.[12]

Although records on smaller loans are fragmentary, it appears that *haciendas* used this credit to bond their sugar workers and rice share tenants to the land. In 1899, for example, an inventory of a Murcia *hacienda* of 259 hectares showed loans totalling P1,147 to seven tenants, eight labourers, and several migratory labour contractors. Most of the loans to *hacienda* workers ranged from P18 to P50, onerous for workers earning wages of P4 to P8 monthly.[13] Similarly, an inventory of a 408 hectares *hacienda* in 1891 showed that P3,833 in loans to rice tenants and sugar labourers accounted for 15 per cent of its total value of P26,376.50.[14] Since these loans were not meant to be repaid in cash, interest rates were usually not recorded. While the minimum interest rate for loans under P200 started at 30 per cent, the region's sugar and rice tenants often repaid in ways that meant an effective interest rate of over 100 per cent.

These loans were not merely passive financial transactions designed to return a simple profit through interest. At each level, the size, terms, and interest rate reinforced a social hierarchy geared to sugar production. For loans of up to P220,000 the largest *mestizo* planters paid only ten per cent, while smaller planters paid 12 to 25 per cent, and peasants 30 to 100 per cent. Paralleling the increasing interest burden, the penalties for default worsened with downward movement on the social scale. A large planter might have to sell a greater share of his sugar crop to the foreign *casa*; a small planter would lose his *hacienda* to the creditor; and a peasant farmer was reduced to a lifetime of plantation labour with his entire family. These credit relations and their contractual sanctions imposed strict controls upon all strata of this sugar society, planter and peasant alike, that were ultimately enforced by the Spanish state, its courts, and constabulary.[15]

Indeed, the industry's intense commercialisation fostered a system of plantation production riven with social tension. After a brief pioneering phase when tenancy was the most cost effective method for clearing land, the planters were sufficiently capitalised by the 1880s to abandon tenancy and bring most of the province's labour force under their direct control on the plantations. Although workers lived in *hacienda* housing and were paid a nominal daily wage, the practice of debt bondage and para-military control restrained workers' mobility and capacity to sell their labour freely, producing severe tensions within Negros plantation society.

THE PHILIPPINE REVOLUTION

During the decade of revolutionary turmoil from 1896 to 1907, the Negros planters demonstrated a remarkable political dexterity that enabled them to survive a chaotic era. In this tumult, the planters' emerging role as the region's political brokers became evident as they maneuvered both the revolutionary Republic and US colonial regime to preserve their local power.

Among all the regional elites, the Negros planters were the last to break with Spain, the first to ally with America, and had the shortest dalliance with the Philippine Republic. While other elites were, at least briefly, swept along in a tide of nationalist sentiment, the Negros planters were, above all else, capitalists concerned with preserving the prerequisites of their plantation economy – labour repression and access to foreign markets. In contrast to the Tagalog rice farmers around Manila who launched a national revolution against the presence of the Spanish friar estates, Negros planters were well served by the colonial order.

Negros planters held clear titles to extensive lands from the colonial state, relied upon the Spanish *Guardia Civil* to control their labour, and had reason to fear the disruption of revolution. Their estates were heavily mortgaged to future crop deliveries under contracts with no exemptions for war or disaster. Any violence that threatened the smooth flow of sugar across the Guimaras Strait into the Iloilo warehouses placed the Negros planters at risk.

In the late nineteenth century the Spanish regime, civil and ecclesiastical, was using its full resources to support the expansion of the Negros plantation economy. Except in cases of blatant abuse that aroused the ire of Spanish planters or municipal officials, the colonial state did little to block the wholesale alienation of land by Negros planters through usury, terror, or purchase. In 1876 and 1880, the Spanish regime promulgated decrees allowing purchase of unoccupied lands at low prices, and colonial authorities in Negros interpreted them in a manner that conceded the local planter class a near monopoly on land acquisition. After filing a land claim through complex legal procedures, planters then bid at public auction. By filing for large blocks of 100 to 350 hectares, the planters were able to exclude small farmers and win the auction with bids of P2.00 per hectare, far lower than the P30 market value for similar land.[16]

The Spanish state also used its colonial constabulary, the *Guardia Civil*, to guarantee plantation security and prevent flight by debt-bonded workers. Iloilo's Spanish-language newspapers of the 1890s are full of planter complaints about absconding workers.[17] To prevent flight planters confined workers to quarters after dark and mounted regular plantation patrols which were coordinated with those of the *Guardia Civil*. Instead of fearing the *Guardia Civil* as an agency of colonial repression as did much of the Tagalog elite, Negros planters petitioned the colonial government to assign these troops to their towns.

Concerned to defend both Spanish and Filipino plantations from brigands active along the length of the Negros mountain spine, the Spanish colonial state often adjusted its array of paramilitary posts to defend the province's plantation frontier as it expanded upward from the plains into foothills frequented by bandits. In a tactical review of 1891–92, for example,[18] the Spanish *Guardia Civil* decided to establish a new post at Pantayanan above the planter towns of Talisay and Silay to protect new *haciendas* there from mountain bandits. Since no government post was available, a local Talisay planter named Aniceto Lacson offered one of his own buildings and was awarded a generous rental of P400 per annum.[19] Showing a similar dexterity in mediating internal and external pressures, it would be the same Aniceto Lacson six years later who would command the Negros forces in their day-long revolution against Spain.

While the *Katipunan*, a secret nationalist society based in Manila, was launching the Philippine Revolution in August 1896, the Negros planters remained loyal to the colonial order. In early September, the Jaro *Ayuntamiento* (city council), then dominated by Filipino families with extensive Negros plantations, condemned the 'ungrateful Filipinos' and thanked Spain for raising them 'to a life of civilization and progress from a state of the most abject savagery and barbarism'.[20] In the northern Negros town of Cadiz Nuevo the Filipino planters passed a similar motion that closed with: 'Viva Alfonso XIII! Viva the Catholic Religion! Viva Spain and the Spanish Philippines!'[21] Encouraged by this outpouring of loyalty, the Spanish church and state organised an Ilongo Volunteer Battalion of 500 Filipino troops from Panay and Negros. With generous contributions from Spanish colonials and Filipino planters such as Eugenio Lopez and Julio Javellana, the Battalion was equipped with modern Mauser rifles and dispatched to Luzon.[22] Representing almost half of Spain's native volunteers, the Ilongo Battalion fought well against General Aguinaldo's revolutionary forces and won Iloilo City honors from the Spanish crown as the 'Most Loyal and Noble City'.[23]

Within Negros itself, the Revolution soon took on the character of class warfare. In marked contrast to Central Luzon, Negros soon produced a remarkable symmetry of opposing forces. On the side of Spain and order were the colonial state, the planters, both Spanish and Filipino, and the Catholic Church. On the side of social revolution were the sugar workers and a leadership cadre in the tradition of the indigenous spirit mediums, the *babaylan*. In November 1896, the Negros revolt began spontaneously on the flanks of a volcano, Mt. Kanlaon, that rises 8,000 feet above the *haciendas* of central Negros.

The movement's leader was Dionisio Sigobela, known as 'Papa Isio', a poor peasant who had been born in southern Iloilo and later migrated to the La Carlota sugar district at the base of Mt. Kanlaon. While working as a labourer on *Hacienda Esperanza*, he made an attempt on the proprietor's life and then fled into the mountains. Although Papa Isio's ideology fused animism with anti-Spanish nationalism, his movement remained a class, rather than racial war, waged by sugar workers determined to destroy the sugar plantations and return the island to peasant rice farming. In testimony before the US Congress and their official correspondence, US Army officers, who fought the rebels several years later, would attribute Papa Isio's popular support to the poor working conditions on the *haciendas*: debt bondage, low and unpaid wages, and whipping for even minor infractions.[24]

War on Negros began in late 1896 when the *Guardia Civil* received reports that workers were fleeing the *haciendas* to join Papa Isio's forces in the mountains above Isabela and Kabankalan. In January 1897, a *Guardia Civil* force of 200 riflemen fought 1,000 *bolo*-wielding peasants in two protracted battles, killing 110 in the first engagement and 85 in the second. For the next two years Papa Isio's forces retreated to the forest fastness of the island's mountain spine.

As Spanish power in the Visayas waned throughout 1898, the Negros planters made a temporary tactical alliance with elite revolutionaries in Luzon and the peasant rebels in the hills above their *haciendas*. Reacting to selective Spanish harassment of suspected subversives in 1897–98, the Negros planters began to reconsider their loyalty to Spain. For example, the wealthy planter Juan Araneta, a future commander of the Negros revolt, was imprisoned for ten months in 1897 on suspicion of sedition and later wrote with great bitterness about the suffering he and his fellow planters had endured.[25] With the declaration of the Philippine Republic in June 1898 and Spain's surrender to America in August, the power equation suddenly changed. On 6 November, the Negros planters, following the example of their peers on Iloilo's Revolutionary Committee, led 7,000 poorly armed *hacienda* workers in a bloodless march on the provincial capital at Bacolod. Assured of good treatment by the Negros planters in command, the colonial garrison of 200 troops surrendered without resistance and local Spanish planters were not molested [*Fuentes*, 1919: 55–63].

Shortly after his capture of Bacolod, General Araneta sought to still the class warfare with a decree on plantation labour and an alliance with Papa Isio. Issued on 9 December, the decree tried to control *hacienda* labour by requiring all planters to register their workers with the local police and issue a travel pass when any left the plantation. Similarly, the treaty with Papa Isio conceded him a localised autonomy in exchange for his promise to return all labourers entering his area without passes indicating payment of plantation debts.[26]

Simultaneous with these revolutionary alliances, the planters sought American protection from the threat of renewed class warfare. Only a week after capturing Bacolod in the name of the Philippine Republic, they wrote the US commander at Manila announcing their decision to 'welcome a protectorate under the great Republic of the United States, placing at once under the safeguard of such a noble ... nation the lives and plantations of this collectivity of Negros Occidental' [*Fuentes*, 1919: 127–9]. After General Araneta raised the American flag at Bacolod and planter delegations met with American commanders to request a US protectorate over Negros, the First Volunteer California Infantry landed at Bacolod on 4 March to the cheers of an enthusiastic crowd.[27] Its commander, Colonel James J. Smith, a future governor general, used his fluency in Spanish to form close personal and political ties to the planters. During his two year term, the Negros Occidental provincial government became a model for America's tutelary collaboration with other Filipino elites.

On 2 March, two weeks after General Araneta's flag raising ceremony, Papa Isio declared class war in a proclamation that attacked the United States and threatened the Negros planters with blood retribution for their treason. Indeed, his subsequent campaign around La Carlota was a fulfilment of that promise: several towns attacked and razed; 70 Spanish planters driven from the district; 12 Filipino planters murdered; and 56 *haciendas*, mainly Filipino-owned, burned out in La Carlota and Isabela.[28] Responding to an appeal from planters in these areas, General Araneta ordered 'extermination of all the *babaylanes*' and Colonel Smith assigned the bulk of his forces to a war against Papa Isio.[29] After losing 115 men in an attack on US forces at Ma-ao and 170 more against the US garrison at Bacolod in July, Papa Isio withdrew to the island's southern plateau and confined his military actions to sporadic raids. For the next eight years Papa Isio continued sporadic warfare, finally surrendering in 1907, five years after the revolution had ended in most areas of the archipelago. Concerned that his execution might inspire a revival of the revolt, the US colonial regime reduced his death sentence to life imprisonment.

THE AMERICAN ERA

Despite the ravages of revolution, the Negros planters preserved enough wealth and power to play an active role in shaping the Philippine state that emerged under American tutelage. As producers of the archipelago's main export crop, the sugar planters of Negros and Central Luzon were more capable than other Filipino elites of exploiting economic relations with the United States and the resources of the nascent Filipino nation. After the start of free trade with the United States in 1913, the colonial government and American investors found the sugar industry, with 80 per cent of its production on Negros, the only sector capable of absorbing a major infusion of capital. A combination of national and foreign capital financed construction of large sugar processing factories that made the Negros mill owners, now allied with their Central Luzon counterparts, the colony's most powerful indigenous interest group.

Once the US Army had driven the rebels into the mountains, the Negros planters turned to winning preferential access to the U.S. sugar market. Although their goal was simple, achieving it involved the planters in complex processes of change: creation of new political institutions for province and nation, and transformation of economic structures at regional, national, and international levels. Of equal importance, the planters' role as political brokers, implicit under Spain, became an explicit, even dominant aspect of their relationship with the new colonial power. In particular, the US decision to administer the islands indirectly through elected officials allowed the planters to elaborate their local power into national influence by becoming intermediaries between Negros and Manila. By mediating between the local and the national, the planters increased their power at both levels throughout the American era.

Restricted by a clause in the Treaty of Paris that placed Spain at commercial parity with the United States for a decade, American officials postponed major changes in the Philippines' export economy until 1909 and devoted their initial efforts to political reform. After establishing elected municipal and provincial governments, the United States transferred a portion of its legislative powers in 1907 to a Philippine Assembly chosen by a restricted electorate. Consequently, the first legislators from Negros Occidental were all prominent planters, including, for example, Agustin Montilla whose family showed 1,293 hectares in the 1897 census.[30] Although Negros Occidental would elect an occasional labour leader or independent over the next 80 years, the planters established a firm control on the province's legislative delegation that they would maintain through the 1987 congressional elections. With their success in the early elections, the planters formalised their role as the region's mediators with Manila, bartering both votes and financial support for constabulary, capital, and favorable customs regulations.

Although the new Assembly eventually became the basis of the Negros planters' national influence, they relied on producer associations and colonial lobbying in their initial push for access to the American market. In the early years of US rule, a combination of weak markets, revolution, and plagues affecting crops and water buffaloes impoverished the Western Visayas. Responding to this economic crisis, the region's planters formed the Agricultural Society of Panay and Negros and, in December 1903, petitioned the US Governor-General for abolition of the Dingley Tariff on imports of Philippine sugar.[31]

When the US Secretary of War led a congressional delegation to the Philippines in August 1905, the Negros planters testified at hearings in Manila and Iloilo to urge a reduction in tariffs. Appearing as the main planter advocate, Esteban de la Rama, son of the Iloilo sugar broker Isidro, claimed that the high cost of working capital and antiquated milling equipment would restrain his island's capacity to flood the US sugar market even if tariff barriers were removed.[32] In alliance with Central Luzon planters and the colonial regime, the Negros planters won duty reductions in 1909 and, finally, unrestricted access to the US market in 1913.

Once the American market was opened, the Negros planters concentrated their efforts on removing barriers to increased production, in particular the critical problem of capital. In 1903, the US Department of Agriculture reported that Negros planters were paying 20 to 56 per cent for crop loans, a sharp rise over pre-revolutionary rates.[33] When the US Secretary of War's inquiry visited Iloilo in 1905, Aniceto Lacson, a former revolutionary general who owned a 200 hectare *hacienda*, testified that loans at comparatively low rates of ten to 15 per cent still accounted for 22 per cent of his annual production costs. In an implicit attack on Illoilo's foreign merchant houses, the *casas*, officials of the Agricultural Association of Panay and Negros told the inquiry that 'the majority of the property owners have their estates eternally mortgaged without hope of freeing them on account of the usurious interest'. The only way to free the planter from the 'bonds of his humiliating condition of dependency' was through 'establishment of mortgage banks by the government itself'.[34]

The industry's antiquated milling technology was not improved until the First World War when a confluence of factors led to the rapid industrialisation of sugar milling: the start of US free trade in 1913, the simultaneous transfer of Hawaiian milling technology, foundation of the Philippine National Bank in 1916, and a wartime boom in world sugar prices. Significantly, the year 1916 brought two new institutions that facilitated the planters' consolidation of national power – the Philippine Senate, an upper house that assumed all colonial legislative prerogatives, and the Philippine National Bank, a major state financial institution controlled by Filipino political appointees. Using the state's legislative and financial power, the Negros planters had, by the mid 1920s, negotiated massive capital investments in new milling technology that simultaneously saved their industry from extinction and elevated them into a national product elite far wealthier than their peers elsewhere in the archipelago. Merging with Central Luzon planters, the Negros elite forged a unified 'sugar lobby' capable of shaping national government policy on tariffs, investment policy, and relations with the United States.

With the start of free trade in 1913, Hawaiian and mainland American investors followed the westward march of the US tariff wall to make some of the first major American investments in the Philippines. Desperate to break the ethnic monopoly of Japanese plantation workers that culminated in the 'Great Strike' of 1909, the Hawaiian Sugar Planter's Association (HSPA) won Philippine government permission to recruit Filipino labourers [*Takaki*, 1983: 145–64]. In 1918, the HSPA financed construction of the Hawaiian-Philippines mill in central Negros 'to show', in the words of its first annual report, 'a reciprocal interest in the development of the Philippines by an investment of capital in appreciation of the assistance which Hawaii and her sugar industry had received from Filipino labour'.[35] In addition to direct investment in three mills, the Hawaiian sugar industry fabricated much of the Philippine milling equipment in Honolulu and dispatched technicians to train Filipino operators.

Some statistics illustrate the scale and significance of the investment in new sugar factories. In 1912, the Bureau of Labour surveyed all 916 steam mills on Philippine plantations and reported a total capital investment of P9.8 million in these low horse-power machines. Significantly, this figure would represent only 6 per cent of the value of the country's 45 modern sugar factories in 1932. Similarly, on Negros 19 centrifugal factories, located at the heart of the island's sugar districts, replaced the 820 steam and animal driven mills operated by individual *haciendas*.[36] In the central Negros district of La Carlota, for example, a P12 million sugar milling factory with 147 kilometers of narrow-gauge rails supplanted 55 hacienda mills worth only P8,000 to P97,000 each. Over 90 per cent of the archipelago's sugar lands were still Filipino-owned, but by 1932 Filipino investors owned only 37 per cent of the mill capital compared to 62 per cent for American and Spanish colonials combined. Paralleling these investments, the colony's sugar exports increased from 157,000 tons (worth 15 per cent of Philippine exports) in 1913 to one million tons (or 62 per cent of exports) in 1930, almost all bound for the US market. The Philippines thus became doubly dependent on a single crop and a single market.[37]

The investment patterns for Negros were, in significant respects, different from those in the rest of the archipelago. Since it produced about 80 per cent of the archipelago's sugar before 1912, it was natural that Negros should have attracted 63 per cent of the Philippines' total mill investment. Since the remaining 37 per cent was divided among ten provinces, these figures indicate a remarkable concentration of capital in a single province. Of the P106 million invested in Negros, 46 per cent was Filipino owned, significantly higher than the national average of 37 per cent. Put differently, the Negros planters had accomplished the remarkable feat of acquiring P48.7 million for mill construction on their island, 29 per cent of the total invested in the entire Philippine sugar industry.[38]

In marshalling these vast resources, the Negros planters had the good fortune to have a well placed ally, General Venancio Concepcion, president of the Philippine National Bank (PNB). One of the four major Filipino merchants enumerated in the Iloilo City tax rolls of the 1890s, Concepcion had long been a local sugar broker and thus had many close friends among the Negros planters.[39] Appointed as second president of the PNB through Assembly Speaker Sergio Osmena's influence in 1918, Concepcion made two major decisions: first, a significant share of the PNB's capital would be invested in Negros crop loans and sugar mills; and, second, local planters would be encouraged to establish the new mills. By the time of his forced resignation in 1920, the PNB had invested P44.1 million in ten Filipino-owned mills, a figure equal to 30.7 per cent of all capital invested in Philippine sugar factories. Significantly, nine of the ten mills favored with PNB capital were in Negros.[40] Before investing, Concepcion approached the dominant planter families in each Negros district to suggest that they organise the local milling factory themselves.[41]

By normal banking standards, Negros planters were not creditworthy for such vast loans. In retrospect, General Concepcion's policy represented both a radical program of economic nationalism and a massive transfer of national resources to a single regional elite. In 1922, a special PNB auditor, Francis Coates, Jr., found that the bank had given these mills 'a capitalization 40 per cent greater than normal', making it unlikely that they would be able to repay the loans. Concepcion had accomplished this capital transfer by accepting the planters' *haciendas* as security on the basis of cursory estimates and then making the loans to Filipino milling firms with low capitalization. In the case of Bacolod-Murcia Milling Company, for example, the PNB had used 'wildcat' financing to extend loans of P2.2 million for 20 years after the planters had invested only P1,039.[42]

By 1921–22, Concepcion's policies had committed 26 per cent of the PNB's P167 million in assets to low-interest, 20 year loans for the Negros sugar mills. When combined with the long-term crop finance required to keep these indebted plantations from defaulting on earlier loans, this policy gave Negros an indefinite claim on a major share of the PNB's resources. Even in 1970, for example, the PNB still loaned P447.5 million for sugar, compared to only P130.4 million for rice and P44.2 million for coconuts. And even among the sugar regions, Negros made a special claim on the PNB. In 1938 its farms secured average annual crop loans more than three times higher those in the second-ranking Central Luzon province of Pampanga.[43] After capturing the PNB in the opening rounds of national politics, the Negros planters would need an effective lobby to defend their prize from poaching by rival elites.

This sudden infusion of capital accentuated differences within the Negros planter class and split a once homogeneous group into competing sectors of planters and millers. Before mill construction, all Negros planters had been united in their conflicts with the foreign *casas*. Once the mills began grinding, however, the wealthier Negros families were transformed into sugar millers whose interests were, at several key points, in direct conflict with those who had remained pure planters. Of the 19 mills contructed in Negros, ten were owned by leading planter families descended from Jaro-Molo merchants. In 1927, for example, the Lopez family pooled their assets to finance the Lopez Sugar Central in an area of north Negros where Eugenio Lopez (b. 1839) had first acquired large *haciendas* in 1879–80.[44] While pure planters remained rooted in local politics and society, the milling families soon entered the ranks of the national financial elite. Moreover, the combination of low-cost PNB finance and high profits enabled Filipino millers to gradually displace the foreign *casas* from the apex of the region's sugar society.

The rise of the Negros milling families was made possible by a new technology and allied contract system that allowed a significantly higher rate of profit and capital accumulation. Under the terms of the 30 year milling contracts signed with all *haciendas* in the districts surrounding their respective factories, the mills processed harvested cane in return for a 50:50 or 55:45 division of the milled sugar, an arrangement that allowed extraordinary profits. In 1925, for example, La Carlota Sugar Central retained 22,777 metric tons of sugar from the 124 *haciendas* in its district. By contrast, Hacienda San Jose, the most productive plantation in La Carlota's mill district, retained only 1,097 tons of sugar. As sugar prices and production increased over the next decade, the mill increased its income far faster than the plantations. By 1933 Hacienda San Jose's retained share had increased by 73 per cent to 1,903 tons, but the mill's had risen by 348 per cent to 79,168 tons. While millers concentrated capital and profit through technology, the planters continued to fragment their wealth through inheritance. Even in La Carlota district, where land ownership was far more concentrated than elsewhere in the province, there was a sharp decrease in the average cane area per planter over the space of a half century – from 186 hectares in 1925, to 154 hectares in 1950, and 64 hectares in 1975.[45]

The sugar mills' capital and profits were unequalled in the processing of the archipelago's other primary industries. Stripping *abaca* (Manila hemp) was a crude, on-farm task; coconut milling was fragmented; and even rice plantation proprietors did not involve themselves in transport or marketing. In short, no other product elite could approach the scale of capital accumulation achieved by the sugar millers, and they soon became a force in Manila's financial markets. Capitalised by a decade of spectacular profits during the 1920s, Negros sugar millers began investing in

Manila during the 1930s, a process that highlighted their continuing prosperity despite the Depression. In the mid-1930s, Jorge Araneta, president of Central Ma-ao, launched his family into the upper echelon of Manila entrepreneurs with a stunning series of acquisitions – the National City Bank Building for P14 million, station KZRM (Radio Manila), and a controlling interest in the nation's largest newspaper chain, the D-M-H-M.[46] Mining stock, timber concessions and Manila property also attracted considerable Negros investment during the 1930s, albeit without the drama of the Araneta acquisitions.

The split of the once homogeneous planter class into rival groups of planters and millers also produced a fundamental political conflict that divided Negros for a quarter century.[47] As production and prices began to rise in the 1920s, planters launched a campaign for legislative revision of milling ratios, culminating in the passage of the Sugar Act of 1952 which provided for a 70:30 division of milled sugar in favour of the planters. For almost 30 years, these ratios were the object of protracted political struggle waged throughout the Philippine political system. Although the mills had far larger financial resources, the planters still held local political leverage and used it as the decisive factor in this struggle.

This implicit conflict of interest first became explicit after the collapse of world sugar prices in 1930. While the mills continued to prosper, planters, who bore the bulk of production costs, were pushed towards bankruptcy and pressured the millers for concessions. In December 1929, the Negros planters formed the Confederation of Planters' Associations and six months later launched a campaign to raise their share of milled sugar by ten per cent to a 60:40 ratio.

Moreover, this internecine conflict allowed organised labour to mount the first challenge to the planters' dominion since the Papa Isio revolt of the 1890s. The industrialisation of milling during the 1920s had further expanded the region's wage labour force and created the conditions for its unionisation. With a work force approaching a thousand men each, the mills became something akin to free industrial towns set down amidst the uniform baronial oppression of the *haciendas*. Moreover, as the volume of sugar transiting through Iloilo City's warehouses rose from 2.1 million *piculs*, or sacks, in 1912 to 10.3 million in 1932, the demand for stevedoring labour kept pace, swelling the size of the urban proletariat. Without adequate wages or social security protection, the region's sugar workers initially formed mutual-aid societies in the 1920s. In Negros, planters and millers patronised rival worker mutual-aid societies, pitting them against each other and encouraging violence that soon brought government suppression. When the Depression lowered workers' wages and brought planter-miller tensions into open conflict, the region's first modern union, the *Federacion Obrera de Filipinas* (FOF), suddenly gained a mass membership among Iloilo City's stevedores. Spreading from the Iloilo docks into the mills and later the *haciendas*, the union played upon divisions among the sugar elites to gather a limited strength.

Led by an Iloilo city newspaper editor, José Nava, the FOF mounted a series of strikes in 1930 and 1931 which represented a serious challenge to the *casas* and mills. In May 1930, some 10,000 stevedores struck for 20 days against the city's sugar exporters and warehouse owner, the *casas*, demanding a restoration of pre-Depression labour rates. With only a temporary holding capacity at the factory sites, the Negros sugar mills could not store their production and the *casas* were forced to capitulate. Emboldened by this victory and another in August against the regional stevedoring monopoly, Vistranco, the FOF launched an ambitious, industry-wide general strike in January 1931. Over a six-week period, the local union branches, loosely coordinated by Nava's central office, managed a series of sporadic stoppages at the mills, their loading docks, and the Iloilo waterfront involving some 50,000 strikers.

Although the union's overt demands for better wages and working conditions were narrowly industrial, the strike itself was essentially political. Simply by establishing an independent workers' organisation, Nava and his FOF were challenging the planters' prerogative of serving as the region's sole intermediary with Manila. If this strike had been successful, the union would have ended the right of planters and millers to play patron, granting wages and benefits as a kindness that drew workers into a restraining reciprocity. In sum, the strike was a direct attack on elite political controls that were the political foundation of the region's sugar society. Not surprisingly, both planters and millers ignored the strike's industrial issues, which required rather small concessions, and responded in political terms.

At the outset of the strikes in January, the millers met in Talisay to plan their response and pleaded with the planters to support the strike-breaking effort. Angered at the mills' refusal to meet their demands for an improved division of milled sugar, the planters decided to remain neutral, in effect shifting their support to the union. Reflecting their political power and the weak position of the union at isolated factories, the mills quickly mobilised Philippine Constabulary troopers to break picket lines and hired scabs to resume milling operations. In Iloilo City, by contrast, the *casas* met much stronger resistance and were ultimately unable to break the FOF hold on the docks. Defeated on the city's waterfront, the *casas* decided to abandon it. By loading their sugar in the open sea along the Negros coast with strike breakers, the *casas* rendered the Iloilo waterfront redundant and broke the FOF's power base. By

1933, only 24 per cent of the Negros sugar crop was transiting through Iloilo City's warehouses. The mills and *casas* had thus deflected a challenge to their control from a militant union by pulling upon the levers of social control to preserve their industry's essential production facilities, the mills and plantations.

Frustrated in his independent bid for industrial power, Nava and the FOF forged a political alliance with the Negros planters. With the union's support in the June 1931 elections, the planter Isaac Lacson, son of revolutionary general Aniceto Lacson, defeated the incumbent Governor Agustin Ramos, an ally of the mills who had been instrumental in breaking the FOF strikes. To defend their interests, the millers contributed to the successful senatorial campaign of Gil Montilla, himself president of the Isabela mill. Although personal and factional alliances complicated political allegiances, the planter – miller cleavage remained the dominant factor in Negros provincial politics until passage of the Sugar Act of 1952.

Planters and millers still, however, maintained a united front at the international level. Convinced that Philippine imports represented a real threat to their markets, Cuban cane and American beet sugar producers combined with other US interests in the early 1930s to advance a Philippine independence bill with limited import quotas through Congress [*Kirk*, 1936: 73–101; *Friend*, 1965: 95–104]. Through an intense lobbying effort, the Philippine Sugar Association (PSA) preserved the industry's immediate access to the US market, albeit at the price of accepting a limited-term quota. After the Philippine Assembly ratified the new independence legislation with its 850,000 ton sugar quota, the US Congress then passed the Jones-Costigan Act which established a comprehensive system of sugar quotas for all exporters [*Friend*, 1965: 142–3, 145–8; *Kirk*, 1936: 115–35]. Although access was supposed to be limited to a finite transitional period, the Jones-Costigan Act in fact initiated a 40-year period, lasting until 1974, in which the Philippines had a sugar quota equivalent to about ten to 15 per cent of the US market. This quota thus became a key feature of Philippine–American relations after independence, and its preservation would oblige the sugar lobby to play an active role in shaping Philippine foreign policy.

THE SECOND WORLD WAR

The Japanese wartime occupation, like the Philippine Revolution, posed a major threat to the survival of the sugar industry. At worst a Japanese victory might mean loss of the US market and the industry's extinction.

Even if America were to win the war, a rigorous implementation of Japanese plans to substitute cotton for sugar would destroy the seed-cane needed for a quick postwar recovery. Finally, the simple destruction of warfare could make postwar recovery financially impractical. Manipulating both the guerrilla resistance and the Japanese military, the Negros sugar elite deftly deflected these threats.

Unlike the Manila political elite who accommodated themselves to the Japanese, the Negros sugar families remained pro-American. After the surrender of the United States Armed Forces in the Far East (USAFFE) in May 1942, the Negros planters played a key role in reorganising its remnants into a guerrilla resistance. The island's civil resistance governor was Alfred Montelibano, Sr., a past president of the Bacolod-Murcia Planter's Association, and all his deputy governors were also leading planters. Although the military resistance was led by young army officers of mixed social backgrounds, Montelibano's planter government managed to subordinate these forces to its larger goal of preserving the sugar industry [*Baclagon*, 1962: 83–97].

Despite their loyalty to the United States, the Negros planters tolerated collaborators among their ranks and co-operated with the Japanese Military Administration (JMA) to preserve key components of the sugar industry. Under its program of autarky for the Philippines, the JMA required a more balanced economy for Negros with a primary emphasis on cotton and a secondary focus on sugar to produce alcohol as a substitute fuel. In March 1943, the JMA appointed Japan's largest sugar milling company, Taiwan Seito Kabushiki Kaisha, owner of 13 mills on Taiwan and 23 on Java, as administrator of three American-owned mills in northern Negros. Several days later, Agriculture Commissioner Rafael Alunan, uncle of resistance governor Montelibano and the prewar president of the Philippine Sugar Association quoted above, met with Negros planters to urge their cooperation. Alunan announced that the Japanese-sponsored government was reducing the country's sugar area by over half, planting cotton on surplus sugar lands, and granting crop loans to sugar planters through Japan's Bank of Taiwan.[48]

Planter opposition to the Japanese cotton program was skillful and successful. Convinced that America would defeat Japan, the Negros planters, working through both anti-Japanese guerillas and nominally pro-Japanese collaborators, positioned their industry for a rapid post-war recovery. As the sugarmen clearly understood, there were two elements to the industry's survival – preservation of sufficient standing cane for seed stocks and protection of milling equipment. Targeting their attacks selectively, the guerrilla army destroyed only those mills that were producing alcohol for the Japanese or were owned by millers who had embittered the local planters [*McCoy*, 1977: 209–69]. In 1946–47, the first full postwar crop year, six of the 16 major Negros mills resumed production, a relatively rapid recovery in comparison with other regions and industries.[49]

The preservation of standing cane was, however, more difficult. Since the Japanese military had initially declared cotton the priority crop, Negros planters were hard pressed to maintain sufficient resources for sugar. By December 1943, there were only 11,580 hectares of sugar cane in Negros Occidental, an 88 per cent reduction in the prewar area. Faced with a loss of the standing cane essential for postwar replanting, the Negros elite, although pro-American, formed a tactical alliance with the Japanese Army to restrain the Japanese cotton companies. As the war brought Allied forces closer to the Philippines and reduced oil imports, the Japanese Army needed alcohol-butanol as a substitute fuel and designated sugar a critical war industry. When the Japanese-client government in Manila finally articulated its new package of financial and market incentives for alcohol production in early 1944, leading Negros sugarmen met in Bacolod to discuss the new program. Since sugar cane grows from a section of the stalk and not from seed, there was a fundamental conflict between chopping the entire cane to maximise replanting or milling it for alcohol. Although they accepted the Japanese incentives, the planters decided to use the finance to chop cane and replant for postwar seed stocks instead of milling to make alcohol for the Japanese Army. The head of the Negros planters' associations, Ildefonso Coscolluela, told the meeting: 'That the case now is not one of planting cane to be milled but planting cane now and keep on planting in order to save the sugar industry from extinction.'[50]

The disruption of enemy occupation, the years of guerrilla resistance, and the final battles of liberation did inflict great damage upon the province's infrastructure and its sugar industry. None the less, Negros Occidental, thanks in large part to planter success in manipulating both the Japanese and the guerillas, emerged from the war with sufficient seed stock and milling equipment for a slow recovery of sugar production.

THE POST-WAR SUGAR INDUSTRY

In the post-war era, Philippine sugar planters negotiated a difficult first decade of reconstruction to move into another two of sustained expansion. In both phases, Negros planters failed to modernise their labour-intensive farming methods or diversify their markets, thereby preserving the colonial character of their industry. By restoring a system that was inherently inefficient, the Philippine planters precluded the possibility of selling sugar profitably on the competitive world market. Instead, after independence in 1946 the Negros sugarmen turned to the United States, as they had after the Revolution of 1896, for capital and markets to finance reconstruction. To recover the inflated production costs, Philippine sugar required access to a protected US market where prices were far higher than international levels. If its sugar industry was to survive, Manila had to forego economic and diplomatic independence to negotiate a continuation of its quasi-colonial relationship with the United States. Although weakened by the war, the planters still had sufficient power to shape national policy to suit the imperatives of their industry.

The Philippine government estimated war damage to the country's sugar mills at P75.1 million, a major investment that could not be recovered without access to the American market. After a period of high inflation, US Rehabilitation Act payments of P26.8 million between 1949 and 1951 provided only ten per cent of the capital for rebuilding, leaving the Philippine Reconstruction Finance Corporation to commit the nation's resources for the bulk of finance. Instead of approving a $500 million package for Philippine reconstruction without conditions, the US Congress, responding to pressure from American interest groups, tied the Rehabilitation Act to the quasi-colonial Bell Trade Act. Passed in April 1946, the Bell Act granted US quotas for sugar and other traditional exports on condition that the Philippines give Americans commercial 'parity' with its own citizens, a concession then barred under the country's Constitution.

Consequently, the sugar industry, the Rehabilitation Act's prime beneficiary, mounted a national campaign for approval of the 'parity amendment' to the Philippine Constitution that would grant Americans this extraordinary privilege. Closely allied with the Elizalde mills and other sugar interests, President Manuel Roxas used the authority of his office to win an overwhelming mandate for the parity clause in a March 1947 plebiscite. Almost simultaneously, Roxas also concluded the Military Bases Agreement with the United States. Stripped of its complexities, the Roxas administration, influenced by the sugar bloc, had struck a post-colonial bargain – conceding territory for U.S. military bases in exchange for war damage funds and access to the American sugar market. Under such terms, it would be difficult for the Philippines to form an industrial base to challenge the influence of the sugar bloc [*Shalom*, 1981: 43–67].

Although ultimately successful, the dislocations of postwar reconstruction accentuated the internal conflicts within Negros of planter versus miller and labour versus capital. As these conflicts intensified, Governor Rafael Lacson, member of a prominent planter family, exploited both the local fissure and the weakness of the new Republic to become an autonomous provincial warlord, the first of this new political type. After forming a private army of 190 'Special Police', Lacson unleashed a terror campaign that disarmed the guards at the Araneta sugar mills and drove the family's political supporters out of the province. As a reward for organising the massive fraud in Negros Occidental that delivered the winning margin for Elpidio Quirino in the 1949 presidential elections, Lacson was given *de facto* control over local units of the Philippine Constabulary – a concession that removed any restraint on the Governor's use of force.[51] Moreover, to strengthen his political base against the mills, Lacson revived the planter-worker alliance by supporting the province's radical union, the FOF, still led by Jose Nava. After repeated failures, the Negros FOF finally won a strike against the La Carlota Central in the late 1940s when Governor Lacson used his Special Police to force concessions from the mill. In the manipulative style of earlier planter politicians, Lacson then used his goons to destroy the FOF and organise a string of municipal unions under the direct control of local planters.

By the early 1950s, however, resolution of the planter–miller conflict allowed reimposition of elite controls. As the original 30 year milling contracts approached expiry, the planters finally raised their share of milled sugar from 50 to 70 per cent with the passage of the Sugar Act of 1952. Using the political strength that came from their spatial control over the province and its plantations, the Negros planters scored a significant political victory over the wealthy millers who were now well established within Manila's financial, political, and social circles. In retrospect, the planters' triumph stands as testimony to the political power of the Negros plantation.

Once the planter-miller breach was healed, the reunified provincial elite, working through the state, dealt harshly with those who had played upon their divisions. In 1952 Governor Lacson was sentenced to death for the murder of a mayoral candidate who had been tortured and executed by his Special Police.[52] Similarly, several FOF leaders were charged with subversion in 1951 and their president Jose Nava eventually died in prison awaiting execution.[53]

After rehabilitation and the resolution of internal conflicts, the sugar industry used the state to remove the last barrier to its long-term prosperity, the declining US sugar quota. Through a major lobbying effort, Philippine sugar won concessions in the Laurel-Langley Trade Agreement of 1955 that modified the Bell Trade Act by changing its limited-term, fixed quota into a more flexible one set at a percentage of the US market.[54] Although the Laurel-Langley Agreement stipulated staged increases in US duties to encourage the Philippines' transition to a world market, the Cuban revolution negated these provisions. After banning Cuban sugar imports, the United States awarded the Philippines a major quota increase, and the Republic's production shot up from 1.3 million tons in 1960 to 2.2 million tons in 1972. However, instead of increasing production by more efficient farming on existing lands, Filipino planters, particularly those in Negros and Panay, extended cane cultivation into marginal uplands and expanded their work force of unskilled labourers. Between 1959 and 1972 the Philippines' sugar area more than doubled from 193,000 hectares to 424,000, while production per hectare dropped from 112 piculs to 68.[55] In marked contrast to the modern technology of the rebuilt mills, the planters were perpetuating their inefficiency, thereby increasing their dependence on the US market.

The decades of post-war prosperity facilitated the emergence of the Negros millers as a major force within the national financial elite. Fuelled by free trade after 1913, several Negros millers, notably Esteban de la Rama and Jorge Araneta, had acquired national assets and influence. But their investments remained rooted in the Western Visayas and they failed to complete their shift to Manila. After the war, however, leading millers made the transition and succeeded in lodging themselves in the upper stratum of the nation's financiers. In 1948, Jose Yulo, a Negrense married into the Araneta family of Central Ma-ao, purchased the Canlubang Sugar Estate, a mill just south of the capital, from the leading Manila financier Vicente Madrigal. After acquiring three Negros mills in the postwar slump, J. Amado Araneta, Jorge Araneta's son-in-law and Jose Yulo's brother-in-law, established himself as a major force in the Manila real estate market. But it was the Lopez brothers who made the most rapid rise, from provincial bus operators in the 1930s to the apex of the nation's financial elite by the 1960s.

After their father's assassination in 1908, the Lopez brothers, still infants, inherited two *haciendas* including one in La Carlota with 200 hectares of cane, an unexceptional endowment by Negros standards.[56] Through a mix of political influence and financial maneuver, Eugenio Lopez built upon this modest financial base to increase the size of the consortia he controlled: Iloilo Transportation Co. was valued at P250,000 in 1937; Panay Autobus was acquired for P565,000 in 1938; Far Eastern Air Transport Inc. was sold in 1947 for P2.8 million; Binalbagan-Isabela Sugar Milling Co. was acquired for P2.7 million in 1951; and Meralco Security Corporation had a total book value of P1.022 *billion* in June 1973.[57] One of the Philippines' largest and most profitable corporations, the latter conglomerate had grown from the Lopez acquisition of the

Manila Electrical Corporation (Meralco) in 1961–62, an American-owned utility that had held a monopoly on Manila's electrical supply during the US colonial era. Although there are no precise statistics on the full scope of the Lopez holdings, it is generally believed that Meralco's acquisition made Eugenio Lopez the Philippines' most powerful financier during the 1960s.

There was, moreover, a symbiosis between Eugenio's rise from provincial merchant to national financier and Fernando's climb from mayor of Iloilo City to vice-president of the Philippine Republic. Their supple fusion of business and politics made the Lopez brothers the region's ultimate political brokers. Backed by family and allies, Fernando was appointed mayor of Iloilo City in 1945 and elected to the Philippine Senate in 1947. Using familial and factional allies, he defeated José Yulo, a leading sugar miller, for nomination as President Elpidio Quirino's running mate at the 1949 Liberal Party convention [*McCoy*, 1977: 491–4]. After finishing his term as vice-president, Lopez returned to the Senate for another two terms. The Lopez brothers then allied themselves with Senator Ferdinand Marcos from Ilocos Norte and Fernando ran successfully as vice-president on his ticket in 1965 and 1969. Although Marcos paid tribute to Fernando as a 'close friend and associate' in 1970,[58] shortly thereafter the two families broke over business conflicts and engaged in a bare-knuckle political brawl that splashed across the front pages of the Manila press. Despite a superficial rapprochement, Eugenio Lopez was still manoeuvring to effect Marcos's political demise when the president's declaration of Martial Law intervened in 1972.

Even after their move to the apex of national power, the Lopez brothers continued to identify with the Negros planters in an effort to maintain political control over their home region. Indeed, the basis of Fernando's political success was a long-term relationship with several leading Negros planters known collectively as 'the sugar bloc'. Led by Oscar Ledesma and Alfredo Montelibano, Sr., the group dominated the Planters' Federation and controlled the region's congressional delegation [*McCoy*, 1977: 441–2, 450–59]. Indicative of their influence within the industry, Montelibano served as president of the National Federation of Sugarcane Planters in 1945–46, as did Oscar Ledesma in 1937–38 and 1950–52. Significantly, in 1951 when the Lopez brothers acquired control over the Binalbagan-Isabela Sugar Milling Company, then the largest in Negros, they did so as leaders of the district planters who were then embittered towards the incumbent milling family.[59] Even through they subsequently purchased another mill in Central Luzon, the Lopez brothers still retained their two Negros *haciendas* and their political identity with the province's planters. Through their leadership of the planters, a strong presence in the legislature, and national financial power, the sugar bloc maintained its dominion over Negros Occidental until President Marcos declared martial law in 1972.

CONCLUSION

The history of the Negros planter class illustrates the importance of plantations in shaping the modern Philippines. By developing capitalist plantations under colonial rule, the Negros planters, in concert with those from Central Luzon, not only transformed their own region, they shaped the character of an emerging economy and nation state. Aside from providing export income for the colony's development, plantations absorbed vast amounts of state capital, a product of the planters' success in promoting government policies essential to their industry's survival. Once fully established in the 1920s, plantation sugar production required a continuing economic dependence on the United States that was ultimately paid for in military base concessions. Internally, the sugar plantations, particularly on Negros, still relied on labour-intensive methods that required low wages, poor conditions, and a concomitant repression to minimise costs within an inherently inefficient farming system. In short, sugar plantations inclined the Philippines toward a path of dependent development that preserved both an inefficient industry and an inequitable society.

In Negros Occidental, sugar plantations formed a provincial society polarized between the country's wealthiest regional elite and its most volatile working class. In a real sense, the Negros plantations had preserved a colonial social system, with a planter elite still relying upon the military to preserve its property and an impoverished working class open to an ideology of radical social change. As the world price of sugar plunged in the early 1980s, Negros suffered a severe economic crisis, rising insurgency, and mass malnutrition. During his 15 years of dictatorship from 1972 to 1986, President Marcos turned the power of the state against the Negros planters, the allies of his enemies the Lopez brothers, thereby breaking their political influence and bringing them to the brink of bankruptcy. Denied both state resources and a role as the province's political brokers, the Negros planters could no longer restrain the spread of the communist New People's Army, whose guerrillas took control of much of the island during the early 1980s. When the Marcos regime's collapse allowed restoration of the Philippine Congress in 1987, however, the Negros planters regained sufficient influence to save their *haciendas* from economic eclipse. As the state dispatched its diplomats to Washington to win new sugar quotas, deployed its troops in the mountains of

Negros, and ordered its banks to provide concessional finance, the province's plantations pulled back from the edge of extinction.

Within Negros, the sugar planters remain an entrenched elite clinging to a plantation system whose rationality is political and to a crop that is fast approaching the end of its product cycle. An impoverished mass of plantation workers still lacks the material, educational, or political means to change their lives through mobility or migration. In the 1990s, Negros remains wedded to an antiquated plantation system, increasingly inefficient by world standards and surviving through a political paralysis that bars any initiative for change.

NOTES

1. *Manila Times*, 6 July 1930.
2. For sources on the emergence of the Chinese *mestizo* elite, see Wickberg [1965: 28–36, 135–45], Larkin [1972: 84–99], Cullinane [1982: 268–71].
3. A more detailed history of the textile industry is found in McCoy [1982b: 301–7].
4. *Estadisticas de los Terrenos Agricolas*, Miagao, Iloilo, 1897, Philippine National Archives (PNA).
5. *Real Audencia Territorial de Filipinas, Tribunal Pleno, Expedience sobre deslinde de los Montes del Estado ... con la hacienda de Sn. Bernardino, Ysla de Negros*, legajo 58, no.1195, 1887, Philippine National Archives (PNA).
6. On tenancy arrangements, see a series of eight contracts between Spanish planter Adolfo Lazarte and local peasants similar to the one contained in 'Reconocimiento de deuda y obligacion al pago a Adolfo Lazare por Rita Miranda', No.218–252, *Protocolos Negros* 1735, 5 May 1871, PNA. The volumes of *protocolos* for the 1880s contain contracts that show a slow shift to wage labour as a requirement instead of tenancy.
7. Tributos Iloilo, 1852, Varias Provincias, PNA; 'Historical Data of Iloilo', Philippine National Library (PNL); *Estadisticas de los Terrenos Agricolas, Negros Occidental*, 1897, PNA.
8. Ibid.
9. *Estado Demonstrative del Numero de Ganaderia*, Ginigaran, 1896, *Estadisticas*, PNA; *Estado del Numero de Habitantes*, Ginigaran, 1896, *Estadisticas*, PNA.
10. Fincas Urbanas, Iloilo City, 1891, PNA.
11. *El Porvenir de Visayas* (Iloilo), 5 June 1896; *Sugar Central and Planters News* (Manila), January 1920.
12. Information on the Ynchausti credit transactions are contained in the company's ledgers, or *Caja Mayor*, for the years 1895–1900, Ynchausti y Cia., Iloilo City. For these loans, see *Protocolos, Negros Occidental*, Vol.1783, pp.1542, 1546, 1552, 1558, 1564; Vol.1787, p.1248; and, Vol.1788, p.1576, PNA.
13. *Arrendamiento de finca rustica por Enrique Jesus Delgado y Beltran a Mateo Henares y Requejo*, Bacolod, 13 June 1899, *Protocolos, Negros Occidental*, Vol.1800, p.568, PNA.
14. *Testamento de Palabra por Balbina Buenaventura y Hernandez de Amenabar*, Bacolod, 12 June 1891, *Protocolos, Negros Occidental*, Vol.1781, p.228, PNA.
15. See, for example, *Adjudicacion pago deuda por el Juzgado de Primera Instancia por Juan Manuel Locsin a Lucio Lacson*, Bacolod, 17 Dec. 1888, Vol.1767, p.566, PNA; Obligacion Simple por Leoncio Costan et al. a Manuel Suarez, Bacolod, 8 Aug. 1872, *Protocolos, Negros Occidental*, Vol.1737, p.350, PNA.
16. See, for example, *Venta terreno baldio realengo por Administracion de Hacienda*

Publica a Vicente Gayona y Grandilla, Bacolod, 14 Nov 1890. *Protocolos, Negros Occidental*, Vol.1773, p.1039, PNA.

17. *El Porvenir de Visayas* (Iloilo City), 4 Oct., 6 Nov., 18 Nov., 20 Nov., 1895.
18. *Expediente sobre distribucion y cambio de puestos de Guardia Civil en la region Occidental de Negros*, 1891, *Guardia Civil*, PNA.
19. *Expediente sobre el establecimiento de un puesta de la Guardia Civil en el Sitio de Pantayanan del pueblo de Minuluan* ... 1892, *Direccion General de Administraction Civil, Guardia Civil*, PNA.
20. *El Porvenir de Bisavas*, 7 Sept. 1896.
21. *El Porvenir de Bisayas*, 12 Oct. 1896.
22. *El Eco de Panay* (Iloilo City), 30 Dec. 1896, 16 March 1897.
23. Monteverde y Sedano [1898: 102–8], *El Eco de Panay*, 4 May 1897.
24. For a fuller account of this revolt, see McCoy [1982a].
25. *La Independencia* (Manila), 28 Dec. 1898.
26. Gen. Araneta, Letter 18 Dec. 1898, R-642, Philippine Insurgent Records (PIR), U.S. National Archives & Records Service (NARS).
27. *La Libertad* (Bacolod), 5 Nov. 1899; Acta, 18 Feb. 1898, No.4201, R-642, PIR.
28. Acta, 2 March 1899, PR-58, PNL; *La Libertad*, 25 July 1899; Telex, to: Smith, 17 June 1899, RG-395, E-2619, NARS; Telex, to: Smith, 21 March 1899, RG-395, E-2619, NARS; Lt. Victor Duboce, Letter, 3 May 1899, RG-395, E-2619, NARS.
29. To: Presidentes Locales de Negros Oriental, 12 April 1899, No.4696, R-643, PIR.
30. Philippine Assembly, *Official Directory: First Philippine Legislature* (Manila: Bureau of Printing, 1908), pp.70, 73, 75; Estadisticas de los Terrenos Agricolas, Negros Occidental, 1897, PNA.
31. *El Porvenir de Bisayas*, 6 Oct. 1896; U.S. Philippine Commission, *Third Annual Report of the Philippine Commission, Pt. I* (Washington, DC: GPO, 1903), pp.353–4.
32. U.S. War Department, *Public Hearings in the Philippine Islands* (Manila: Bureau of Printing, 1905), pp.30–35.
33. U.S. Senate, 60th Congress, 1st Session, *Senate and House Documents and Reports* (Washington, DC: GPO, 1907), pp.14–15.
34. US War Department, *Public Hearings in the Philippine Islands*, pp.150–52, 161–3.
35. *First Report for the Hawaiian-Philippine Co., Fiscal Years Ended September 30, 1921, and September 30, 1922* (Manila), pp.14–19.
36. Philippine Islands, Department of Commerce and Police, Bureau of Labor, *Third Annual Report of the Bureau of Labor* (Manila: Bureau of Printing, 1912), pp.57–79; Philippine Sugar Association, *Summary of Information about the Philippine Sugar Industry* (Manila: 17 Nov. 1933), pp.25–6.
37. Bureau of Labor, *Third Annual Report*, pp.57–79; Philippine Sugar Association, *Summary of Information*, pp.12–15, 25–6; Association de Agricultores de La Carlota & Pontevedra, *Memorial Anual Ejercicio 1924–1925* (Iloilo, 1925), pp.1–22.
38. Philippine Sugar Association, *Summary of Information*, pp.24–6.
39. *Contribuciones Industriales*, Iloilo, PNA.
40. Letter from V.C. Conception, to: Board of Directors, Philippine National Bank, 21 June 1920, File: Philippine National Bank – 1920, Box 169, Quezon Papers, PNL.
41. Ramon M. Lacsamana (ed.), *Negros Occidental 1938–1938 Year Book* (Bacolod, 1939), pp.F1–3.
42. Francis Coates, Jr., 'Report of an Examination of the Philippine National Bank as of at the close of business November 30, 1919', Bureau of Insular Affairs, War Department, NARS.
43. Bureau of Census, *1971 Philippine Yearbook* (Manila: Bureau of Census, 1971), p.301; Commonwealth of the Philippines, Commission of the Census, *Census of the Philippines, 1939: Summary for the Philippines, II* (Manila: Bureau of Printing, 1941), p.995.
44. Lacsamana, *Negros Occidental Year Book*, pp.F3–31; Lopez (ed.) (1982): Vol.I, pp.xliv–xlvi.
45. Asociacion de Agricultores, *Memoria Annual 1924–1925*, pp.33–6; Asociacion de Agricultores, *Memoria Annual 1933–1934*, pp.37–43; Asociacion de Agricultores,

Memoria Annual 1949–1950, pp.8–12; Asociacion de Agricultores, *Annual Report Crop 1974–75*, pp.12–48; Lacsamana, *Negros Occidental Year Book*, p.G3.

46. Lacsamana, *Negros Occidental Year Book*, p.C11.
47. For an analysis of this period in the region's history, see McCoy [1984].
48. Minutes of the Meeting of Sugar Planters at the Cosmos Building, Bacolod City, on 24 May 1943, at 10 a.m., Box 11, RG 16, MacArthur Memorial, Norfolk, Virginia.
49. *Sugar Central and Planters News*, 19 Dec. 1947.
50. Philippine Sugar Association, Visayas Branch, Minutes of the Meeting of the Managers of Sugar Centrals Held in the Conference Room of the Philippine Sugar Association on 24 May 1944, File: CAC 9144, Ynchausti y Compania, Iloilo City.
51. *El Tiempo* (Iloilo), 28 Jan. 1946, 13 Oct. 1947; *People v. Rafael Lacson et al.*, Criminal Case 3220, Court of First Instance, Bacolod; Letter from Rafael Lacson to President Quirino, 31 July 1950, as Sept. 1950, Quirino Papers, Vigan, Ilocos Sur; *El Civismo*, 23 July 1950, 20 Aug. 1950.
52. *People vs. Lacson*, Criminal Case 3220; *Philippines Free Press*, 28 Aug. 1954.
53. *People vs. Guillermo Capadocia*, Criminal Case 2878, Court of First Instance, Iloilo City, Decision, 5 April 1952; For a fuller discussion of this case, see McCoy [1977: 717–38].
54. Shalom [1981: 94–98]; Jose E. Romero *et al.*, 'The PSA Record of Promoting and Protecting the Sugar Industry', *Sugar News* (Manila) 49, Nos.2–3 (Feb.–March) 1973, pp.58–66.
55. *Sugarland* (Bacolod) 11, No. 4, p.25.
56. Asociacion de Agricultores de La Carlota y Pontevedra, Inc., *Memoria Anual de la Junta directive y Comite* (La Carlota, 1941), p.14.
57. *The Times*, 12 Jan. 1939; *Makinaugalingon* (Iloilo City), 2 March 1938; Andres Soriano, Eugenio Lopez, Memorandum Agreement, n.d., Quirino Papers; Letter from Pio Pedrosa to President Manuel Roxas, 3 May 1947, File: Memorandum, Misc., Roxas Papers, PNL; *Sugar News*, Jan. 1952, p.31; *Isabela Sugar Co., Inc. vs. Eugenio Lopez et al.*, Order, Court of First Instance of Manila, Branch VII, Civil Case No. 14831, Quirino Papers; Meralco Securities Corporation, *Annual Report 1973* (Manila), pp.28–35.
58. Honorable Fernando Lopez, Vice President of the Philippines, *Silver Jubilee in Public Service, 1945–1970* (Iloilo City: Yuhum Press, 1970).
59. Jacinto Montilla, 'Squandering of the People's Money by Influential Politicians and Businessmen Denounced to Integrity Board', 12 Sept. 1952, p.6, Quirino Papers.

REFERENCES

Baclagon, Uldarico S., 1962, *They Chose to Fight: The Story of the Resistance Movement in Negros and Siquijor Islands*, Manila: Capital Publishing House.

Cullinane, Michael, 1982, 'The Changing Nature of the Cebu Urban Elite in the 19th Century' in A.W. McCoy and C. de Jesus (eds.), *Philippine Social History: Global Trade and Local Transformations*, Quezon City: Ateneo de Manila University Press.

Elson, R.E., 1984, *Javanese Peasants and the Colonial Sugar Industry: Impact and Change in an East Java Residency, 1830–1940*, Singapore: Oxford University Press

Friend, Theodore, 1965, *Between Two Empires: The Ordeal of the Philippines, 1929–1946*, New Haven, CT: Yale University.

Fuentes, C.R., 1919, *Apuntes Documentados de la Revolucion en Toda La Isla de Negros*, Iloilo: El Centinela Inc.

Geertz, Clifford, 1963, *Agricultural Involution: The Processes of Ecological Change in Indonesia*, Berkeley, CA: University of California Press.

Kirk, Grayson L., 1936, *Philippine Independence: Motives, Problems, and Prospects*, New York: Farrar & Rinehart.

Larkin, J.A., 1972, *The Pampangans: Colonial Society in a Philippine Province*, Berkeley, CA: University of California Press

Lopez, Oscar M. (ed.), 1982, *The Lopez Family: Its Origins and Genealogy, I*, Manila: Eugenio Lopez Foundation.

McCoy, Alfred W., 1977, 'Yloilo: Factional Conflict in a Colonial Economy, Iloilo Province, Philippines, 1937–1955', Ph.D. dissertation, Yale University.

McCoy, Alfred W., 1982a, '*Baylan*: Animist Religion and Philippine Peasant Ideology', *Philippine Quarterly of Culture and Society*, Vol.10.

McCoy, Alfred W., 1982b, 'A Queen Dies Slowly: The Rise and Decline of Iloilo City', in *Philippine Social History: Global Trade & Local Transformations* (eds. A.W. McCoy and E.C. de Jesus), Honolulu, HI: University of Hawaii Press.

McCoy, Alfred W., 1984, 'The Iloilo General Strike: Defeat of the Proletariat in a Philippine Colonial City', *Journal of Southeast Asian Studies*, Vol.15, No.2.

Monteverde y Sedano, Federico 1898, *Campana en Filipinas: La Division La Chambre*, Madrid: Hernanda y Compania.

Moynagh, Michael, 1981, *Brown or White?: A History of the Fiji Sugar Industry, 1873–1973*, Canberra: Pacific Research Monograph Number 5, Australian National University.

Rivera, Temario C., 1982, 'Rethinking Philippine Social Formation: Some Problematic Concepts and Issues', in *Feudalism and Capitalism in the Philippines: Trends and Implications* (ed. T.C. Rivera), Quezon City: Foundation for Nationalist Studies.

Shalom, Stephen R., 1981, *The United States and the Philippines*, Philadelphia, PA: Institute for the Study of Human Issues.

Takaki, Ronald, 1983, *Pau Hana: Plantation Life and Labor in Hawaii*, Honolulu HI: University of Hawaii.

Wickborg, Edward, 1965, *The Chinese in Philippine Life 1850–1898*, New Haven, CT: Yale University Press.

'Tea and Money versus Human Life': The Rise and Fall of the Indenture System in the Assam Tea Plantations 1840–1908

RANA P. BEHAL and PRABHU P. MOHAPATRA

INTRODUCTION

The plantation complex in tropical Asia in the nineteenth century differed from the much older Atlantic plantations in one important respect: they did not have a history of slave based production. The main similarity between the two plantation systems lay in the export-oriented nature and industrial organisation of agricultural production. Nineteenth century Asian plantations were also similar to their Atlantic counterparts in that both utilised indentured labour, transported to the plantations over a long distance and employed under contractually-specific conditions. The two most important aspects of indentured servitude were, first, the provision that a breach of contract resulted in criminal prosecution, and second, the widespread penal santions granted to planters. It was the unfreedom inherent in indenture, namely the inability of the worker to withdraw his/her labour-power to bargain over the terms of the contract or for higher wages, that led to it being viewed as a 'new system of slavery' (see, for example: Tinker [1974]; Rodney [1981]; Gillion [1962]).

Recent literature on this subject (see, for example, Galenson [1984], Emmer [1986; 1990]) has challenged many of these arguments. In contrast to the contention of the older school, therefore, revisionists underplay the element of unfreedom involved in indenture, and emphasise the economic rationality of a system which they argue benefits employers and labourers alike. Galenson maintains that indenture was essentially a facilitating device that enabled labour to flow from low labour produc-

---

Rana P. Behal, Deshbandhu College, University of Delhi, New Delhi – 110 019, India; Prabhu P. Mohapatra, Nehru Memorial Museum and Library, New Delhi 110 011. This is a much revised version of the papers presented by the authors at the Workshop on 'Capitalist Plantations in Colonial Asia' held at CASA, Amsterdam, in September 1990. They are grateful to the participants of the workshop, Sumit Guha, and to an anonymous referee of *The Journal of Peasant Studies* for their useful comments. The article in its present form owes much to the editorial comments of Tom Brass.

tivity areas (Asia) to high labour productivity areas (Carribbean or North America). It allowed the cost of migration to be worked out over the contract period 'in the form of advances against future labour' [*Galenson*, 1984: 16]. The powerful attraction of economic opportunity was sufficient to 'prompt many migrants over the course of three centuries to enter indentures giving up much of their freedom for a period of years in hope thereby to amend their estate' [*Galenson*, 1984: 24].

Emmer, another leading revisionist, similarly argues that indentured emigration from India to overseas colonies was the result of rational and deliberate choice on the part of migrants, prompted by hopes of bettering their future, an 'escape hatch' from social and economic oppression at home [*Emmer*, 1986: 204]. Although the revisionist view has modified the traditional notion of indentured emigration as a system based largely on the ignorance of the migrant, it has little to say about the conditions of unfreedom structuring the indenture contract. Moreover, the revisionist argument emphasising the benefits of indenture is suspiciously similar to that advanced by contemporary defenders of indentured system, notably planters and colonial officials. For example, during the fierce debate over a nominal wage increase for indentured labour in Assam tea gardens, a planters' representative in the Central Legislative Council of India in 1901 said: 'When the coolie goes to a garden he begins to receive a wage and begins to live, whereas in Chotanagpur (the home district of the labourer) he only exists . . .'. Another planter was even more emphatic in saying that the indentured contract was 'an effort to give to the coolie in the only possible way an ability which he did not before possess; the ability to make a contract to better his condition . . .' [*GOI*, Leg. 48–102, March 1901].

That the indentured system was designed not so much to provide economic opportunities for the labourer as to secure for the planters a labour force whose wages were determined completely outside the labour market is a point that revisionists overlook. Nothing illustrates this better than the stagnant nominal wage rates paid to contract labour on both Caribbean and Asian plantations throughout the period when the indenture system was in force [*Rodney*, 1981: 34]. Low wage indentured labour was utilised effectively to curtail the bargaining power of free wage labourers, and thus functioned to 'guarantee planter *control* over the entire labour process' [*Rodney*, 1981: 38]. Accordingly, the analogy between slavery and indenture is nearer the truth than the attempt by revisionists [*Emmer*, 1986: 187–8; *Emmer*, 1990: 246–8; *Galenson*, 1984] to equate it with the free emigration of Europeans to America in the nineteenth century.

A related aspect of the revisionist argument concerns the causes given for the decline of the indenture system. Drawing an analogy with the debate about the abolition of slavery, the end of indenture is attributed by revisionists to political reasons and not economic considerations. Such an argument is a corollary of assumptions regarding the economic rationality of the indenture system: since the system benefitted all, it could not have been in the interest of any to demand its abolition [*Emmer*, 1986: 204–5; *Galenson*, 1984: 25].

The transition from indentured to free wage labour has not received the attention it deserves from the historians of the plantation system, especially from critics of indenture, partly because the nominal change in the legal status of workers seemed of much less consequence than the continuity of the coercive apparatus of plantation labour system. Thus, Ann Stoler in her study of North Sumatran tobacco plantation concludes:

> The distinction drawn between 'indentured' and 'free wage labour' marked neither the exit from a 'pre-capitalist' system nor the entry into a capitalist one; there was virtually no change in the working conditions or the production relations that prevailed in Deli estates. The penal sanction was simply the legal component of a much more widespread system of manipulation, whose corpus permeated within and outside the labour process – long after the penal sanction itself was abolished [*Stoler*, 1985: 45].

However, the assumption of an unchanging plantation system implicit in the above statement is difficult to sustain. Even if the essential aspect of production relations did not change with the abolition of indenture, it had important consequences for both planters and workers. To the former it meant adoption of different strategies of labour control in the absence of penal powers, and to labourers it opened up the possibility of greater control over the conditions of life and work in the plantations. In the Assam tea gardens during the period of indenture, a labourer fleeing the harsh work regime of an unhealthy garden could be arrested even by a ferry man who transported him across the rivers in the valley. After the withdrawal of penal powers, he perhaps had a much better chance of crossing that river.

The object of this essay is to examine in some detail the issues discussed above: namely, the emergence, operation and the eventual decline of the indentured labour system on the tea plantations in Assam during the colonial period. The focus will be on four specific aspects of the indenture system: the role of the colonial state through its various legislative and executive actions, the problem of labour force recruitment, the conditions of production on the plantation and the reproduction of its labour force, and the resistance of the latter to the indenture system. Finally, the reason for its decline is located in the contradictions generated by the working of indenture system itself.

THE STRUCTURE OF THE TEA INDUSTRY

The British annexation of Assam during the 1830s expressed the East India Company's need to sustain its tea trade in the wake of the Opium War with China, and the Assam Company was formed in London with a capital of £500,000 in 1839. The colonial state provided British capital with vast tracts of land on extremely favourable terms, special rules on land grants being introduced from time to time [*Behal*, 1983: 8–12]. However, only six companies were registered and 51 tea gardens in operation in 1860, when a highly speculative boom that lasted until 1865 was marked by the doubling of tea prices, soaring profits, and the registering of a further 86 tea producing companies [*Griffiths*, 1967: 61–99].

After 1865, the growth of the tea industry was characterized by three major features. First, rationalization, which took the form of increasing transfer of control over the management of tea gardens to the British managing agency houses, a process initiated by the Assam Company placing its gardens under the management of Schoene, Kilburn and Company. The operations of the bigger companies stimulated amalgamation of a number of smaller gardens into large-scale enterprises under managing agents, and by the end of the century, the seven top managing agency houses controlled nearly 61 per cent of all tea gardens. Second, most of the expansion of the tea industry after 1870 was financed not by Britain's home savings but rather by investments by British residents in India, and the undistributed surplus and ploughed back dividends of older companies already in the field [*Bagchi*, 1972: 161–2, 176; *Gladstone*, 1910: 93–4]. And third, in contrast to the pre-1865 expansion, a large part of the expansion of the tea industry during the period until the end of the century took place in a context of steadily falling prices; between 1880 and 1900, tea prices fell by half [*GOI*, 1901 (b)]. All these features had an important bearing on the structure of recruitment for the development of a labour system in Assam.

THE ORIGIN AND DEVELOPMENT OF INDENTURED LABOUR

While the availability of land and capital presented no difficulty, the mobilisation of labour for the growth of the tea industry was a critical problem. Assam was sparsely populated, and the absence of a substantial class of agricultural labourers, as well as the alleged reluctance of Assamese peasants to work on the plantations, led to the search for a source of labour outside the province. The Assamese peasant was stigmatised as being naturally 'indolent', 'lazy', and largely addicted to the use of opium [*ACR*, Ms. 9925: 154; *Griffiths*, 1967: 101], and the corresponding absence of a labour market was frequently cited as the reason for long distance recruiting. However, there is no doubt that the high wages demanded by local labourers was an important consideration in the resort to long distance recruitment.

In 1861, a manager of the Jorhat Tea Company reported to his superior that large scale importation of labour was a *sine qua non* for expansion, considering that local labourers were demanding higher wages and would not agree to work at existing rates, given the high standard of *nerick* (tasks) set in plantations. The Commission of Enquiry reported in 1868 that local wages had doubled since the 1850's and 'even if local labour was now largely available, the high price which it commands would materially affect the profits of tea cultivation' (cited in Griffiths [1967: 73, 103]). Thus the need for cheap and pliant labour in a strict work regime weighed more in the consideration of the planters than the availability of local labour *per se*. However, the argument of the absence of a labour market would be a constant refrain of the planters in their efforts to secure the active support of the colonial state for establishing and maintaining indentured recruitment along with a vigorous penal contract system in Assam.

Colonial state intervention in the recruitment and labour system of the Assam tea plantations dated back to 1863, when the first Act to control the recruitment and conveyance of labour was passed in the Bengal Council. This was occasioned by massive mortality in the transportation of labour to Assam, reaching figures of fifty percent in some cases. This Act stipulated that all recruiters be licensed, that every intending emigrant be medically examined, and that adequate sanitary arrangements for the conveyance of labour were made by recruiters. The penal contract system was introduced in Act VI of the Bengal Council in 1865, which stipulated minimum monthly wages [Rs5 for men and Rs4 for women], a three-year contract, a nine-hour work day, and a government inspector of labour empowered to cancel the contract of labourers on complaints of ill treatment.

The main provision of the Act lay in the sanctions for breach of contract by labourers: planters were given powers to arrest labourers who 'absconded' without warrant, and imprisonment was the penalty for refusal to work. However, the Act applied only to newly recruited labour, and not to time expired labourers or locally recruited labourers, who were now increasingly contracted under the Workman's Breach of Contract Act XIII of 1859 (extended to Assam in 1864) [*GOI*, 1906: 136]. Subsequently, all labour contracted under the Labour Immigration Acts was referred to as Act labour, while the rest of the labour force, including those contracted under the 1859 Act, was termed non-Act labour in official reports.

The origin of colonial state intervention in the establishment of a penal contract system has to be placed in the context of the rapid expansion during the early 1860s, when the appalling conditions of fraudulent recruitment and insanitary transportation led to high mortality rates and large scale desertions from the plantations: of the 85,000 labourers imported into Assam between 1863 and 1866, no less that 35,000 were reported to have died or deserted [*GOB*, 1868]. The Bengal Act VII of 1873 introduced for the first time provision for recruitment by licensed garden *sirdars*, and also for recruitment outside the Act on short term one year civil contracts. However, given the overwhelming preference for penal contract by the planters, the innovation of the 1873 Act was left practically unused.

Between 1865 and 1882, the tea industry grew steadily, and recruitment continued to be under the supervision of the government while a penal contract system came into being in Assam. In the course of this period, official enquiries carried out in 1868 and 1873 revealed that statutory minimum wage levels for indentured labour in Assam were not paid, recruitment abuse flourished, and appalling transportation, living and working conditions continued to result in high mortality rates.

The trigger for the next round of changes in the legislation and labour system was provided by the slow but steady drift downwards of prices from 1878 onwards. The planters now launched a campaign for drastic cost-cutting measures combined with rapid expansion.

Since labour was the main component of production costs, the Indian Tea Districts Association (ITDA) in a memorandum to the Government of India in 1880, stated that the 'future of the tea industry hinged on the maintenance of an adequate supply of coolie labour at a cost calculated to leave a fair margin of profit on the capital invested'. The Association argued further that the most important hinderance to the expansion of the tea industry was the 'excessive cost of recruitment largely due to the stringency of labour laws'. In the opinion of the planters, the three year contract period was too short to allow them to recover the cost of recruitment and outlays on health care; they also complained that time-expired coolies not subject to penal sanctions had become 'complete masters of the situation' due to competition for their labour-power, and went in search of the highest remuneration. The planters concluded by demanding the abolition of all governmental regulation on recruitment, the extension of the term of penal contract from three to five years, and greater control over time-expired labourers.

Opinion was sought from district officials on the ITDA demands for suitable amendment to the existing Labour Emigration Act of 1873. Recognising the contradictory nature of the demand for deregulated emigration alongside a vigorous penal system in the labour districts, the Government of India suggested that the true remedies of the difficulties of tea industry lay in 'free emigration and free labour' [*GOI*, Emigration, A, 20–23, June 1880]. However, the Assam planters strongly opposed the suggestion of a non-penal labour system: a leading planter from Assam wrote that the absence of any labour market in Assam placed the planters so completely in the hands of his coolies that the majority would prefer to work under the unamended 1873 Act than allow any change in the penal system. In fact, the planters demanded *further* penal powers, aimed largely against time expired labourers engaged under short-term contracts of the 1859 Act, which planters felt had inadequate penal provisions. For instance, under the 1859 Act there was no right to private arrest without warrant, and prosecution automatically terminated the contract.

The reason for planters' demands for more penal powers over time-expired labourers was the growth of a labour market and competition amongst planters for time-expired labourers: by 1880 a bonus system had developed, such workers being paid between Rs12 and Rs24 to re-engage. Absence of legal control over time expired labourers also enabled them to seek preferred employment in favourably situated and healthy gardens rather than in remote and unhealthy gardens of upper Assam. Furthermore, planters were afraid that lack of adequate control over time-expired labourers would – and often did – lead to mass refusal to work: The Mangaldai Tea Planters Association, for example, observed that 'old coolies who are in the know ... will not only meet the demand for labour by a distinct refusal, but will incite other coolies to do the same'. Another planter argued for the whole labour force to be put under penal contract, citing the need for absolute control over labourers during the rainy season when the heaviest tasks (hoeing, planting and plucking) were performed, and when refusal to work by old coolies was most common.

There were, however, a few dissenting voices amongst local officials and even some planters, who were against the extension of the term of contract. They pointed out that it would result in inefficient work by a disheartened labour force and an increased death rate due to the greater application of the Act in unhealthy gardens. Consequently, there would be a greater incentive for the coolie to desert, and the managers would be tempted to overexploit the coolie. All this, the dissenters argued, would lead to greater rather than lower labour costs for the planters. The Chief Commissioner of Assam, however, fully supported the planters' demand for more powers of control, and suggested that they be freed from any restriction on their private power to arrest without warrant [*GOI*, Emig. A, 68–72, Dec., 1880].

The Commission appointed by the Government of Bengal agreed with the planters in their demand for deregulation of recruitment, but 'in the interest of "gardens" which cannot have the garden *sirdar* recruiting and must depend on professional assistance', it insisted that the system of licensed contractors and recruiters should continue under government supervision as before. On the question of penal contracts, the planters' demands were met in full. 'In deference to the almost unanimous wish of the tea planters', and against the objection of some of the district officials, the Commission proposed the extension of the period of contract to five years. For local labour contracts, planters now had merely to file the papers of the contract within one month, without having to present the labourer in front of a magistrate, and instead of the cumbersome procedure for endorsement of days of unlawful absence, to provide only a list of defaulters to the Inspector of Labour on their annual visits. All restrictions on the power of arrest, such as requiring a warrant if the magistrate's court was within ten miles of the plantation, were lifted; moreover, planters could now delegate the power of arresting deserters to others.

As regards labour, the Commission re-introduced the statutory minimum wage of Rs5 for men and Rs4 for women for the first three years of the period of contract, with the provision for an increase of Rs 1 over the last two years. However, the minimum wage was made conditional on completion of the daily task, thus formally recognising the piece work system which had become standard practice on the plantations. The inspectors were permitted to reschedule the daily workload if this was judged to be too heavy [*GOI*, Emig. A 1–15, Aug. 1881], and the government was given powers to close down a garden if mortality rates were high and could be shown to be the result of inadequate sanitary arrangements.

With a few minor changes, all the suggestions of the Commission were accepted by the Government of India and accordingly made into law: the 1882 Labour Districts Emigration Act I was passed without much difficulty in the Governor General's Council. The 1882 Act remained the major legislative device for the regulation of recruitment and the labour system in Assam until 1901, being modified slightly in 1893 when the maximum term of contract was reduced to four years. The major features of the Act, namely deregulated emigration and freedom for local contracts, were clearly designed to served the need of planters to regulate an expanding labour force. By strengthening planters' control over time-expired labour, therefore, such legislation thwarted the development of an embryonic labour market.

It is crucial to locate Act I of 1882 in the context of the strategy adopted by planters in the face of two important developments in the world market: first, the need to acquire the pre-eminent position in exporting tea to the UK market, where the main rival was China, and second, the steady yet fluctuating decline of tea prices from 1878 onwards. The planters' twofold response was to expand production at an unprecedented scale while simultaneously attempting to cut costs of production, a strategy described in the Indian Tea Association Report of 1884 as the reduction of expenditure 'by a more economic use of the labour available and by getting better work from the coolies' [*Griffiths*, 1967: 123]. Under Act I of 1882, the tea industry experienced phenomenal growth: between 1880 and 1900 acreage under tea increased by 120 per cent, while production of tea quadrupled. Indian tea acquired a leading position in the world market, outstripping China tea in 1888 for the first time, and its share of the UK market increased from 27 per cent in 1880 to 54 per cent by 1901 [*Griffiths*, 1967: 120–29]. In the phase of a downward drift of prices, such growth was made possible by importing no less than one million persons into Assam during the period in which Act I was in operation [*ALR*, 1880–1900]. By keeping wages below even the statutory minimum level, the strengthened penal contract system aided the intensification of labour that contributed to reduced production costs. We now turn to the history of emigration and the structure of recruitment during the period of indenture.

LABOUR RECRUITMENT UNDER INDENTURE

Long distance recruiting was practically coterminous with the inception of plantations in Assam. The Assam Company initially relied on European contractors to supply them with labour, these contractors working through paid native agents who did the actual recruiting in districts of Bengal and Bihar [*ACR*, Ms., 9925; *Griffiths*, 1967: 64]. The phenomenal expansion of tea gardens during 1860–65 period saw the proliferation of native Indian recruiting contractors who supplied tea planters with thousands of labourers from Chotanagpur, Bihar and eastern U.P. During the 1860s the planters also sent out coolies from their gardens to recruit in their home districts. This system is known as garden *sirdari*. After the 1863 and 1865 Acts, both forms of recruitment came under government supervision: recruits had to be registered in their districts of recruitment, medically examined, and their conveyance to Assam had to follow health regulations laid down by the government.

Although the 1882 Act, which introduced the system of free or deregulated emigration, was designed to encourage private recruitment through garden *sirdars*, an immediate result, however, was the increase of unregistered recruitment by unlicensed contractors which outstripped registered recruitment by a ratio of 4 to 1 [*GOB*, 1883: 3]. This development was facilitated by the emergence of the 'Dhubri contract' system. Instead of registering emigrants at the district headquarters where the labourer also signed a contract, as was done earlier, the unregistered emigrants were taken straight to Dhubri in the Goalpara district of Assam, from where a regular steamer service carried the coolie recruits to the tea gardens. Goalpara, which did not have a single tea garden, had been declared a labour district under the new Act; it was here that a vast number of recruits under the unlicensed system were made to sign their contracts nearly 400 miles away from their homes in front of a magistrate who, by all accounts, made only perfunctory inquiries as to the willingness of the coolie to emigrate to Assam [*ALR*, 1887: 2–3; *Griffiths*, 1967: 276].

Unregistered emigration and Dhubri contracts became the standard practice for the recruitment of indentured labour in the era of deregulated emigration. The first casualties of the new recruiting system were the old agencies of licensed contractors and recruiters. In 1882 there were about 20 licensed contractors and 1,000 licensed recruiters in the field, but by 1890 they had vanished from the recruiting districts of Chotanagpur and Santhal Parganas: many licensed contractors and recruiters were reported to have gone over to unlicensed recruiting [*GOI*, Emig. A, 6–14, Nov. 1891].

There was a similar decline in the number of licensed tea garden recruiters (*Sirdars*); the latter were recruiting only about two coolies per head in 1895, compared to about ten per head in 1882 [*GOB*, 1896: Para 66]. The manager of the Jorhat Company complained in 1889 'We have to send away 100 garden *sirdars* for 300 or 400 coolies required. Formerly we calculated on getting 15 to 20 coolies for each *sirdar*' [*GOI*, 1891: Para 24]. In many cases, garden *sirdars* went to work for the local agents of the Calcutta managing agency houses, who were also unlicensed contractors. Thus many *sirdari* recruits were in reality unlicensed contractors' coolies, and the garden *sirdar* often acted as a chaperone for these coolies on their journey to a specific tea garden. This system, known as 'mixed sirdari', was very common in the Assam Valley and Cachar [*GOI*, 1891: Para 19; *ALR*, 1884: 3].

The deregulated recruitment system had at its base the *arkatti*, or the village recruiter, and at its apex the great coolie recruiting firms of Calcutta controlled by managing agency houses (Begg and Dunlop, Balmer and Lawrie) who also had a great many tea gardens under their control [*Griffiths*, 1967: 295]. The village *arkatti* was not in any sense a creation of the 1882 Act, and there is much evidence to show that he played a crucial role even in the phase of controlled emigration. In 1877, the magistrate of Ranchi, the largest coolie recruiting district, noted that most of the recruitment under the licensed recruiting system and garden *sirdari* was actually done by *arkattis*, who received Rs 2 to Rs 5 per coolie supplied [GOI, Gen. A, 22–23: Jan. 1878]. The village *arkatti* was variously described as 'belonging to the very class of people who emigrate as labourers and live and move among them', or in more uncharitable terms as a 'notoriously bad character', or being 'as a rule the scum of the country, ex-convicts, burglars, thieves, dacoits and notorious badmashes' [*GOI*, Emig. A, 6–31: Sept. 1889].

The sphere of operation of the *arkatti* was usually a small group of villages around a weekly market or *haat*. His success lay in being in close touch with the village, or as a contractor put it 'to know the wants and inclinations of the people' [*GOI*, Emig. A, 6–14, Nov. 1891]. Above the village *arkatti* was the *sudder arkatti*, who normally resided in major towns of the recruiting districts and often fixed deals with the local agents of the coolie contractors. The function of the local agent was to finance the *sudder arkatti* through cash advances, and to confirm the suitability of the recruits. The coolies would then be sent to the depots which dotted the major towns of these districts or railheads, whence they were sent with contractors' escorts to the forwarding depot at Dhubri. Accordingly, this system involved three to four distinct levels of transaction and officials often disparagingly referred to the *arkatti* recruitment and transfer of coolies through various intermediaries as the 'sale system' [*GOI*, Emig. A, 4–5, Oct. 1883].

The boom in unlicensed recruitment led to the proliferation of these professional recruiters on an unprecedented scale. In 1890, by a conservative estimate, no less than 5,000 *arkattis* were said to be operating in Ranchi district alone [GOI, Emig. A. 6–14, Nov. 1891]. By 1895, the Bengal Labour Commission [1896: Appendix] found that coolie contractors in Purulia, the major recruiting station in Chotanagpur, each had about 500 to 1500 *arkattis* on their books, and there were no less than 14 coolie contractors in the town.

The high density of *arkattis* could only be explained by the increasing profits of the recruitment business. In 1877 *arkattis* received between Rs 2 and Rs 5 per head for each coolie supplied to the contractors. In 1883, immediately after the promulgation of the 1882 Act, the price paid to village *arkattis* had nearly doubled to Rs 10 per head. By 1888 it was reported that *arkattis* got Rs 20 per coolie supplied to the *sudder arkattis*, who in turn were receiving Rs 50 to Rs 60 per head from the depot owners and local agents of the contractor [*GOI*, Emig. A, 6–14, Nov. 1891]. The high rates paid to *arkattis* also affected the price of garden *sirdari* coolies. Many tea gardens and their managing agents complained bitterly in 1889 about the practice of garden *sirdars* selling their coolies to *arkattis*, since the price they received was invariably much higher than the commission paid by the tea gardens [GOI, Emig. A. 6–31, Sept. 1889].

The profits of recruitment at the lowest level was replicated at all stages of coolie transactions, the net result being a steady increase in the cost of recruitment to the planters. The rising cost of recruitment had been cited as the primary reason for the planters' demand for deregulated emigration, yet the results were contrary to their expectation. In 1875, the cost of importing coolies in Sibsagar district of Assam was reported to be Rs 35 per head [*GOI*, Emig. A, 68–72: Dec. 1880]. By 1889, evidence suggests that the cost of recruitment had more than doubled, a leading contractor estimating the cost to be about Rs 90–100 per head [*GOI*, Emig. A, 6–14, Nov. 1891]. In 1895, the best coolies – the tribal labourers from Chotanagpur – were costing planters no less than Rs 150 per head, the average cost for an average coolie being estimated at Rs 130 [*GOB*, 1896: Para 76].

The massive famines which occurred in the recruiting districts during 1897 and 1900 led to a slight decline in the price of coolies; yet even in famine years the price did not fall below Rs 100 per head. The cost of coolies otherwise varied according to region and caste, the overwhelming favourite with the planters being tribal workers from Chotanagpur, referred to in the recruiting parlance as '1st class jungley' [*GOB*, 1896: Para 73]. The next in terms of preference were the so-called 'semi-aboriginal castes', and the lowest were the coolies from Bihar and the United Provinces, who were to be had from contractors at half the rate of tribal coolies from Chotanagpur. Between 1880 and 1900, of the 710,000 adult coolies recruited for the tea gardens of Assam, no less than 46 per cent were from Chotanagpur, while only 21 per cent were from the congested plains of the United Provinces [*ALR*, 1877–1902].

The cost of recruitment in Assam contrasted unfavourably with that in the Dooars gardens of North Bengal, where in 1895 unlicensed garden *sirdars* recruited coolies at a cost of Rs 6 per head, the same labourer costing Assam planters Rs 60. The host of middlemen spawned by the deregulated emigration system shared roughly Rs 40 to Rs 120 between them, depending on the class of coolies supplied, and the Bengal Enquiry

Commission did not mince words in describing them as 'doing the work of hunting and offering the spoils of the chase to the planters but at an enormous cost' [*GOB*, 1896: para 73, 85].

In a sense, the metaphor of hunting is not inappropriate for the *arkatti* system of recruitment: 'The essence of hunting is snaring and violence, and these two are the favourites of the *arkattis*' observed an official in 1895 [*GOI*, Emig. A, 9–10: April 1897]. Thus, while the immediate results of deregulated emigration were welcomed unabashedly by planters and officials in Assam as having made 'emigration simpler and more popular', they evoked an instant cry of disapproval from officials in the recruiting districts. The first reports concerning the increased abuses associated with deregulated emigration came from the Chotanagpur division, where the Commissioner reported in 1882 a great increase in enticing minors from the protection of their guardians and the abduction of married women. Missionaries in Chotanagpur supplied the Commissioner with 50 documented cases of kidnapping and abduction in 1882, and the Deputy Commissioner of Ranchi and Manbhum reported another 100 cases of abduction of women [*GOI*, Emig. A, 2–6, Nov. 1883; *GOI*, Emig. A, 4–5: Oct. 1883]. The Government of Bengal reported in 1885 'the unanimous condemnation of free emigration system by the district officers as affording the greatest facility of wrong-doing' [GOI, Emig. A, 15–17, Aug. 1886]. Similarly, in Chotanagpur, 262 cases of kidnapping, cheating and wrongful confinement were filed in 1888, while the Commissioner received some 500 complaints of missing persons.

Yet these cases were said to represent gross underestimates of actual abuses since 'in the majority of the cases the complainants are themselves in the dark about the *arkattis* who caused such trouble and therefore consider it futile to undertake the cost and journey to the police station or magistrate' [*GOI*, Emig. A, 9–19, Nov. 1899]. There were reports of unwilling victims being kept under lock and key in secret depots and then being spirited away to Dhubri where they had no choice but to sign a labour contract. In a note submitted to the Lt. Governor of Bengal in 1899, Father Hoffman, a Jesuit missionary stationed at Chotanagpur, listed the typical methods of *arkatti* recruitment. These included loans advanced to potential victims, tempting young men with liquor and women, and kidnapping young girls and marrying them off in the depots, a system known as 'depot marriages' [*GOI*, Emig. A, 9–19, Nov. 1899].

The high profits of early recruiting lured a significant number of freelance Europeans and Indians. An example is John Henry Lawton, a dismissed soldier, who with his posse of armed peons set up the business of 'coolie raiding' in the highways along which gangs of coolies recruited by *sirdars* of the Dooars gardens travelled to Jalpaiguri. Lawton posed as a government official with his retinue, and ordered the coolies into specially constructed temporary depots from which they were taken to Assam. He was prosecuted for five such cases of coolie raiding in 1888 alone, a year in which a special police officer was appointed to investigate 32 such cases of coolie raiding [*GOI*, Emig. A, 4–7, Nov. 1888; 3–4 July 1889; 6–14, July 1890; Nov. 1891]. The Bengal Labour Enquiry Commission [1896: para 76] noticed the existence in 1895 of a new set of men called 'daffadars', armed musclemen from the United Provinces, who had inserted themselves in the recruiting business between the *arkattis* and contractors.

The depredations of *arkatti* recruiting caused near panic in many recruiting districts. A magistrate in Hazaribagh district noted in 1889 that: 'the fear of recruiting in this district is very great, and it is not too much to call it a terror. People even fear to go to their several *haats* and *bazars* for fear of them. Hardly since the days of *thugee* has there been such widespread fear of travelling' [*GOI*, Emig., 6–14, Nov. 1891]. Similar reports of widespread fear of Assam recruiting were received from Santhal Parganas and Ranchi [*GOI*, Emig. A, 9–19; Nov. 1899].

By making emigration to Assam unpopular, *arkatti* recruitment and its related host of intermediaries no doubt contributed directly to the increasing cost of recruitment. However, the most important reason for the existence of the *arkatti* system, as well as the rising cost of recruitment, was the ever increasing demand for labour that emanated from the operation of the penal contract system in Assam. The question to be asked is: despite the migration of one million persons to Assam over the last two decades of the nineteenth century, why did the tea garden labour force amount to only 400,000 workers in 1900? Planters and government alike justified the existence of the penal contract system in terms of a chronic labour shortage in Assam. Is it not pertinent to ask rather to what extent the penal contract system itself was a factor in causing the shortage of labour?

THE LABOUR SYSTEM UNDER INDENTURE

The characteristic feature of the labour system in Assam was the extensive use of the penal contract. In this the 1882 Act marked the watershed. Prior to this only 22 per cent of the labour force were under penal contract. By 1891, at its height, 55 per cent of the labour force were contracted under the 1882 Act [*ALR*, 1891], and another 40 per cent of the labour force were under the 1859 Act [*GOI*, 1891: Para 111]. The widespread use of the 1882 Act during the period in which the tea industry expanded immensely strengthened planters' power in relation to their work force. This expansion was carried out under the statutory minimum wage provision of the Act (in effect the maximum wage paid), since the major function of penal sanctions was to keep down wages, by preventing the operation of a labour market. It did so by checking competition among planters for scarce labour, and by preventing workers from taking advantage of this situation.

The passing of the 1882 Act was followed by the opening of new gardens in remote and unhealthy areas of Assam (such as the Balisera gardens in Sylhet). Thousands of *arkatti* labourers were hastily assembled in newly opened gardens without adequate health protection, availability of provisions, or safe drinking water, which resulted in increased mortality rates within the year. As the civil surgeon of Sylhet noted, 'Anybody of experience who saw the ground would know before the garden opened that it was a question of tea and money versus human life'. But other officials were certain that it was a direct consequence of the application of the new Act and its stringent provisions. 'The Emigration Act' wrote an official, 'no doubt operates to keep down the scale of wages. Unhealthy gardens could not be opened out without its aid. It would be too expensive. Free coolies would demand the wages sufficient to compensate them for the risk they run or would desert' [*ALR*, 1883: Para 15].

The working of the penal labour system had often come under criticism from the emerging nationalist press. In 1887 D. N. Ganguly, the Assistant Secretary of the Indian Association, published a series of articles in the *Bengalee* depicting the abysmally low wage rates and bad working conditions in the Assam gardens, a situation which he compared to slavery [*Ganguly*, 1972]. The Government's reaction was to dismiss these criticisms as orchestrated by 'perambulating philanthropic Babus and garden sea (sic) lawyers' and 'a mere handful of obscure and disreputable men representing nobody but themselves' [*GOI*, 1891: 239].

An interesting insight into the working of the penal labour system at this time is provided by a confidential report on Assam written by a group of Santhal tribal headmen in 1894. They had been taken on a tour of Assam tea gardens by the President of the Indian Tea Association, J. Buckingham, the purpose being to create a favourable impression amongst Santhals and to encourage them to migrate to Assam. The tour lasted 17 days, during which they visited 27 gardens and personally interviewed 216 labourers, visited the coolie lines, and inspected the working conditions in the gardens. The Santhal headman and an accompanying British official submitted separate reports of their visit to the Government of Bengal [*GOI*, Emig. B, 1–3, Sept. 1894]. Predictably, the official report described the working conditions of the gardens in glowing terms; by contrast the report of the Santhals was highly critical of the labour system in the tea gardens. It is almost certain that they had been taken to 27 of the best run gardens in Assam, and their criticism of the labour system is thus all the more incisive. The first point they made in their report concerned the insufficient wages in the tea gardens. The monthly rate of Rs 5 per man and Rs 4 per woman was considered by the workers themselves to be too low, given the high price of rice in the gardens. The daily wage of three annas (3/16th Rupee) was similarly considered inadequate in relation to the heavy workload which left no opportunity to earn overtime. Wages were only about half of what Santhal migrants could earn by working on road construction in Chittagong (where the monthly rate was Rs 9) and lower than those paid in the Dooars gardens of Bengal.

The second point in their report dealt with the general working conditions in the gardens. They found conditions of work unacceptable in 23 of the 27 gardens they visited. In these gardens coolies were not allowed any sick leave, and prolonged illness was penalised by a cut in pay. In most gardens non-completion of tough work schedules was punished with fines and substantial pay cuts. Flogging of men and women was common in every garden, either for non-completion of work or for disobedience and desertion. Women were not allowed any rest after giving birth to a child, and were 'required to work within six days of delivery'. In contrast to the claim made by the accompanying British official that Santhal labourers did not return home because they were 'too well off', the tribal headmen reported that labourers were unable to return home even after 15–20 years because they did not have enough money and their wives and children were still working out the contract [*GOI*, Emig. B, 1–3, Sept. 1894].

In short, the report indicted the penal contract system in Assam for not providing workers with a living wage, arguing that low wage levels were maintained by the use of force, both legal and extra-legal. Many of the claims made by the Santhal headmen were subsequently confirmed by the detailed investigation into wages in Assam during the famous controversy of 1899–1900, initiated by Henry Cotton, the Chief Commissioner of Assam. Cotton showed that throughout the whole period of indenture to 1900, the average wage of 1882 Act labour remained well below the statutory minimum, the highest being only about Rs 4.75 for men and Rs 3.75 for women. More importantly, while wages in tea gardens had stagnated for 35 years (since 1865), between 1875 and 1900, the price of foodgrains increased by 100 per cent. Wage levels in the tea gardens also compared extremely unfavourably with those of agricultural labourers outside the gardens, amounting to only about half the latter rate [*GOI*, 1901a, 94–96].

Non-Reproduction of the Plantation Labour Force

Along with a low living wage and also a consequence of it, the plantation system in Assam was characterised by a persistent problem of work force reproduction. Throughout the period during which the indenture and penal contract system was in operation, labour force mortality constantly outstripped the birth rate, with the result that the net reproduction rate was negative. Between 1882 and 1901 the average annual death rate for adult workers was 3.93 per cent. More significantly, over the same period, the death rate of workers recruited under the 1882 Act was 5.4 per cent [*ALR*, 1882–1901]. In other words, one in every four of the adult work force recruited under the 1882 Act did not survive the full period of the contract. Labourers contracted under the 1859 Workman's Breach of Contract Act had a much lower death rate, the annual average for the period being 3.07 per cent while that for the non-tea-garden population of Assam of a comparable age group was only 2.4 per cent [*GOI*, 1901: 97].

The high death rate among the tea plantation work force in Assam drew repeated criticism, both from the nationalist Indian press and from independent missionaries like Charles Dowding [*Ganguly*, 1972; *Dowding*, 1894]. It was evident that the local government had failed abysmally to enforce the sanitary measures provided under the Labour Immigration Act: of the 725 gardens which had been officially declared unhealthy (those with more than seven per cent mortality of the labour force) between 1882 and 1892, only 50 gardens had been inspected by the officials; of these only ten gardens were mildly penalised. The standard response by the tea planters to such criticism was to blame the high mortality rate on 'bad batches' or 'unsuitable coolies' supplied by the contractors [*ALR*, 1884: 29], the fault being attributed to the coolies themselves for bringing with them diseases like anchylostomasis or the infamous 'coolies' anaemia' and cholera contracted in their home districts [*Dowding*, 1894: 9–11], explanations designed to delink mortality and the penal contract labour system. A special enquiry conducted by the Assam Government in 1889 into the causes of mortality in 107 gardens concluded that 'unacclimatised' new coolies were most liable to suffer from diseases and death. This new explanation now became a stock argument for planters and officials alike in defence of the penal contract system, since they could claim that the death rate was independent of the system under which the coolies worked.

The problem of high death rate was thus sanitised and made a natural condition rather than a direct consequence of the penal contract system.

Replying to Dowding's criticism of the Assam labour system, Henry Cotton, the Chief Commissioner of Assam wrote in 1897:

> Unfortunately, the fight of civilisation against nature demands its victims no less than war against a human enemy and land cannot be reclaimed from jungle except at the price of comparatively high mortality amongst the pioneers of cultivation. . . . But the cost of life and treasure has never been allowed to count in the balance and the victories of peaceful industry must continue to claim their victims [*GOI*, Emig. A, 28–29, Sept. 1897: Para 6].

The 'victories of peaceful industry' were, however, gained at a cost; ironically, even here the official sympathy lay not with the coolies who died but rather with those who had paid for them – the planters. Thus William Ward, Cotton's predecessor in Assam justified penal powers for the 'hapless' planters because 'it seems right to give him (the planter) something more for the cost he incurs, the *inferior article* he has to put up with, and the consequent risks he runs in losing by death or desertion the coolies he has paid for . . .' [GOI, Emig. A, 3–11, April, 1893]. Hence, workforce mortality was attributed to the coolie being an 'inferior article': as a planter wrote in reply to Charles Dowding's criticism, 'They do not die from diseases engendered by local conditions of gardens. They die from diseases which are induced by their own inherent faults . . .' [*Dowding*, 1894: 51]. Accordingly, attempts to explain away the high death rate on the tea gardens in Assam avoided any reference to the operation of the penal contract system, and how this enabled planters to intensify the labour process while paying workers below subsistence wages.

There is much evidence to show that the mortality pattern was intimately connected with the rigours of the penal contract system operating under the 1882 Act. The special enquiry of 1889 pointed out that the 1882 Act was utilised mainly in the unhealthy gardens (with a mortality rate above seven per cent of their total labour force), noting that 'many unhealthy gardens owe their unhealthiness, in a great measure, to the insufficiency of labour force and to the fact that the greatest pressure on the coolies to work is exercised during the rains, which is also the most unhealthy time of the year' [*GOI*, 1891: 112, 253, 264].

It also showed that, due to these labour shortages, the enforcement of penal sanctions in unhealthy gardens coincided with a stricter work regime, and death rates mounted when coolies were not allowed sick leave [*GOI*, 1891: 252]. A large number of deaths attributed to anchyolostomasis in 1888, the Civil Surgeon of Lakhimpur reported, were because the 'number of coolies palpably suffering from it are allowed to go on working or trying to work, until it develops into an incurable stage' [*GOI*, 1891: 280]. In 1889, when the death rate was as high as 6.4 per cent among 1882 Act labourers only five per cent of total working days in Assam were granted as leave of absence due to illness. New gardens opened in unhealthy and remote sites invariably utilised 1882 Act labourers for their ventures. When there was a proposal to repeal the 1882 Act in 1897, the planters unanimously opposed this, as 'without the penal contract the labour required for starting new ventures cannot be obtained' [*GOI*, Emig. A, 11–13 April 1897: Appendix D].

Overwork and undernourishment remained the most common cause of death. Non-Act labour invariably obtained a higher wage, which in no small measure contributed to the differential mortality levels between the two groups. Thus the Deputy Commissioner of Sibsagar noted in 1894, that: 'Most of the new coolies have little or nothing to live on besides their bare wages. Bearing in mind the fact that the wages they earn are usually below the average, I think, it is impossible to avoid the conclusion that coolies for some time after they arrive must find it difficult to subsist' [*ALR*, 1894: 35] Similarly, by 1899 Henry Cotton was already coming round to the view that a considerable number of deaths among the 1882 Act coolies were due to the abysmally low wages they earned and to consequent malnourishment. During the controversy over the wage question in 1900–1901, Cotton showed from the wage data of a number of gardens that the high death rate correlated with low average wages and high daily tasks set for the 1882 Act coolies [*Behal*, 1983: 206–10; *GOI*, 1901a, 94–6]. It was only in the context of a penal contract system characterised by low wages, an intensified work regime, and grossly inadequate health care, therefore, that the impact of the so-called hostile influence, which the coolies brought with them, was magnified to epidemic proportions.

The penal contract system systematically militated against a normal rate of reproduction by the labour force. Thus the birth rate for women employed in the tea gardens over the period 1880–1901 averaged only 86 per 1000, compared with an average of 127 births per 1000 women in the non tea garden population of Assam [*ALR*, 1882–1901]. Moreover, during the last two decades of the nineteenth century the birth rate declined steadily, the overall effect being a persistent negative rate of net reproduction among the tea garden work force. The low birth rate was attributed by medical opinion to the widespread practice of abortion amongst women labourers [*GOI*, 1891: 273–4]: for example the medical officer of Nazira observed in 1906 that, 'the amount of abortion on some gardens is appalling; on some gardens 65 per cent of pregnant women do not give birth to living children' [*GOI*, 1906a: 343]. The reason for abortion was ascribed to the women 'not wanting her earning interfered with' or, as the Indian Tea Association concluded, to the 'weakness of marriage ties amongst the coolies' [*GOI*, 1906a: 6]. It was impossible to escape the conclusion that abortion was due to the absence of maternity leave coupled with low wage levels that compelled women to work throughout pregnancy and immediately after childbirth. Women workers under the 1882 Act contract suffered specially, since they 'were kept regularly at work during the plucking season and malingering was not allowed' [*GOI*, 1891: 74].

Low birth rate and high death rates were both outcomes of the low wage penal contract labour system. The labour force in the Assam tea gardens was as a result constantly depleted, necessitating much greater recruitment of fresh labour each year. The planters' strategy of intensification and expansion of production was based largely on coercion of fresh emigrants, which in turn militated against a positive rate of reproduction of the labour force. It is in this contradiction that we must seek the causes of the tremendous pressure exerted on the structure of recruitment to supply an ever increasing number of workers for the Assam plantations.

POWER AND PROTEST

The various penal Acts, culminating in that of 1882, armed the planters with immense legal powers over labour; these included the power to prosecute defaulting and malingering labourers and – the keystone of the whole penal contract system – the power of private arrest of deserters. Yet these legal powers were not enough; the penal system was buttressed by a variety of extra-legal powers which the planters used to control their labour force. Through mergers of small gardens, large units averaging 1200 acres in size had emerged as the typical plantation in the late nineteenth century. Most of these gardens, specially in the Upper Assam Valley, were physically isolated units, separated and often fenced off from neighbouring villages. These huge private estates, with compulsorily resident indentured labour in the coolie lines, provided the milieu for the exercise of virtually unlimited powers by planters over their workers. A large number of *chowkidars* (security guards), *sirdars* and garden *mohurirs* (field staff) supervised by European managers and their assistants constituted the apparatus that controlled the labour force [*Barker*, 1884: 134].

The drive for intensification of the labour process, and the constant supervision to prevent coolies from running away, reinforced the strong element of violence and the tendency to overstep the limits of legality. An elaborate system of tracking down deserting coolies developed, and in 1873 Ware Edgar reported that planters often employed 'savage hillmen' to catch absconders with the promise of a reward of Rs 5 per head. Flogging was fairly common for defaulting and absconding, and it was reported in 1873 that frequently 'runaways enfeebled by their sufferings in the jungle died under or from the effects of floggings they received when caught' [*GOB*, 1873: XXI]. The typical planters' attitude towards the recalcitrant or deserting coolie was that displayed by W. A. Stoddard, the manager of Maphock tea estate in 1872, who wrote to the government demanding whipping of the offending labourers: 'The stick has a greater terror for these innate thieves and scamps for whom flogging is useful; especially without hurting the man much, the quiet, firm, systematic way the government floggings are conducted' [*GOB*, 1873: 87]. J. Ware Edgar reported in 1873 that the 'systematic and gross ill treatment' of the coolies was due to the planters' attitude of 'aversion deepened into a mingling of hatred and contempt' for them [*GOB*, 1873: XXI].

Extra-legal actions on the part of planters were confined neither to the work place nor to the prevention of desertions, but permeated the whole social life of a labour force isolated in the coolie lines. In his memoirs, Barker fondly remembered sitting in judgement every morning on cases of coolie 'delinquency' reported to him by the *jamadar* (headmen in the coolie lines), and meting out punishment to 'evildoers' [*Barker*, 1884: 171]. Similarly, a Bengal Government official was shocked to find in 1888 a planter organising a polyandrous marital union between five time-expired coolies and a single woman brought in under the Act; in return each man had agreed to re-engage in the same garden for a five-year period. 'The disposal in marriage of all imported female coolies', the official noted, 'is regarded as a matter entirely within the jurisdiction of the manager' [*GOI*, Emig. A, 2–9, Feb. 1889]. This exercise of power by the planters without explicit legal sanction was regularly condoned by the colonial government on the ground that the 'tea planter as master of a large and irregular labour staff must enforce discipline by occasionally severe measures which need not be looked into too closely, because they are substantially just and for the good of the general body of coolies' [*GOI*, Emig. B, 1–3, Sept. 1893: Para 4]. The coolies themselves no doubt thought differently. The common term by which they designated the penal contract system was '*phatak*' (literally jail), and the Deputy Commissioner of Lakhimpur noted ruefully that 'Every form of punishment however mild and whether really done for the coolie's good or not is designated *phatak*' [*GOI*, 1891: 240]. J.B. Fuller noted in 1903 that amongst the coolies the whole plantation system was commonly regarded as a *phatak* [*GOI*, Emig. A, 12–14, Dec. 1904].

Given the rapid increase in production and productivity in Assam, planters seem at first sight to have succeeded in utilising the penal contract system to intensify the labour process and restrict competition for scarce labour. Yet this apparently invincible power structure was very abruptly transformed in the first decade of this century: to what extent, therefore, were planters able to impose their will on labour, and how did the latter react? In the first place, the 1882 Act, the core legal device of the penal contract system, could not be applied to the whole labour force. The proportion of 1882 Act labour in the Surma Valley declined from 1885 onwards, amounting to only three per cent of the workforce over the 1892–1901 period. Why is it that the 1882 Act, which gave such wide ranging powers over new migrants as well as time-expired coolies, was not applied on a larger scale?

Between 1882 and 1885, the proportion of labour under the 1882 Act in the Surma Valley increased from about 15 per cent to 34 per cent [*ALR*, 1882, 1885], and many time-expired labourers as well as those serving under the Masters' and Workmens' Act were put under the 1882 Act. This in turn led to a rise in desertions by plantation labour; overall some ten per cent of workers absconded, and in Sylhet district 15 per cent of Act labour escaped during the 1882–85 period [*ALR*, 1882, 1885]. High desertion rates thus acted as an important deterrent to the large application of 1882 Act contracts in the Surma Valley.

Despite the rigid enforcement of the power of private arrest granted under the 1882 Act, nearly five per cent of Act labour deserted each year over the period 1882–1901, most escapes being from the unhealthy tea gardens [*ALR*, 1882–1901]. However, these are figures for successful desertions; a far larger percentage of labour tried unsuccessfully to leave the gardens, none of which was recorded in the official returns. New coolies, brought under the *arkatti* system often by force or fraud, reacted strongly to the harsh work regime and the low wages of the plantations. They were more prone to desert, an official noted, because the 'system and discipline under which garden work must necessarily be carried out, the muster, the gong to return, etc., the general coercion in short, mild as it may be made, is a condition to which he cannot sometimes reconcile himself' [*GOI*, 1891: 185]. Desertion, sometimes meant a transfer of labour-power from a particularly unhealthy, low wage, or highly undermanned garden to a more favourable one, which to some extent mitigated the high mortality rate in such contexts. On the other hand, it put additional pressure on the unhealthy gardens to recruit more labourers under the *arkatti* system, and in these gardens the tendency to use the 1882 Act contracts correspondingly increased [*GOI*, 1891: 187]. On the whole, the steady rate of desertion from the tea gardens throughout the period in which the penal contract system was in operation raised the cost of supervision generally, and – in the case of unhealthy gardens – gave rise to increased demand which raised the cost of recruitment.

The major reason for the decline in the proportion of 1882 Act Labour in Assam was the inability of planters to force long-term contracts on time-expired coolies. In 1884 about half of time-expired coolies were on five year contracts, but by 1891 only about five per cent of such workers renewed their engagement for the full five-year period, most renewals being for only one year [*ALR*, 1884; 1891]. From 1893 onwards, when the 1882 Act was amended and local contracts without the intervention of magistrates were limited to one year, planters found no advantage in putting time-expired coolies under Act I. Instead renewals were generally under the 1859 Act, which had far less stringent penal powers than its 1882 counterpart. This suggests that one of the main purposes of the 1882 Act – the suppression of the labour market for skilled, time-expired labour – was to some extent negated by the resistance of the labour force. The absence of any legal hold over time-expired coolies often led to a mass exodus from the remote and stricter work regimes of unhealthy tea gardens to more favourable ones [*GOI*, 1891: 207]. By the end of the nineteenth century the workforce on many gardens in Assam Valley was composed of time-expired coolies attracted by a larger bonus, or land to cultivate. In these gardens, coolies were allowed 'ample leave and unusual liberty to work or absent themselves' [*GOI*, 1891: 207]. From the 1890s onwards, there was also a much greater movement of time-expired labour to settle outside the gardens in *bastis*, renting land from either government or neighbouring villages. In 1906 it was reported that *basti* labour, consisting largely of time-expired coolies, regularly got about 25 per cent higher wages on daily task compared to labour residing in the coolie lines [*GOI*, 1906b: 335].

These tendencies grew on the one hand because of the workers' desire to commodify their labour-power, and on the other in spite of – and also because of – the planters' attempt to negate the operation of the labour market as well as their efforts to intensify the labour process. Planters in the Assam Valley had successfully opposed a government scheme for land colonisation on the grounds that while it might encourage labour migration 'it would also encourage time-expired labour to go away and become cultivators instead of re-engaging as labourers in the gardens' [*Dowding*, 1894: 36]. The planters put immense pressure on time-expired coolies to re-engage, often by force: holding wives as surety, confinement in *phataks*, and physical assault or deception were regularly practised by planters to prevent time-expired labour from leaving. Many of the cases of overt conflict referred to in annual labour reports involved planters' pressure on the coolies to re-engage [*GOI*, Emig. A, 11, Nov. 1903].

The existence of endemic resistance in the tea gardens was deliberately suppressed in official reporting so as to present a picture of well-being and harmony, and reports often invoked the low level of prosecution under the penal law as a sign of good relations prevailing on the plantations. Yet a careful scrutiny of the labour reports indicates a growing number of incidents of 'rioting, mobbing, assaults and unlawful assembly'. The introduction of the 1882 Act led to an increase in such incidents, specially in the Surma Valley. The intensification of work norms and the labour process led to a rise in the number of prosecutions, cases of 'rioting' and 'unlawful assembly', reaching 272 prosecutions per year by the end of the century [*ALR*, 1896: 46]. In 1903, instances of serious conflict between European planters and the labour force were numerous enough to merit the attention of the Governor-General Curzon, who wrote that 'the number of assault cases are steadily increasing in Assam and the relations between masters and coolies in many plantations are becoming a public danger' [*GOI*, Emig. A, 11, Nov. 1903]. It was noted that the immediate cause in many cases was the resentment of coolies about increased workloads.

The violent clash between coolies and the management of Rowmari Garden in Lakhimpur District during 1903 was a typical example of overt conflict in the plantations. In July 1903, at the height of the plucking season, the coolies of the garden rose as a body and assaulted the European managers and their Indian underlings, causing grievous injury to them (the manager had to return to England for treatment). The next day they all marched to the magistrate's court as Dibrugarh to complain about the management. Subsequently, 42 coolies were prosecuted and many of them sentenced to long terms of imprisonment. The ostensible cause of the violence was the garden management confiscating umbrellas which the coolies used while plucking tea leaves during rains, and forcing them instead to wear broad-rimmed hats (*jhampi*). The ferocity of the assault for an apparently trivial reason surprised the investigating official, and initially it was ascribed to their natural 'excitability' and to a few 'malcontents' who made the reversion from umbrella to a *jhampi* a matter of prestige. However, further investigations showed that the garden had been subject to a fairly strict work regimen: leave was not granted freely, and there was considerable resentment over production speed-ups. Unlike the broad-rimmed hat, which allowed both hands of the coolie to be used for plucking leaves, the umbrella had to be held with one hand, and this affected the efficiency of plucking. The use of the umbrella by the coolies was accordingly a form of resistance to the speed-up of production demanded by the garden, and its deprivation acted as a flash-point for the built-up resentment against the drive for work intensification. The incident in Rowmari was not an isolated one: it was reported that in many gardens the managers' attempt to enforce the use of *jhampi* instead of umbrella was met with stiff resistance. 'The coolies will use umbrellas and the managers have generally had to give in' [*GOI*, Emig. A, 11, Nov. 1903].

Violent clashes between planters and coolies in Assam became more frequent at the turn of the century, and caused the government increasing concern. At Curzon's insistence, a special investigation into their causes was initiated, and an Assam officer was deputed to study the labour system in the Dooars, Madras and Ceylon plantations. On the basis of this report, the Government of Assam concluded that violent conflicts between planters and coolies were far more common than in any other plantation system throughout the sub-continent due to the widespread use in Assam of penal power to import and regulate labour. The *arkatti* system of recruitment was also unique to Assam, plantations in other parts of the sub-continent utilising various forms of *sirdari* to recruit their labour.

Two features of Assam plantations, argued the Chief Commissioner of Assam, J.B. Fuller, made them extremely prone to violent conflicts. First, the absence of a garden *sirdari* system of labour recruitment and management put the European management directly in contact with the work force. Without an authoritative spokesman to give vent to their grievances, the conflicts over minor grievances assumed major proportions. Second and more importantly, the power of private arrest in particular and the penal contract system in general made employment in the plantations extremely unpopular. The costly system of *arkatti* recruitment put such a premium on the coolie that penal powers were utilised much too severely. By allowing the private power of arrest of deserters, the Chief Commissioner argued, the Government had closed the 'most effective safety valve against the occurence of labour disturbances and rioting'. Against the defenders of penal powers for planters, the Chief Commissioner noted that the penal system was not just the result of labour shortage in Assam, but 'also its most potent cause'. The widespread use of penal powers turned Assam into a land of prisons, thus materially hampering the natural flow of labour to Assam [*GOI*. Emig. A, 12–14, Dec. 1904].

On these grounds, the Government of Assam sought the complete withdrawal of the 1882 Act from the Surma Valley, and withdrawal of the provision of power of private arrest from Assam Valley. These recommendations were reiterated by the Assam Labour Enquiry Committee of 1906, which led to the enactment of 1908 that struck a blow at the penal contract system in Assam.

DISMANTLING OF THE INDENTURE SYSTEM

The penal contract labour system and *arkatti* recruitment which grew up alongside it, were substantially transformed during the first decade of this century. The first component to be affected was the system of 'free' or deregulated emigration; by closing about 90 per cent of the chief recruitment areas to unlicensed recruiting, Act VI of 1901 drastically curbed the *arkatti* system [*ALR*, 1902: 2]. Under section three of the Act, the government reverted to the previous system of supervision and registration, an important innovation being the special concessions granted to centralised recruiting agencies like the Tea District Labour Supply Association. The planters were now forced to take on the responsiblity of recruitment directly, in order to avail themselves of these concessions.

The second important piece of legislation was Act XI of 1908, enacted on the recommendation of the Assam Labour Enquiry Committee's report of 1906. The requirement for local contracts and penal provisions of Act I were withdrawn completely from the Surma Valley, and the power of private arrest was similarly withdrawn from the Assam Valley. Unlicensed recruiting was also abolished. These two legislative actions of the colonial state effectively disarmed much of the penal contract labour system, complete abolition of recruitment by licensed contractors and special penal legislation taking place only in 1915.

The last vestige of the indentured labour system, namely Act XIII of 1859, was finally abolished in 1926. Despite the staggered nature of the demise of the penal contract and indentured labour, the critical importance of the legislation of 1901 and 1908 should not be minimised. The number of 1882 Act labourers in Assam declined rapidly after 1901, indicating an abrupt shift in the pattern of labour recruitment and contract system. 133,000 Act workers in 1900 had by 1910 declined to only 23,000, a mere five per cent of the total labour force. Similarly, though restricted unlicensed recruiting through *arkattis* continued till 1908, the major burden of recruiting was taken over in 1902 by the Tea District Labour Supply Association, under whose auspices *sirdars* recruited for their respective gardens.

The transformation in recruitment for, and the labour system on, the tea plantations in Assam was undoubtedly due to a drastic reversal in the policy of the colonial state, beginning with Act VI of 1901. Until then, state policy towards the tea industry was essentially set in a gradualist framework. Special labour legislation was considered as a transitory measure, necessitated by the absence of a labour market in Assam, and designed to promote the ultimate objective of 'genuinely free emigration and free labour'. In the sphere of recruitment the transition was conceived as one from state supervised emigration to privately assisted emigration, and finally to completely unassisted emigration. The penal powers of the planters were justified as necessary to the phase of assisted emigration and an insufficiently developed labour market; they would be rendered unnecessary by a fully developed labour market and completely free emigration. In 1891, these policy assumptions were clarified by the Government of India in the following words: '... there is a tendency towards the development of free emigration from the assisted to the unassisted form and when this system has reached a sufficiently advanced form, free labour properly so called shall be rendered practicable' [*GOI*. Emig. A 2–8: June 1892]. The Chief Commissioner of Assam was more specific as to conditions under which penal contract could be abolished when he wrote in 1886:

> The time when the planters in the upper districts of Assam Valley can do without such a contract can only come when ten million acres of cultivable land in that valley are taken up and the population has increased from 2½ million to 7½ millions of people [*GOI*, Emig. A: 2–8 June 1886].

Thus the ultimate goal of the colonial state policy, 'really free emigration and absolute liberty of contract' obviously lay very much in the distant future. Given this perspective, the reversal of colonial state policy in the first decade of this century appears all the more abrupt.

CONCLUSIONS

Within the narrative of colonial discourse, this reversal of policy is generally ascribed to the realisation by an increasing number of colonial state personnel that special labour legislation had failed in its objective of developing genuinely free emigration and free labour in Assam. For example, Charles Eliott and Stewart Bailey, who as the heads of the Assam administration had ardently defended the 1882 Act, became its bitter critics on becoming aware of the enormity of abuses under the 'free' emigration system when they came to head the Bengal administration. Similarly, Henry Cotton who began his tenure as Chief Commissioner of Assam by sending troops to apprehend runaway coolies from the Assam Bengal railway constructions, became extremely critical of the labour system in Assam, specially the tendency to pay abysmally low wages to contract workers. The findings of the Bengal Labour Enquiry Commission of 1896 and the Assam Labour Enquiry Commission of 1906 were in this regard crucial. The former conclusively demonstrated that *arkatti* recruitment under the free emigration system of Act I was extremely unpopular in recruiting districts and a major factor in the rising cost of recruitment. The Assam Labour Enquiry Committee supplied the arguments from the point of origin, 28 of the 31 witnesses it examined in the recruiting districts citing the existence of penal contract as the most important cause for the absence of a free flow of labour into Assam [*GOI*, 1906a: 27]. These findings then became the basis of the 1901 and 1908 legislation.

However, the realisation by the colonial state of the inceasing distance between policy goals and practical results is not entirely adequate as an explanation, and fails to account for the timing of policy changes in the first decade of this century. Moreover, it merely begs the question: why were the practical results of the colonial state policy so different from its avowed objectives? The real nature of colonial state interventions, and the changing relation between plantation economy and the colonial state, cannot be fully comprehended purely within the parameters of colonial policy.

The legislation of 1901 and 1908 was enacted when the tea industry was undergoing a severe economic crisis, prices having plunged to an all-time low between 1895 and 1906. Much of the crisis was caused by overproduction due to acreage expansion in the previous decades. Profits and dividends were under threat for the first time since the depression of 1865–70. Intervention by the colonial state was thus intended to rationalise the structure of recruitment and the labour system in order to overcome the inbuilt contradictions of the plantation system in Assam, the most acute form of which was the tendency towards overproduction. The legislation of 1901 and 1908 thus indicated a shift in the strategy of the colonial state, from the provision of a constant supply of cheap labour to ensuring the long-term viability of the plantation system as a whole.

The major contradiction affecting the plantation system arose from the necessity of expanding production by increasing the acreage planted under tea, in response to the declining tea prices of the last two decades of the nineteenth century. The exceptional labour legislation aided the planters by providing a vast number of labourers who were paid wages much below the market rate. Yet precisely because of the penal contract and its accompanying low wages, the labour supply remained largely independent of the wage rate and dependent on the capacity of the *arkatti* recruitment structure. While penal labour legislation held down wages by not allowing the operation of the labour market, competition for labour amongst planters gave rise to a 'recruiters' market'. What the planters gained by paying below market wage rates might have been offset to a great extent by the rapid rise in the premium paid to the recruiters.

Planters' responses to the rising cost of recruitment reinforced the intensification of the labour process, in effect pushing wages below the level at which labour power could be reproduced. This had a direct effect in increased mortality and desertion rates, which raised the real cost of labour further and induced greater dependence on *arkatti* recruitment. This basic contradiction became apparent when the rise in the output per unit of labour employed was far outstripped by the rise in the cost of recruitment. Between 1880 and 1900, output per unit of labour rose from 276 lb to 350 lb, or by 32 per cent, while the cost of recruitment more than doubled. This contradiction of the plantation system, in an expansionary phase based on the payment of low living wages, was expressed paradoxically in the *high cost of cheap labour*.

At another level, the same contradiction was manifested in the delinking of the social reproduction of plantation labour-power from the physical reproduction of the workforce. Low wages and the intensification of the labour process accordingly contributed significantly to the combination of high mortality and low fertility. Throughout the phase of indenture, and specially in its last two decades, the planters' main concern was ensuring an adequate supply of labour for expansion of production rather than the long term goal of building a resident workforce. Until the end of the century, planters actively discouraged the grant of plots of land to time-expired labour, fearing a reduction in the supply of labour. Low wages and high work norms enforced by the penal sanctions led to the exhaustion of fresh emigrant labour, and the negative rate of workforce reproduction contributed directly to planter dependence on the *arkatti* system and the high cost of recruitment it entailed.

Resistance on the part of the labour force to the indenture system was also an important factor in hastening its demise. In the recruiting districts, resistance manifested itself in the immense unpopularity of the *arkattis* and in the growing opposition to the penal contract, while in the plantations resistance took the form of desertion and other modes of opposition to intensification. Despite the best efforts of the planters, therefore, labourers – and especially time-expired coolies – persisted in operating a fledgling labour market.

In the end, the indenture system fell largely because of its inherent contradictions. It became economically untenable and politically costly to maintain. It was altered substantially in a phase of acute crisis, when the colonial state stepped in to rationalise the labour system. The gradual dismantling of indenture marked the transition from a system based on the physical coercion of labour to one based on its rational exploitation. In one sense it was an immensely successful system, in so far as it laid the foundation of the tea industry by permitting a large expansion in the acreage under tea; yet this success was achieved at the expense of no less that half a million deaths, and in the end this proved a little too much even for a system that looked at coolie deaths merely as 'depreciation *en bloque*' [*Dowding*, 1894: vi].

REFERENCES

Abbreviations

ACR	Assam Company Records
ALR	Assam Labour Reports
GOB	Government of Bengal
GOI	Government of India
GOI Emig.	Government of India, Emigration Department Proceedings
GOI Leg.	Government of India, Legislative Department Proceedings
GOB Gen.	Government of Bengal, General Department

Antrobus, H.A., 1957, *A History of the Assam Company, 1839–1953*, Edinburgh: T & A Constable.

Assam Company Records, 1840, Proceedings of Bengal Committee, India Committee Minute Books, MS No.9925, Guildhall Library, London.

Assam Labour Reports, 1877–1902, *Annual Report on Labour Immigration into Assam*, Calcutta.

Bagchi, A.K., 1972, *Private Investment In India*, London: Cambridge University Press.

Barker, George M., 1884, *Tea Planter's Life in Assam*, Calcutta.

Behal, Rana Partap, 1983, 'Some Aspects of the Growth of the Tea Plantation Labour Force and Labour Movements in Assam Valley Districts (Lakhimpur, Sibsagar and Darrang), 1900–47', unpublished Ph.D. Thesis, Jawaharlal Nehru University, New Delhi.

Dowding, Charles, 1894, *Tea Garden Coolies in Assam*, London: Spink & Co.

Emmer, P.C., 1986, 'The Meek Hindu: The Recruitment of Indian Indentured Labourers for Service Overseas, 1870–1916', in P.C. Emmer (ed) *Colonialism and Migration: Indentured Labour before and after Slavery*, Dordrecht: Martinus Nijhoff Publishers.

Emmer, P.C., 1990, 'The Great Escape: The Migration of Female Indentured Servants from British India to Surinam 1873–1916', in David Richardson (ed.), *Abolition and Its Aftermath*, London: Frank Cass.

Galenson, Donald, 1984, 'The Rise and Fall of Indentured Servitude: An Economic Analysis', *Journal of Economic History*, Vol.XLIV, No.1.

Ganguly, Dwarkanath, 1972, *Slavery in British Dominion* (edited by Sris Kumar Kunda; reprint of 13 articles published in the newspaper *Bengalee* between September 1886 and April 1887), Calcutta: Jignasa.

Gillion, K.L., 1962, *Fiji's Indian Migrants*, Melbourne: Oxford University Press.

Gladstone, J.S., 1910, *History of Gillanders, Arbhuthnot and Company and Ogilvy, Gilanders and Company*, London: Part of Gladstone Family Records, Clwyd Record Office, Howarden, UK.

Government of Bengal, 1868, *Report of Commissioners Appointed to Enquire into the State and Prospects of Tea Cultivation in Assam, Cachar and Sylhet*, Calcutta.

Government of Bengal, 1873, *Papers Regarding the Tea Industry in Bengal*, Calcutta.
Government of Bengal, 1883, *Report on Inland Emigration from Bengal*, Calcutta.
Government of Bengal, 1896, *Report of the Labour Enquiry Commission of Bengal*, Calcutta.
Government of India, 1891, *Special Report on the Working of Act I of 1882, for the years 1886–89.*
Government of India, 1901a, *Proceedings of the Central Legislative Council, Volume XL*, Calcutta.
Government of India, 1901b, *Prices and Wages*, Calcutta.
Government of India, 1906a, *Report of the Assam Labour Enquiry Committee*, Calcutta.
Government of India, 1906b, *Proceedings of the Assam Labour Enquiry Committee*, Calcutta.
Griffiths, P., 1967, *A History of the Indian Tea Industry*, London: Weidenfeld & Nicolson.
Rodney, Walter, 1981, *A History of the Guyanese Working People, 1881–1905*, London: Heinemann Educational Books.
Stoler, Ann, 1985, *Capitalism and Confrontation in Sumatra's Plantation Belt, 1870–1979*, New Haven, CT: Yale University Press.

Unpublished Records

Government of Bengal, 1878, Monthly 'A' Proceedings of the General Department, National Archives of India.
Government of India, 1880–1904 Monthly 'A' and 'B' Proceedings of the Emigration Department, National Archives of India.
Government of India, 1901, Monthly 'A' Proceedings of the Legislative Department, National Archives of India.

Plantation Labour in Colonial India

RANAJIT DAS GUPTA

I. INTRODUCTION

This article makes an attempt to analyse different types of capitalist plantations in colonial India. While there is considerable controversy over the definition and essential characteristics of plantations,[1] in the present paper plantations in colonial conditions are viewed as a particular type of 'capitalist' enterprise with the following basic features: an agro-industrial enterprise raising one or several crops on a large scale under tropical or semi-tropical climatic conditions; an international market orientation; the launching and subsequent maintenance of plantations under the ownership and control of foreign capital with the backing of the colonial state; the employment of a large number of producers and labourers (not necessarily wage workers) doing hard manual work under conditions of a primitive labour process; the use of a migrant and/or immigrant labour system; and the mobilisation and control of direct producers through economic and extra-economic coercive methods with the direct and indirect support of the colonial state. The class structure thus created through the use of coercion, open or concealed, was sharply divided between white or *sahib* owners, managers and supervisors on one hand and non-white labour on the other [*Bagchi*, 1982: 42–4, 77]. Another major consequence was the creation of a low wage economy [*Bagchi*, 1982: 22–3].[2]

To reiterate, the use of unfree labour, in one form or another, with a highly authoritarian structure of management of labour [*De Silva*, 1982: 185] is viewed here as a distinguishing characteristic of the colonial plantation system. Unfreedom lay in the use of non-market power, and even physical coercion, as well as market power entailing the threat of hunger and denial of employment in conditions of endemic unemploy-

Ranajit Das Gupta, Regional Development Group, Indian Institute of Management Calcutta (IIMC), Joka, Calcutta, 700 077, West Bengal, India. I am grateful to the participants in the Amsterdam conference on 'Plantation Labour in Colonial Asia' held at the Centre for Asian Studies Amsterdam for their helpful observations, to an anonymous referee for the *Journal of Peasant Studies*, and also to Bhubanes Misra, Amiya Kumar Bagchi and Raghabendra Chattopadhyay for comments on earlier drafts of the paper. The research for this study has been funded by the Centre for Management Development Studies, IIMC.

ment in which vast sections of direct producers faced significant risks of starvation and death [*Bagchi*, 1973; *Chakravarty* 1978: 264–6; *Bagchi*, 1988].[3] This results in the deprivation of freedom, partly or fully, to enter into or withdraw from the labour process and labour market, and to bargain over such conditions as working period, work rhythm, and wages. Unfreedom is viewed here as a crucial dimension of class struggle between capital and labour in the plantation context.[4] Two more features of colonial plantations were the maintenance of links between the pre-capitalist/non-capitalist subsistence economy and the plantation sector, and the externalisation of at least part of the cost of reproduction of labour power.[5] These features of plantation society originated in, and were maintained as a part of, the wider colonial economy.

Proceeding from the above understanding, the article presents a study of plantation labour in India during the period of British rule. The country came to have several types of plantations exhibiting great diversities over space and time, crops, ownership patterns, and methods of mobilisation and control of labour both 'on work' and 'off work'. Considerable diversities were also noticeable in the relationships between plantations and peasant and/or subsistence economies.

Indigo, tea, coffee and rubber were the four principal plantation crops (see Appendix). Of these, the indigo plantation was launched in the middle of a peasant society in some of the densely populated districts of Bengal Presidency in the 1790s, flourished till the Indigo Revolt of 1859–60, and thereafter had a chequered career until its end in the mid-1920s. Tea plantations were started in several widely separated locations in the 1830s, all of which were sparsely populated. Further, while in the areas of plantation areas of North-East India (Assam, Dooars and Darjeeling) tea was the only crop, plantation areas in South India (Mysore, Coorg, Nilgiris, the Shevaroy hills, the Wynaad region, Anamalai, Travancore and Malabar) included tea, coffee and rubber.

The article makes an empirical comparison of capital–labour relations in different plantations and plantation areas in India, and on this basis seeks to address some wider conceptual and analytical issues.[6] Section II discusses aspects of mobilisation and control of labour in indigo plantation; section III deals with labour in tea plantations in North-East India; section IV examines the recruitment and disciplining of labour in the tea, coffee and rubber plantations of South India; section V contains a summary and concluding observations.

II. INDIGO PLANTATIONS

Indigo figured prominently in the trade conducted by the English East India Company in the eighteenth century [*Watt*, 1972: 393–5; *Chaudhuri*,

1964; *Kling*, 1977], and its production as an important commercial crop dates from the late 1790s. The growth of the cotton textile industry in England, coupled with the decline in the supply of indigo from traditional sources like the West Indies and Santo Domingo, stimulated its production in India. A factor of critical importance was that indigo became a principal means of remitting part of the tribute from the company's territories in India, and the earnings of individual European officials and traders, to England. This led the Directors of the Company to promote not only trade but also cultivation and manufacture of indigo. Persons with experience in the slave plantations of the West Indies were brought to India. The Company's officers were encouraged to engage in the indigo trade, and individual Europeans were licensed to move into the interior and set up plantations [*Chaudhuri*, 1964: 81–4; *Bagchi*, 1987: 133–4]. Thus, the very origin of European enterprise in indigo lay in international networks. This link was maintained and came to influence how indigo producers were mobilised and controlled.

By the early nineteenth century, there were two major centres of indigo production: the deltaic region of Bengal, and north and central Bihar, both with extensive peasant agriculture and densely populated. For reasons discussed below, the planters, mostly Europeans, encountered widespread and strong reluctance by peasants to engage in indigo cultivation. As the British government did not encourage colonisation by Europeans in India, planters pressed for *zamindari* and other tenurial land rights to enable them to overcome peasant reluctance and to control peasants. By a regulation of 1829 they were given these rights and the resultant system of sub-infeudation proved to be of critical help to them [*Chaudhuri*, 1964: 165–81; *Kling*, 1977: 19, 42; *Sen*, 1982: 49–59].

In the early nineteenth century European enterprise in indigo remained limited primarily to several forms of trade with little or no involvement in production,[7] but this was soon to change.

Variants of Production Organisation

The organisation of production consisted broadly of two types: (1) cultivation of indigo by *raiyats* (tenant producers) under contract with the planters with some input from the indigo concern, known as *raiyati* or *assamiwar* cultivation; and (2) hired or wage labour-based cultivation under complete control of the planters known as *nij* or *zerat* cultivation.

Under *raiyati* or *assamiwar*, the holding of a *raiyat* was the basic unit of cultivation, with two categories of tenants: *be-ilaka raiyats* on lands over which planters lacked any *zamindari* or tenurial rights; and tenants resident in villages comprising the estate or farm of a planter-*zamindar*, who exercised the power and authority of a *zamindar* or tenure-holder

[*Chaudhuri*, 1964: 124–5]. All categories of *raiyats* under the *raiyati* type raised indigo on their tenanted lands with their own family-labour, cattle and implements at their own expense and risk, but all operations from preparation of the soil through selection of the moment for sowing to harvesting and delivery to the indigo factory had to be carried out under the close supervision of the European planters and their Eurasian or Indian *amlas* or supervisory personnel.

There were several sub-types of *raiyati* or *assimiwar* cultivation. The main ones were the following.[8] A major variant, particularly till the late 1820s, was cultivation by a *be-ilaka raiyat* who had entered into a contract with a planter, usually a European, through inducements on advances, to grow and sell indigo at a fixed price per unit of weight. Prepayment of cash advances, however, was likely to be a permanent liability. Not infrequently planters found it difficult to induce peasants who were *raiyats* of an Indian *zamindar* to raise indigo. From the European planters' point of view the difficulty was that legally virtually all cultivation was *be-ilaka* and all cultivators *be-ilaka raiyats* over whom planters had no formal power [*Kling*, 1977: 19, 53; *Chaudhuri*, 1964: 170–72; *Sen*, 1982: 50–51].

Another major variant was cultivation by an *ilaka raiyat* of a planter who under the 1829 Regulation had acquired *zamindari* or other tenurial rights over land on which indigo was grown, or held a long lease, or had secured occupancy rights from another *zamindar*. While the relationship between *ilaka raiyat* and planter was basically similar to that described above, the tenant was subject to greater control by the planter-*zamindar*. By the time of the Indigo Revolt the extent of *ilaka* cultivation was far greater than *be-ilaka* cultivation.[9]

Yet another variant was the *tinkathia* system, prevalent in Bihar, in which a *raiyat* had to agree with a planter-*thikadar* (or temporary leaseholder of land from a *zamindar*) to grow indigo on a specified portion of his lands, usually three *kathas* per *bigha* (one *bigha* = 20 *kathas*). At the time of the agreement, an interest-free advance was given and at the end of the indigo year the *raiyat* was paid according to the extent of land sown with indigo rather than by the quantity of output [*Hunter*, 1976: 102, 268–9; *Henningham*, 1982: 38; *Fisher*, 1978: 115].

Under *nij* or *zerat* cultivation a planter held land as a *zamindar* or took a terminable lease (*patta*) from a *zamindar*, but did not enter into any agreement with the peasants nor pay any advances, production being organised entirely by him at his own expense and risk. The practice was to hire poor peasants and labourers residing in the locality, along with their bullocks, ploughs and carts for cultivating and transporting the crop, and to pay them by the day. A sub-variant was cultivation by labour, often non-local labour (largely tribals) hired in the 'market' and generally settled on the factory grounds, working with cattle, ploughs and carts owned by the indigo concern.[10]

Till around 1860 the *raiyati* system was the predominant one in both Bengal and Bihar, and in Bengal proper the extent of *raiyati* cultivation was at least three times *nij* cultivation [*Chaudhuri*, 1964: 125–6]. While after the Indigo Revolt indigo plantations in Bengal suffered a drastic decline, in the major indigo districts of Bihar the *nij* or *zerat* gradually began to replace the *raiyati* or *assamiwar* [*Chaudhuri*, 1970].

In sum, indigo estates in Bengal and Bihar exhibited features of both tenant labour-based and hired labour-based plantations. Under the former indigo cultivation, though done on peasant holdings, was subject to the highly authoritarian control and power of the planters. In the case of *nij* or *zerat* cultivation, the planters used the powers obtained from the *zamindars* to dispossess tenants and then engage them as hired labourers, also under stringent control.

One common basic feature, however, was that the entire indigo plantation system, together with the deployment and mobilisation of labour, was built primarily upon extra-economic compulsion and non-market mechanisms. It merits repeating that indigo was cultivated in areas which were highly populous, yet indigo planters were confronted with the problem of mobilising labour. Peasants were highly reluctant to grow indigo for many reasons. First, given the cost to the *raiyat* of indigo cultivation, the price paid by the planter was unremunerative.[11] Second, since the lands on which indigo was grown were very fertile, suited for the cultivation of rice and other crops, indigo could be cultivated only at the cost of these crops. Third, being dependent on good and well-distributed rainfall it was a highly uncertain and speculative crop. Fourth, the wage paid to labourers under the *zerat* system was much below the prevailing market rate [*Chaudhuri*, 1970: 215]. Fifth, and importantly, the indigo growers – tenant cultivators as well as hired labourers – resented the rigorous and arbitrary control exercised by planters and their *amlas*. Sixth, the labour process, particularly at the manufacturing stage, involved most unpleasant hard labour. All this combined to make indigo production extremely unattractive for almost all categories of indigo producers, and to create considerable tensions between them and the planters. These were intensified by the character of the *sahib* planters who were, generally speaking, rough, uneducated and arrogant racists. Not infrequently, they also indulged in sexual violence.[12] These tensions generated a strong aversion to raising indigo and resistance to incorporation into the production relations installed by the planters.

This resistance was the principal reason why planters resorted to various forms of extra-economic coercion. In the early decades of the nineteenth century these were mostly direct and overt: outright violence, confinement and flogging, sometimes resulting in death; plunder and affrays; enforcing of contracts on peasants by *lathials*, or planters' retainers armed with long bamboo sticks; and rent increases for tenants unwilling to cultivate indigo, forcible and illegal seizure of their lands, and their eviction [*Buckland*, 1976: 238–9; *Chaudhuri*, 1964: 190–91]; *Kling*, 1977: 175–6]. In later decades, because of public outcry, and because of official reluctance to support such coercion that might threaten the long term viability of the indigo plantations, and, not least, in the interest of the stability of colonial rule, the methods of the planters became less crude and somewhat indirect. But 'force' in one form or another remained a characteristic feature of indigo plantations not only in Bengal prior to the Indigo Revolt but also in Bihar as late as the 1910s [*Buckland*, 1976: 102: 240; *Henningham*, 1982].

Apart from the frequent recourse to overt violence, the single most powerful means of compulsion under the *raiyati* or *assamiwar* system was debt servitude. Planters forced advances on the peasants to make them agree to enter into contracts to grow indigo. After the crop was delivered to the factory, payments were adjusted against the advances made. Because of unremunerative prices and occasional bad crops, peasants soon accumulated arrears. Such debts were passed from generation to generation, and the peasant became bonded to the factory. The planters were not interested in debt repayment, since this 'never-to-be liquidated balance' came to represent 'an inexhaustible reservoir of labour' that 'the planters could tap at anytime' [*Chaudhuri*, 1970: 221].

The *nij* or *zerat* system which spread in Bihar later involved 'the expropriation of peasant producers and the employment of their labour' for wages [*Chaudhuri*, 1970: 215]. The coercive power and influence acquired by planters who leased large areas from the big *zamindars* of Bihar, such as the Bettiah *raj* or Hatwah *raj*, on *thika* (temporary lease) and *mukarari* (permanent) tenures, meant that they could pay less than the prevailing wage rate [*Yang*, 1989]. What were the reasons for the transition from the *raiyati* system to the *nij* system? Prior to the Indigo Revolt the *raiyati* system was preferred to *nij* cultivation because of planters' difficulty in securing sufficient land with a secure right of occupancy, and in organising cultivation by ensuring a large enough supply of labour and mobilising an adequate number of ploughs. The fact that under the *zerat* system the entire risk was borne by planters was another reason why they preferred the *raiyati* system. But the following factors contributed to the beginning of a change-over to the *zerat* system: the intensification of peasant discontent under the *raiyati* system culminating in the Indigo Revolt in Bengal and the agitation of 1867–69 in Bihar; the need felt for establishing greater control over land, labour and production; and the realisation that the *zerat* was the best method of increasing the profitability of the indigo system [*Chaudhuri*, 1964: 124–66]. But the change-over was never total, and the *raiyati* system with some variations remained prevalent in early twentieth-century Bihar [*Henningham*, 1982: 43–4].

How did the indigo plantation affect agrarian society and particularly the peasant economy? Briefly, the indigo plantation was a major component of the process of commercialisation of agriculture during the nineteenth century, as a classic case of 'forced commercialisation' bearing the mark of domination by foreign capital [*Bagchi*, 1988: PE 40].

In relation to the internal social structure it needs to be stressed that indigo production came to be closely linked with the system of subinfeudation. From the planters' viewpoint, profitable indigo production required extensive acquisition of intermediary tenures and sub-tenures. Throughout the first three decades of the nineteenth century they pressured the government into granting them this right and eventually obtained it. Thus, indigo production gave impetus to the spread of subinfeudation a major factor in the stagnation of peasant agriculture in eastern India, even after the end of indigo cultivation (among others, Sen [1982: 4–5]).

All this indicated the basic contradictions of a production system which combined features of capitalist enterprise with coercive feudal power exercised by indigo planters with *zamindari* rights, or holding leases under *zamindars* along with police and judicial authority. These contradictions and their consequent social tensions manifested themselves in a series of extensive anti-planter peasant struggles that broke out in Bihar between 1907 and 1923 [*Henningham*, 1982: 44–65]. But it was an external factor – the marketing of a competitive artificial substitute produced from a more advanced capitalist labour process elsewhere, and the consequent disastrous fall in the price of the natural dye in the international market – that eventually brought the extinction of the indigo plantation by the mid-1920s.

III. LABOUR IN TEA PLANTATIONS IN NORTH-EAST INDIA

Have we consider some significant aspects of the capital–labour relation in the tea plantations in North-East India, more specifically in Assam, comprising the Brahmaputra and Surma Valleys, with reference also to the relations that emerged in the Dooars in Jalpaiguri district and the

Darjeeling hills in North Bengal. Tea had some special requirements with regard to labour.[13] It needed an abundant supply of cheap, docile and highly malleable workers capable of hard labour under severe and difficult conditions. These requirements led to the imposition of a particular mode of recruitment, control and authority over tea plantation labour. The tea plantation sector, though capitalist, was launched and maintained on the basis of wage labour that was unfree.[14]

The regions in which tea gardens were opened were most sparsely populated,[15] and large tracts of cultivable land could be obtained for ownership or rent at relatively low cost. Given the legacy of a semi-tribal, semi-feudal society and the constraints of a colonial economy, social differentiation had not proceeded far in Assam and planters had great difficulty in obtaining adequate labour. Dooars and Darjeeling, despite some differences, had similar problems [*Arbuthnot* 1904: 2]. Also, because of the low wage policy of the planters, and the harsh conditions of work and life in the plantations, local peasants and agricultural labourers had a strong aversion to plantation work.[16]

Mobilisation of Labour

Faced with an acute shortage of labour (among others, Sircar [1986b]), and following the failure of attempts to depeasantisise Assam,[17] the planters had to seek labour from sources some hundreds of miles away. They resorted to a policy of organised recruitment of labour from tribal/semi-tribal, as well as non-tribal low caste peasant, communities inhabiting an extensive area spread over the Bengal Presidency or the latter day provinces of Bengal, Bihar and Orissa (particularly Chhotanagpur and Santal Parganas), the eastern part of the United Provinces, certain areas of the Central Provinces, and the districts of Ganjam and Vishakapatnam in the Madras Presidency. These recruits belonged to diverse peasant, labouring, artisan and service groups. Because of the operation of a host of socio-economic forces [*Mohapatra*, 1985: 259–61], they were subjected to a complex process of depeasantisation, deskilling and proletarianisation by which this extensive region was reduced to a vast reservoir of cheap labour or labour catchment area (LCA).

Labour under penal contract: From the very beginning not only men, but also women and children were recruited though a system based on coercion. Instead of the market mechanism, force – both overt and concealed – and politico-legal mechanisms became of crucial importance in mobilising labour.

Apart from the labour shortage mentioned above, planters were confronted quite early on with labour protests in the forms of desertion and assaults, as well as informal collective bargaining and even strikes [*Guha*: 1981]. In response, and following the model of indentured labour recruitment in India for the British colonial sugar islands of the West Indies, Mauritius and Fiji which came into operation in 1834,[18] the indentured system with penal provisions was introduced for the Assam plantations. The Workmen's Breach of Contract Act (or Act III) of 1859, reinforced by Section 492 of the Indian Penal Code, introduced provisions for arrest and punishment by the government under civil laws of workers considered responsible for any breach of contract. The Act VI of 1865 (an amended version of the Transport of Native Labourers Act of 1863), along with certain so-called welfare provisions relating to work period, working hours, wages and sanitation, not only contained penal clauses and made deserting or 'absconding' workers criminally punishable, but also gave planters the power to arrest such workers without any warrant. The Inland Emigration Act of 1882, which introduced 'free' emigration and unlicensed recruitment, also strengthened the penal contract system.[19] Contemporary accounts testify to the ruthlessness with which such power was exercised by the planters [*Ganguli*, 1972]. The amending legislations of 1893 and 1901 retained the system with all the penal power already conferred on planters. The planters' private power of arrest was abolished in 1908. But under the Act XIII of 1859, and the Assam Labour and Emigration Act of 1915, recruitment of labour under penal contract continued. Further, time-expired labourers were frequently induced under duress to re-engage themselves. All this reduced indentured wage labour employed on Assam plantations to forced labour or unfree wage labour.

This forced nature of indentured labour was considerably enhanced by how labourers were recruited. Under the 1863 Act recruitment was assigned to licensed contractors and their agents or *arkattis*, and alongside them, under the Amendment Act of 1870, to garden *sardars*.[20] Methods of recruitment allowed hardly any element of free choice on the part of labourers. Being concerned solely with rounding up as many people as possible, the contractors and their men resorted to all forms of deceit and trickery, intimidation, violence, and even abduction of married women, to secure labour. Once men and women were recruited by such means, they were kept in transit depots and later despatched to plantation districts under the strictest control. Planters soon found recruitment through licensed contractors too expensive and inconvenient. 'Free' emigration, or the right of a worker to enter a contractual agreement in Assam rather than in his/her own home district, and deregulated recruitment by unlicensed contractors introduced under Act I of 1882, sustained and exacerbated previous abuses. The entire system of recruitment came to resemble 'the slave-running in Africa and the global slave trade' [*Guha*, 1988: 45]. One may wonder how long recruitment based on deception, fraudulent practices and misrepresentation can continue. But even when those recruited became aware of the deceptive practices, and of the conditions prevailing in the plantations, they had few options. Successive official reports show the persistence of this manner of securing labour for more than four decades.[21]

Recruitment involved diverse forms of extra-legal, non-market power as well as market power. Often poverty-stricken potential migrant workers – poor peasants, agricultural labourers and ruined artisans – were given loans by garden *sardars* to clear their debt to *zamindars* and *sahukars*, and to meet their travel expenses. In this way they entered the grip of the *sardars*. It is hardly meaningful to speak of free choice and optimising decisions by people faced with the prospect of persistent unemployment, disguised and open, and starvation [*Chakravarty*, 1978: 264–6.]

Control of Labour

It was only in 1926, as a result of bursts of labour protests in myriad forms – individual and mass desertions, most strikingly demonstrated by the Chargola exodus of 1921 [*Guha*, 1988: 129–33], evasion of assigned tasks, litigation, assaults on planters, and strikes [*Guha*, 1981: 87; *Guha*, 1988: 133–4; *Behal*, 1985: PE 20] – and growing nationalist criticisms, that the 1859 Act and indentured labour were finally abolished. With the repeal of all penal legislation there was undoubtedly a mitigation of some of the most blatant coercive practices, but the consequences of abolition of the legal power of planters must not be exaggerated.

It deserves emphasis that in colonial India in general and on plantations in particular, there were both legal and extra-legal sanctions and power, with no strict separation between the two in practice. It was because of various extra-legal non-market mechanisms that even labourers who were juridically free were subjected to diverse forms of bondage. In Assam, as noted above, the planters' private power of arrest was abolished in 1908. In North Bengal gardens labour was never under the indentured system [*Arbuthnot*, 1904: 2, 4–5]. But the *mystique* of *sahib* planters, enjoying the support of the government, was as a law unto themselves [*Das Gupta*, forthcoming a: 65–9]. Forcible capture of 'absconding' coolies and corporal punishment were viewed by *sahibs* as their rights; beating, flogging and other kinds of coercion were considered the most effective way of dealing with labourers regarded as recalcitrants. Rape and other forms of sexual coercion were also elements in the brutal regime of the plantation system. Contemporary evidence shows that even in cases of rape or murder there was virtually no recourse to law. Labourers could complain to government officials and could, in theory, even go to the court, but in most cases these measures were of no avail, the racial bias of white colonial officials playing its part [*Ganguli*, 1972: 45–96]. Later important changes began to take place, largely in response to labour protests [*Guha*, 1981; *Behal*, 1985; *Das Gupta*, forthcoming a], but to the last days of the colonial period the labour system remained essentially based on coercion.

On arriving at a garden, labourers were put under legal and extra-legal control. The practice of paying a 'bonus', which originated under the indentured system, continued after indenture was abolished, and was described as *girmit* (agreement) or contract money. This practice gave labourers the impression that, as before, they had no freedom to leave their existing employment, an impression reinforced by the practice of taking thumbprints when so-called bonuses were paid [*RCLI*, 1931: 376].

On arrival, workers were confined to the tea gardens. The *chowkidar* visited each 'coolie' line every night to keep watch on the movement of 'coolies'. Contact with the outside world, even with neighbouring tea gardens and villages, was virtually prohibited. For all practical purposes, the coolies were prisoners under guard, day and night [*RCLI*, 1931: 15–16]. The coolie lines and most of the roads and pathways through the sprawling tea gardens were the private property of the estates, and outsiders were strictly excluded from them. In addition, private armed police forces were organised by planters in all the major plantation regions to ensure that labour remained under control.[22]

In many tea gardens plots of land, usually less than one-third of an acre for each adult labourer, were given to the workers for cultivation, either free or at a nominal rent, as a means of binding them more strongly to the plantation [*Guha*, 1988: 45]. Further, agreements among employers not to entice away each other's labourers, or to employ a labourer coming from another garden, effectively restrained the mobility of labourers and their ability to bargain [*RCLI*, 1931: 377; *LIC*, 1946: 28].

It also deserves mention that poor peasants, agricultural labourers and displaced artisans who became plantation workers were subjected to a regimen quite unlike pre-industrial rhythms of work dependent on variations of weather or seasons. Plantation work imposed on them the 'tyranny' of the gong. The day began with a gong at sunrise and ended with a gong at sunset. Every worker whose name was on the garden roll had to report for work regularly and to work hard. Illness was no reason for not reporting for work. Kadamoni, a woman worker in a Jorehat garden, told the Royal Commission on Labour in India [RCLI] (1930): 'I am even made to work even when I am ill' [*RCLI* 6(2): 115].

In sum, whether under the indentured system or not, the mechanisms of a free labour market without any restraint were never allowed to operate. In resorting to the penal contract, employers disregarded the difference between a labourer and labour power, that is, the labour capacity owned as a commodity by the labourer who is free to exchange it in the labour market. In the case of non-indentured labour this difference was, no doubt, formally and juridically recognised. But all through planters resorted to non-market as well as market mechanisms of various kinds to keep tea plantation workers under regimented control in both the work-place and the 'cooly' lines, indeed, in a state of virtual captivity.

Twin Dependency: Capitalist Sector and Subsistence Sector

To fully understand the labour system of the tea plantations of the North-East, it is also necessary to take into account certain features of the wage system. Plantation workers, both indentured and juridically free, were paid *below-subsistence* wages, resulting in high mortality and low fertility.[23] Moreover, this fixing of wage levels was the *basic mechanism* for compelling whole families, including children of four and five years of age, to work in the gardens [*RCLI*, 1931: 415].

But even the plantation earnings of the entire family amounted to only 62 to 78 per cent of family expenditure.[24] In other words, total family earnings from plantation wage work failed to meet the requirements of reproduction. Under the circumstances, workers had also to depend on production and consumption outside the wage economy, that is, on non-capitalist subsistence sectors. Thus, the plantation system encompassed two different but closely interrelated and interacting sectors: the dominant capitalist plantation sector and the subsidiary non-capitalist subsistence sector.

The latter had several components. Most significantly, among plantation workers in Assam, Dooars, Terai and Darjeeling – regions in which there had been no significant peasant economy – there emerged elements of tenancy and subsistence farming. A fairly large proportion of the tea garden workers were given plots of land by the planters to raise paddy and vegetables towards their subsistence requirements. Surma Valley planters introduced this practice in the late nineteenth century and enjoyed the additional hold over workers settled on garden lands.[25] In the Brahmaputra Valley, in the early years planters were reluctant to allot garden plots to workers for paddy cultivation, fearing that this would keep them away from tea-garden work. Later Brahmaputra Valley planters also came to recognise the advantage of allotting garden rice land to workers as a new instrument for control. By 1906 allotment of garden lands for private cultivation was a fairly widespread practice, and one encouraged by the colonial government.

This does not mean that all tea garden labourers in all estates had plots of rice land. Even those given land usually got minuscule plots amounting to no more than two-thirds of an acre.[26] Given that in Assam and Dooars paddy production was around 20 mds. (1 md. = 37.5 kg) per acre [*LIC*, 1946: 48; *Mitra*, 1951: lxviii] and that the usual annual paddy intake for a family of husband, wife and two minor children was around 33.5 mds. (approximately), garden rice land at best could provide no more than 40 per cent of a working family's staple food requirement.[27] This does not mean that the paddy raised on a garden plot fully covered the deficit in the cereal budget of a worker's family caused by low wage income. It only helped to mitigate the deficit to some extent.[28]

A second component was setting ex-tea garden or time-expired workers on government *khas* lands or the land of private landholders on the plantation's periphery, who continued to work part-time on the plantation. In the later years of the nineteenth century and early years of the twentieth, there was much controversy within government and among planters about the desirability of encouraging settlement on government land. However, the Assam government was keen to promote the colonisation of the sparsely populated province [*ALECR*, 1906: 89–92; *RCLI*: 361] and by 1906 many ex-coolies held land received directly from the government in both valleys and the number of such land-holders continued to increase.[29] A separate but related form, much more limited in extent, was access to land as tenants of private landlords.

A variety of other activities and occupations constituted the third component, including growing vegetables and fruits, rearing and grazing cattle, raising poultry, collecting forest produce, wood-cutting, fishing, and other activities within the garden grant and in its periphery [*RCLI*, 1931: 384–5]. Though these were perhaps of lesser importance than subsistence farming, they had a noticeable place in the life of the workers.

In sum, the colonial authority and the capitalist planters pursued an active policy of creating and supporting a subsistence sector within the plantation areas and their peripheries. This sector had several important economic and non-economic implications. In view of the inadequate and unsatisfactory nature of available data it is difficult to calculate the precise contribution from the subsistence economy, but a rough idea is possible. Thus the nine detailed family budgets collected by the 1921–22 Assam Labour Enquiry Committee suggest that income other than wages income, that is, the value of produce of the various activities listed constituted from 28 to 40 per cent of total family income [*ALECR*, 1921–22: 123–37]. Also, in tea gardens which could not provide rice lands to their workers, wage rates were higher than in those which settled workers on garden lands [*ALECR*, 1906: 73; *RCLI* 6(2): 150; *Das Gupta*, 1986: PE7]. In short, other activities met a large part of the subsistence requirements of plantation workers and thus supplemented their wage earnings, suggesting their twin dependency on the capitalist sector and a non-capitalist subsistence sector. The cost of production and reproduction of labour power was at least partly externalised to a non-capitalist economy, and borne by the workers themselves through self and familial exploitation [*Burawoy*, 1980: 138–73].

In addition, the practice of allotting garden plots served as a constraint of a non-market kind on plantation workers' freedom to enter and withdraw from the labour market.[30] It was an instrument for tying and disciplining labour; any worker accused or even suspected of insubordination was under the threat of being dispossessed of the plot and driven out of the garden. The 'willingness' of the workers to accept this sort of tie-in allotment was not independent of the state of the labour market.

Extremely low cash wages and the need to supplement them led workers to accept land allotment. The necessity for survival constrained all their choices [*Bagchi*, 1973: 1507].

IV. LABOUR IN PLANTATIONS IN SOUTH INDIA

Around the time tea was launched in Assam, tea, coffee and rubber plantations were introduced in South India. Even though, plantations were located in densely populated regions and districts, their specific sites were virtually unexplored and/or very sparsely populated hill and forest tracts.[31] These areas were situated in both British India and areas of indirect colonial rule, that is, in the Princely States of Mysore, Travancore and Cochin. This political geography had an important bearing on the formation of the plantation labour force, particularly its recruitment and, from the planters' point of view, disciplining, given the relative proximity of the plantations to their LCAs.

In tea, and in large coffee and rubber estates, Europeans held dominant positions, while there was also significant Indian ownership of coffee and rubber plantations. Differences in crop and area notwithstanding, employers organised in one single body, the United Planters' Association of Southern India (UPASI) founded in 1893.

In the south most plantation labourers came from the Tamil districts of the Madras Presidency.[32] Some labourers resided within walking distance of the estates and travelled to work each day, but for the majority the distance between their home areas and the plantations was 100 to 300 miles.[33] The majority of workers were Adi-Dravidas or low Hindu castes, drawn from landless or land-poor and oppressed agricultural labouring groups.[34] As in plantations of north-eastern India, (but in contrast to the largely male-dominated Tamil labour migration to Ceylon, Malaya and Straits Settlements, and Burma) entire families worked in the coffee and tea plantations of southern India while in the rubber estates most workers were single males.

Mobilisation and Disciplining

For a fairly long period control was exercised by putting labour under politico-legal forms of compulsion and contract. The Act XIII of 1859 was, as in Assam, a major instrument of control. In addition, the Madras Planters' Act of 1903 (repealed in 1929) and the Coorg Labour Act of 1926 (withdrawn in 1931) both had penal provisions.[35] However, from the planters' points of view there were two problems. It was usually the *kanganies* who received an advance from the planters under penal contract to secure a stated number of workers, also under contract, and bring them to the estates, but often *kanganies* failed to meet the contract obligations. In such cases the courts sometimes took the view that even if *kanganies* had not recruited the number of workers agreed, the penal provisions of the 1859 Act were not applicable.[36] The second problem arose from the fact that the plantations and their labour catchment areas extended over British Indian territories and three Princely States. The planters often complained that taking advantage of different jurisdictions a *kangani*, a *maistry*, or a labourer – after accepting an advance – slipped away from one territory to another, and that there was no method of extraditing the defaulter and compelling him to meet the contract.[37]

The UPASI, the Government of Madras and the Government of India from time to time engaged in protracted negotiations to work out the most effective measures for enforcing labour contracts. The 1903 Madras Planters' Act, which strengthened the system of penal contract, was one result of these negotiations,[38] and was useful primarily as a threat to recalcitrant workers.

There were additional problems for the planters. The LCAs for plantations in the South were also the catchment areas for Ceylon, Malaya and Burma, and the early years of the present century witnessed increased competition from these countries. Furthermore, crimping of labour by different estates within South India itself was common. The relatively short distances between plantation areas and the recruitment grounds made any strict application of the indenture system difficult. It was probably due to these factors that, compared to the North-East, the indenture system was of lesser importance and its abuses less blatant in the south.

Debt Bondage and the Kangani *System*

In those conditions a variant of debt bondage enforced by the planters through the *kanganies* came to play a critical role in the mobilisation and disciplining of labour. Labour contractors, known as *kanganies* or *maistries*, were often drawn from the ranks of the plantation workers themselves. They did not, however, work as labourers, and many of them belonged to higher castes than the labourers [*George et al.*; 1986: 210]. Head *kanganies* received interest-free loans directly from the planters, from which cash advances were given either by themselves or through junior *kanganies* or *kole maistries* to low caste, land-poor, and landless agricultural labourers to induce them to move to the estates. Head *kanganies* also arranged for labourers' transport to the estates, and supervised their work with the assistance of the *kole-maistries*, in return for a commission on the earnings of the workers in their respective gangs. This commission amounted to 'unearned income', squeezed out of the labourers' legitimate dues.[39]

Some evidence suggests that mobilising labour through advances had its origin in the need felt by planters to 'emancipate' potential plantation workers from their state of 'hereditary serfdom' and indebtedness to local moneylenders.[40] For many workers, however, this meant just a transfer of 'debt from the hereditary landholder [and moneylender] to the planter or to the planter's agent', that is, the *kangani*.[41] But whatever its origin, the explanation for the persistence of debt bondage over several decades was the planters' attempt to overcome the resistance of labourers to plantation work, as well as to curb or even eliminate the mobility of labour within the labour market.

Once labourers accepted an advance, for all practical purposes they became permanently tied to the recruiting estate. Though they may not have been under any formal politico-legal constraint, they could not leave the estate till the advance was fully repaid.[42] Even if full repayment was made, in a context of intense misery and persistent and substantial unemployment they remained under the control of the *kanganies*, and could not withdraw from plantation work or even move from one estate to another. Furthermore, politico-legal constraint, though of an informal nature, was not altogether absent, as one of the rare pieces of evidence from a worker shows. In response to a query from the RCLI, Maranney, a labourer in an Anamalai tea garden, stated (1930): 'I had to sign a bond when I took Rs.45. I have not taken back that bond from the *maistri* though I have repaid the amount'.[43] Continuing he said: 'If the Sahib gives me permission, I can leave the garden; otherwise not'. Further: 'If I run away without paying back the advance, I will get a warrant, that is to say, the *maistri* will inform the agent and the agent will send policemen to bring me back'.

The hold on workers was also maintained through various devices resorted to by *kanganies* with the support of planters. These included striking off the names of recalcitrant workers, underscoring the pounds of leaf plucked, beating and kicking, and sexual violence against women workers.[44] The near-universal practices of paying wages through *kanganies*, and of withholding wage payments, were particularly effective methods of intensifying dependence as well as disciplining labour. Moreover, as in Assam and Dooars, labourers were kept segregated and outsiders were prevented from any contact with them.[45] In fact, an interplay of indebtedness, caste hierarchy, personal dependency relations, and *sahib-power* backed by the state, helped to keep workers under permanent control and in a state of perpetual fear, and significantly to restrict the operation of the labour market.

Work Regime and Rural Connections

The labour process involved long hours of hard, unpleasant manual labour under the strict supervision of the *kangani*. Palani Ammal, a woman worker of an Anamalai tea estate, testified before the RCLI (1930): 'We work during the rains. Even if it rains very heavily, we have to go out to work. Even the children will have to go out and work. They [*kanganies*] will come in search of us . . .'.[46] Workers unwilling to go out to work were physically coerced to do so.

The planters, though interested in organising a labour force tied to the plantation system in general and individual estates in particular, did not want a permanently settled residential estate labour force. They recruited with the understanding that labourers would work on a tea estate for about ten months and on a coffee or rubber estate for six or seven months, and would then return to their villages for two to five months, provided no arrears were due. This pattern was partly due to the nature of the plantation crops, particularly coffee and rubber, which did not require round the year work, and partly to the inadequacy of wage earnings, which made it necessary for labourers to obtain a portion of their subsistence requirements from their rural links.[47] In tea work was of a perennial nature, hence there was less oscillatory movement between the estates and village homes.

V. CONCLUSION

The creation and organisation of plantation sectors, dominated and controlled by British capital, led to the growth of a labour force with several distinctive features. It was composed of several segments, originating from impoverished or small peasant households, landless agricultural labourers and displaced artisans belonging to tribal, semi-tribal and low caste people.

This capital–labour relation was characterised by varying degrees and diverse forms of unfreedom. Different segments of the plantation work-force were subjected to different degrees of subordination through the use of both non-market and market powers. During the long period of colonial capitalism the methods and forms of control went through many changes, but no strict separation between non-market and market relations took place and the deprivation of freedom remained a condition of the social existence of plantation labour. It has also been seen that unfreedom occurred in both labour-surplus and labour-scarce situations.

It should be stressed that unfreedom was a central element in the struggles between planters and plantation workers. These struggles took varied forms ranging from indirect and subterranean ones such as desertion, feigning of obedience or of illness, avoidance of assigned work, individual or small group violence against planters or their accomplices, and so forth, to overt and concerted collective action like mass desertion, 'unlawful' assembly, submission of demands, violent protests or riots and strikes. Some anti-planter protests converged with wider anti-colonial conflict. But till the end of the Second World War most labour struggles on the plantations were independent of nationalist and other political organisations.[48]

A related critical aspect was that proletarianisation was complex, partial and uneven, with many different, overlapping levels: proletarianisation of sections of the peasantry already dispossessed, fully or partly, of land; partial commoditisation of the workers' labour-power and differing extent of dependence on wage earnings; incomplete separation or alienation from means of subsistence production; participation of large sections of the wage-earning work force, often under 'inducement' by the planter, in peasant-like work and subsistence production; participation of ex-tea garden workers settled on land on the periphery of the gardens in part-time plantation wage work; and, as in the case of southern plantations, oscillatory movements between village based peasant production and wage work on the plantations. Significantly, processes of proletarianisation encompassed both struggles between capital and labour, and resistance on the part of peasants, agricultural labourers and artisans to incorporation in plantation production relations and their specific conditions of capitalist accumulation.

Significantly, in the context of colonialism capitalist transformation was marked by ambiguities, pluralities and enclave formation with effects for the formation and growth of plantation labour. The forms of labour within capitalist plantations were extraordinarily diverse and plural: tenant labour; indentured labour, that is, labour under overt politico-legal constraints; labour bonded by indebtedness and relations of personal dependency; labour not subject to any explicit or formal politico-legal constraints but still under diverse forms of non-market control; labour under mediated control; wage-labour tied by housing and the allotment of tiny plots of land; and wage-labour receiving payment partly in kind. Moreover, it is not always possible to make any strict separation between these different forms.

In sum, the functioning of plantations in colonial India was based on both extra-economic and economic methods of compulsion, and the unfreedom for labour that resulted. It also depended on the maintenance of a particular kind of relationship between the capitalist sector and the non-capitalist subsistence economy which never freed labourers from significant dependence on the latter.

NOTES

1. See, among others, Jones [1968]; Beckford [1972]; Bernstein and Pitt [1974]; De Silva [1982]; Bagchi [1982]; Graves [1987].
2. These features are indicated in the studies referred to in note 1 with varying emphases.
3. This chronic excess labour situation caused by landlessness and loss of livelihood as a result of colonialism is distinctly different from situations of seasonally excess labour and cyclical unemployment or Marx's reserve army of labour.
4. The literature on unfree labour is large and growing. See, among others, Kloosterboer [1960], Marx [1973], Corrigan [1977]; Brass [1986]; Miles [1987]; Patnaik and Dingwanay [1985]; Breman [1989].
5. See Deere [1979: 133–48]; Nichols [1980: 107–73].
6. The abbreviations used to indicate source references in this article are the following: *ALECR 1906: Report of the Assam Labour Enquiry Committee 1906*; *ALECR 1921–22: Report of the Assam Labour Enquiry Committee 1921–22*; *LECR: Report of the Labour Enquiry Commission 1985*; *LICR*: Government of India, Labour Investigation Committee, *Report on an Enquiry into Conditions of Labour in the Plantations in India* by D.V. Rege, 1946; *RIC: Report of the Indigo Commission, 1860*; *RCLIR: Report of the Royal Commission on Labour in India*, 1931; *RCLI* 6 (2): Royal Commission on Labour in India, *Minutes of Evidence*, Vol.VI, *Assam and the Dooars*, Pt.2 Oral Evidence; *RCLI* 7 (1): Royal Commission on Labour in India, *Minutes of Evidence*, Vol.VII, *Madras Presidency and Coorg*, Pt.1, Written Evidence; *RCLI* 7(2): Royal Commission on Labour in India, *Minutes of Evidence*, Vol.VII, Madras Presidency and Coorg, Pt.2, Oral Evidence, 1931; *SIPECR: Report of the South of India Planters' Enquiry Committee*, 1896.
7. For details see *RIC* [paras 32, 35]; Buchanan [1966: 43].
8. For all the variants see, among others, *RIC* [paras 31–32, 34]; Buchanan [1966: 43–4].
9. For details of land area under *ilaka* and *bi-ilaka* cultivation in some of the largest indigo concerns see Kling [1977: 53–4].
10. See *RIC* [paras 23–4]; Buchanan [1966: 50–51]; Chaudhuri [1964: 124–5].

11. In 1860 the Lieutenant-Governor of Bengal estimated that the loss to the peasant was no less than Rs 7 a *bigha* (approximately one-third of an acre), which was 'equivalent at least to 7 times the rent of the land' (cited in Buckland [1902: 249]). See also Chaudhuri [1964: 130–32]; Kling [1977: 34–6].
12. For contemporary evidence of violence by a civilian officer see, among others, Beames [1984: 172–3]. While Beames does not refer to sexual violence, Rev. Lincke and Rev. Long told the Indigo Commission that they had heard about outrages against women [*RIC*: A894–96, A1663–64], and it was a popular belief that such outrages were committed. It has been suggested that, as publicising such incidents would have brought shame on the victim and her family, no formal complaints of this nature were made before any official body [*Kling*, 1977: 198–9 n.8].
13. For the early history of tea in India and particularly in Assam see Antrobus [1957]; Buchanan [1966: 54–5]; Bose [1954]; Griffiths [1967: 33–75]; Guha [1981]; Guha [1988: 12–13]; Sircar [1986a; 1986b].
14. See Miles [1987: 32–4]. It should be mentioned that Miles speaks of unfree wage labour in the sense of wage labour under politico-legal constraints.
15. For population density in the different districts of the Brahmaputra Valley see Guha [1988: 339].
16. It is noteworthy that the plantation wage was lower than the locally prevalent wage see Guha [1988: 15–16, 43].
17. To obtain an adequate labour supply the planters pressured the government to raise land revenue to force peasants to work for wages on the plantations. In 1860 land revenue rates on the dry lands in the Brahmaputra Valley districts were raised by 15 to 30 per cent and in 1868 were doubled throughout the valley [*Guha*, 1988: 9–10]. But this did not have the result the planters desired.
18. See, among others, Saha [1970: 29–30]; Tinker [1974]; Chakravarty [1978].
19. See Das [1931]; Bose [1954: 73]; Guha [1988: 16–17]; Das Gupta [1981: 1785]; Chatterjee [1987].
20. Garden *sardars* were labourers employed on tea estates who were licensed to recruit other labourers in the recruiting districts for their gardens. They were remunerated on a commission basis. For details see *ALECR* [paras 90, 97]; *RCLIR* [363–7]; *LICR* [21, 24–5]; Behal and Mohapatra [this volume]. While in addition to recruitment the garden *sardars* at times performed supervisory functions, there was also a separate category of supervisory *sardars* [*RCLIR*: 383]. The place and role of the *sardars* in the Dooars differed to some extent from those in Assam. See Arbuthnot [1904: 2–3, 5–6, 26–7].
21. With reference to the blatant abuses of recruitment as late as 1901 Sir Charles Rivaz, a top government official, observed that a horde of unlicensed and uncontrolled labour recruiters made large gains 'by inducing, under false pretences, ignorant men and women, chiefly from the most backward districts of Bengal [and Bihar and Orissa], to allow themselves to be conveyed to Assam, and by practically selling these people to the planters for the purpose of being placed under labour contracts in that Province'. Cited in ALECR [1906: Ch.XVII, para 48]. For evidence relating to the early 1890s, see LECR [1895: paras 67–71, 79, 82, 90].
22. For example, the Dooars and Darjeeling planters organised the North Bengal Mounted Rifles [Dash, 1947: 146].
23. See the contribution to this volume by Behal and Mohapatra.
24. Computed from budgets presented in the ALECR [1921–22: 123–37].
25. The 1906 Assam Labour Enquiry Committee observed that this hold was 'comparable to that which a strong *zamindar* has over his *ryots* in the backward parts of Bengal' [ALECR, 1906: pp.66–7, para 143].
26. One enquiry shows that in 1944 the average holding per adult worker in the Assam gardens was 0.66 acres, and in the Dooars and Darjeeling gardens 0.78 and 0.64 acres, respectively [LICR: 48, 82].
27. As no systematic data are available, only a rough estimate has been made here.
28. It merits mention here that the usual diet of plantation workers consisted of rice and some vegetables and occasionally pulses. Normally it did not have any fruit, milk, eggs,

fish or meat. It was a highly unbalanced diet and the nutritional value was extremely low.

29. See ALECR [1906: 89, 92]; ALECR [1921–22: 23–4].
30. For details see Das Gupta [1986: PE6]. This is, however, not a new point. See, among others, Guha [1981].
31. In 1901 population density per square mile was 116 in the tea-producing Nilgiris, 114 in the coffee-producing province of Coorg, and less than 100 in the tea and coffee producing Wynaad highlands in Malabar.
32. For 1896 data see SIPECR [94], and for 1943 LICR [128].
33. For details see SIPECR [15]; RCLI [7(1): 250].
34. See RCLIR [355–6]; Das [1941: 28–32]; Chatterjee [1987: 245–74].
35. See RCLIR [355–6]; Das [1941: 28–32]; Chatterjee [1987: 245–74].
36. Ibid., p.252.
37. Ibid.
38. Ibid., pp.261–7.
39. See LICR [123].
40. See SIPECR [18].
41. Ibid. See in this connection also Patnaik and Dingwayney [1985: 5]; Sarkar [1985: 113–14].
42. See, among others, RCLIR [237; LICR [123–5].
43. RCLI [7(2): 337]. The next two citations are also from the same source.
44. RCLI [7(1): 201].
45. For an example of a prohibitory order on holding a meeting to make representation to the RCLI see *RCLI* [7 (1): 248].
46. See RCLI [7(2): 376].
47. For details see RCLI [7(1): 346]; LICR: [5]. The Labour Investigation Committee stated that the period of absence was usually either from December to February or July to September in tea areas, from April to July in coffee areas and from January to March in rubber areas [LICR: 5].
48. See, among others, Guha [1988]; Behal [1985]; Das Gupta [1989]; Das Gupta [forthcoming b]. All these studies are on labour protests on Assam and Dooars plantations. Any systematic study of protests on plantations in South India is yet to be done.

REFERENCES

Antrobus, H.T., 1957, *A History of the Assam Company 1839–1953*, Edinburgh.

Arbuthnot, J.C., 1904, *Report on the Conditions of Tea Garden Labour in the Duars of Bengal, in Madras, and in Ceylon*, Shillong: Assam Secretariat Printing Office.

Bagchi, A.K., 1973, 'Some Implications of Unemployment in Rural Areas', *Economic and Political Weekly*, Vol.VIII, Nos.31–33, Nov.

Bagchi, A.K., 1982, *The Political Economy of Underdevelopment*, Cambridge: Cambridge University Press.

Bagchi, A.K., 1987, *The Evoluation of the State Bank of India: The Roots 1806–1876*, Bombay: Oxford University Press.

Bagchi, A.K., 1988, 'Colonialism and the Nature of "Capitalist" Enterprise in India', *Economic and Political Weekly*, Vol.XXIII, No.31, 30 July.

Beames, J., 1984 (first published 1896), *Memoirs of a Bengal Civilian*, Delhi: Manohar.

Beckford, L., 1972, *Persistent Poverty: Underdevelopment in Plantation Economies of the Third World*, New York: Oxford University Press.

Behal, R.P., 1985, Forms of Labour Protest in Assam Valley Tea Plantations 1900–1930, *Economic and Political Weekly*, Vol.XX, No.4, 26 Jan.

Bernstein, H. and M. Pitt, 1974, 'Plantations and Modes of Exploitation', *Journal of Peasant Studies*, Vol.1, No.4.

Bhowmik, S., 1981, *Class Formation in the Plantation System*, New Delhi: People's Publishing House.

Bose, S., 1954, *Capital and Labour in the Indian Tea Industry*, Bombay: All India Trade Union Congress.

Brass, T., 1986, 'Unfree Labour and Capitalist Restructuring in the Agrarian Sector: Peru and India', *Journal of Peasant Studies*, Vol.14, No.1.

Breman, J., 1989, *Taming the Coolie Beast: Plantation Society and the Colonial Order in Southeast Asia*, New Delhi: Oxford University Press.

Buchanan, D.H., 1966 (first published 1902), *Development of Capitalistic Enterprises in India*, London: Frank Cass.

Buckland, C.E. (ed.), 1976 (first published 1902), *Bengal under the Lieutanant Governors*, Vol.I, New Delhi: Deep & Deep.

Burawoy, M., 1980, 'Migrant Labour in South Africa and the United States', in T. Nichols (ed.).

Chakravarty, L., 1978, 'Emergence of an Industrial Labour Force in a Dual Economy: British India 1880–1920', *Indian Economic and Social History Review*, Vol.XV, No.3, July–Sept.

Chatterjee, R., 1987, 'Social Perspective of Labour Legislation in India 1859–1932: As Applied to Tea Plantation', Ph.D. dissertation, University of Calcutta.

Chaudhuri, B.B., 1964, *Growth of Commercial Agriculture in Bengal (1757–1900)*, Calcutta: Indian Studies Past & Present.

Chaudhuri, B.B., 1970, 'Growth of Commercial Agriculture in Bengal 1859–1885', *Indian Economic and Social History Review*, Vol.7, Nos.1 and 2, Jan. and April.

Cohen, R., Gutkind, P.C.W. and P. Brazier (eds.), 1979, *Peasants and Proletarians: The Struggles of Third World Workers*, London: Hutchinson.

Corrigan, P., 1977, 'Feudal Relics or Capitalist Monuments? Notes on the Sociology of Unfree Labour', *Sociology*, Vol.11, No.3.

Das, R., 1931, *Plantation Labour in India*, Calcutta: R. Chatterjee.

Das, R., 1941, *History on Indian Labour Legislation*, Calcutta: University of Calcutta.

Das Gupta, R., 1981, 'Structure of the Labour Market in Colonial India', *Economic and Political Weekly*, Vol.XVI, Nos.44, 45 and 46.

Das Gupta, R., 1986, 'From Peasants and Tribesmen to Plantation Workers: Colonial Capitalism, Reproduction of Labour Power and Proletarianisation in North East India, 1850s to 1947', *Economic and Political Weekly*, Vol.XXI, No.4, 25 Jan.

Das Gupta, R., 1989, 'Oraon Labour Agitation: Duars in Jalpaiguri District, 1915–16', *Economic and Political Weekly*, Vol.XXIV, No.39, 30 Sept.

Das Gupta, R. (forthcoming a), *Economy, Society and Politics in Bengal: Jalpaiguri 1868–1947*, Delhi: Oxford University Press.

Das Gupta, R. (forthcoming b), 'Plantation Workers and Collective Action in Jalpaiguri Dooars, 1890s–1947', in Das Gupta, *Labour and Working Class in Eastern India: Studies in Colonial History*, Calcutta: K.P. Bagchi & Co.

Dash, A.J., 1947, *Bengal District Gazetteers: Darjeeling*, Alipore: Bengal Government Press.

De Silva, S.B.D., 1982, *The Political Economy of Underdevelopment*, London: Routledge & Kegan Paul.

Deere, C.D., 1979, 'Rural Women's Subsistence Production in the Capitalist Periphery', in Robin Cohen *et al.* (eds.).

Dewey, C. and A.G. Hopkins (eds.), *The Imperial Impact: Studies in the Economic History of India and Africa*, London: University of London.

Fisher, C., 1978, 'Planters and Peasants: The Ecological Context of Agrarian Unrest on the Indigo Plantations of North Bihar 1820–1920', in C. Dewey and A.G. Hopkins (eds.).

Ganguli, D., 1972, *Slavery in British Dominion*, Calcutta: Jijnasa (reprint of articles from Bengalee, September 1886 to April 1887).

George K., Tharian, and P.K. Michael Tharakan, 1986, 'Penetration of Capital into a Traditional Economy: The Case of Tea Plantations in Kerala, 1880–1950', *Studies in History*, Vol.2, No.2 (New Series).

George K., Tharian, Haridasan, V. and B. Sreekumar, 1988, 'Role of Government and Structural Changes in Rubber Plantation Industry', *Economic and Political Weekly*, Vol.XXIII, No.48, 26 Nov.

Grant, J.P., 1976, 'Minute of the Lieutanant Governor of Bengal on the Report of the Indigo Commission', in C.E. Buckland (ed.).

Graves, A. 1987, 'Plantations', *New Palgrave Dictionary of Economics*, London: Macmillan.

Griffiths, P., 1967, *History of the Indian Tea Industry*, London: Weidenfeld & Nicolson.

Guha, A., 1981, 'Formation of a Working Class in Assam Plantations: A Study in Retrospect', in North East India Council for Social Science Research, *Problems of Tea Industry in North East India*, Calcutta: Research India Publications.

Guha, A., 1988, *Planter Raj to Swaraj: Freedom Struggle and Electoral Politics in Assam 1862–1947*, New Delhi: People's Publishing House.

Henningham, S., 1982, *Peasant Movements in Colonial India: North Bihar 1917–1942*, Canberra: Australian National University.

Hunter, W.W., 1976 (first published 1877), *A Statistical Acount of Bengal*, Vol.XIII, *Tirhut and Champaran*, New Delhi: Concept.

Jones, W.O., 1968, 'Plantations', in D.L. Sills (ed.).

Kling, B., 1977, *The Blue Mutiny*, Calcutta: Firma KLM Pvt. Ltd.

Kloosterboer, W., 1960, *Involuntary Labour Since the Abolition of Slavery: A Survey of Compulsory Labour throughout the World*, Leiden: E.J. Brill.

Marx, K., 1973, *Grundrisse*, Harmondsworth: Penguin.

Labour Investigation Committee (LIC), Government of India, 1946, *Report on an Enquiry into Conditions of Labour in Plantations in India* by D.V. Rege, Delhi: Manager of Publications.

Miles, R., 1987, *Capitalism and Unfree Labour: Anomaly or Necessity*?, London: Tavistock Publications.

Mishra, G., 1978, *Agrarian Problems of Permanent Settlement: A Case Study of Champaran*, New Delhi: People's Publishing House.

Misra, B., 1985, The Growth of Industries in Eastern India 1880–1910, Ph. D. dissertation, University of Calcutta.

Mitra, A., 1951, *Census 1951: West Bengal: District Handbooks; Jalpaiguri* [Census Publications for West Bengal, Sikkim and Chandernagore, Calcutta.

Mohapatra, P.P., 1985, 'Coolies and Colliers: A Study of the Agrarian Context of Labour Migration from Chotanagpur, 1880–1920', *Studies in History*, Vol 1, No.2.

Myint, H., 1973, *The Economics of the Developing Countries*, London: Hutchinson.

Nichols, T. (ed.), 1980, *Capital and Labour: A Marxist Primer*, Glasgow: Fontana Paperbacks.

Patnaik, U. and Manjari Dingwanay (eds.), 1985, *Chains of Servitude: Bondage and Slavery in India*, Madras: Sangam.

Royal Commission on Labour in India (RCLI), 1931, *Report of the Royal Commission on Labour in India*, Calcutta: Government of India Central Publications Branch.

Saha, P., 1970, *Emigration of Indian Labour 1834–1940*, New Delhi: People's Publishing House.

Sarkar, T., 1985, 'Bondage in the Colonial Context', in Utsa Patnaik and M. Dingwanay (eds.).

Sen, A., Chatterjee, P. and S. Mukherjee, 1982, *Perspectives in Social Sciences: Three Studies on the Agrarian Structure in Bengal 1850–1947*, Calcutta: Oxford University Press.

Sen, A., 1982, 'Agrarian Structure and Tenancy Laws in Bengal 1850–1900', in Sen *et al.*

Sills, D.L. (ed.), 1968, *International Encyclopedia of Social Sciences*, New York: Macmillan and Free Press.

Sircar, K.K., 1986a, 'A Tale of Two Boards: Some Early Management Problems of Assam Company Limited, 1839–1864', *Economic and Political Weekly*, Vol.XXI, Nos.10 and 11, 8–15 March.

Sircar, K.K., 1986b, 'Labour and Management: First Twenty Years of Assam Company

Limited (1839–59)', *Economic and Political Weekly*, Vol.XXI, No.22, 31 May.
Tinker, H., 1974, *A New System of Slavery*, Delhi: Oxford University Press.
Watt, G., 1972 [first published 1889], *A Dictionary of Economic Products of India*, Vol.IV, Delhi: Periodical Experts.
Yang, A.A., 1989, *The Limited Raj: Agrarian Relations in Colonial India, Saran District, 1793–1920*, Delhi: Oxford University Press.

APPENDIX TABLE
PLANTATIONS IN INDIA: ACREAGE, YIELD AND DAILY WORKING STRENGTH, 1929–30

Province or Area	Total area of plantat-ations(000 acres)	Area under crop(000 acres)	Production (000 lbs.)	Average daily working strength (Permanent and Temporary)
Tea (1929).				
British India				
Assam				
Surma Valle	609	145	73,784	156,489
Assam Valle	1,039	285	185,157	400,995
Total	1,648	430	258,941	557,484
Bengal				
Darjeeling	158	61	23,009	65,522
Jalpaiguri	288	128	85,427	125,632
Chittagong	28	6	1,517	5,745
Total	474	195	109,953	196,899
Madras				
Nilgiris	73	32	11,403	30,759
Malabar	49	13	6,493	12,832
Coimbatore	36	22	9,700	27,217
Others	1	*	34	44
Total	159	67	27,630	70,852
Coorg	1	*	169	620
Others	53	20	4,272	17,768
Total British India	2,335	712	400,965	843,623
Princely States[b]	190	77	32,033	86,849
Total India	2,525	789	432,998	930,472
Coffee (1929–30)				
British India				
Madras				
Nilgiris	41	16	3,655	12,424
Malabar	15	4	548	2,302
Coimbatore	4	3	1,080	3,005
Others	27	15	2,786	8,889
Total	87	38	8,069	26,620
Coorg	61	37	14,668	21,886

APPENDIX TABLE (cont.)

PLANTATIONS IN INDIA: ACREAGE, YIELD AND DAILY WORKING STRENGTH, 1929–30

Province or Area	Total area of plantat-ations(000 acres)	Area under crop(000 acres)	Production (000 lbs.)	Average daily working strength (Permanent and Temporary)
Total British India	148	75	22,737	48,506
Princely States[b]	130	88	16,687	43,998
Total India	278	163	39,424	92,504
Rubber (1929)				
British India				
Madras				
Nilgiris	3	1	213	377
Malabar	44	13	2,301	4,634
Others	2	1	166	343
Total	49	15	2,680	5,354
Coorg	5	3	529	1,115
Total British India	54	18	3,209	6,469
Princely States[b]	86	61	13,644	25,219
Total India	140	79	16,853	31,688

Source: Report of the Royal Commission on Labour in India, 1931.

Notes: [a]The table does not present data on indigo plantations, for by mid-1920s indigo plantation had suffered total extinction.

[b]Princely States refer to Mysore, Travancore and Cochin.

'Enclave' Plantations, 'Hemmed-In' Villages and Dualistic Representations in Colonial Ceylon

ERIC MEYER

Several authors have undertaken the critical task of exposing the inadequacies of dualistic theories for the analysis of the village/estate relationship in colonial Ceylon. The first section of this article critically reviews their work. In a second section, popular representations of that duality are analysed, and the historical genesis and possible meanings of the 'enclave' myth so common in Sri Lanka examined. The debate on the 'positive' or 'negative' impact of the plantation system on villages will not be tackled as such, but cannot be totally avoided as it lies behind most interpretations of the village/estate relationship.

The term 'dualism' covers a series of theories about economic (and socio-cultural) configurations in which two contiguous entities, supposedly foreign to each other, non-symbiotic, and possibly contradictory, function side by side according to their own logics. Even if the term does not exclude relations (generally unequal) between the two entities, and domination of one by the other, its emphasis is on separation rather than hierarchy.

'Enclave', when used inside a dualistic conceptual framework, qualifies it by adding the idea of territorial configuration and possibly power relations between the two sectors, but if relations of domination – dependence are given the foremost place, the term implicitly or explicitly loses its 'dualistic' connotation to fit into marxist theories of political economy, and/or nationalist interpretations of unequal development. Then 'enclave' is commonly used to characterise the plantation system of production as forcefully inserted into a pre-existing socio-economic fabric, to emphasise its primordial links with an external capitalist system controlled by the metropolis, and to minimise its links if any with the local system of production (see the contributions by Behal and Mohapatra, and Murray in this volume); while other authors (for example, Das Gupta

---

Eric Meyer, Centre d'Etude de l'Inde et de l'Asie du Sud, 54 Bld Raspail, 75006 Paris.
Participants in the Conference and editors of the present issue of *The Journal of Peasant Studies* are thanked for their stimulating criticism and editorial help.

in this volume) use a somewhat different image of superimposition of the plantation upon the traditional economy.

'Enclave' images can be ambiguous: they may imply that an enclaved entity ignores its surrounding environment or that it dominates and ultimately possibly absorbs it. In the latter case, it is the village which eventually figures as an enclaved, or, to use a metaphor common in Sri Lankan anti-colonialist literature, a 'hemmed-in' entity, suggesting that the planter's *raj* has reduced the majority sector and population to a minority status.

I do not pretend that the 'enclave' and 'hemmed-in' metaphors are meaningless. On the contrary, they are historically meaningful; but my point is that when uncritically applied to village-estate relations, they might obscure important issues. The genesis and historical meaning of these representations must be kept in mind when examining evidence regarding the estate/village relationship. 'Enclave'/'hemmed-in' representations will therefore be treated as powerful reinforcing myths, the political implications of which, in the present crisis in Sri Lanka, are highly significant.

DECONSTRUCTING THE DUALISTIC MODEL

The dualistic interpretation of the Sri Lankan economy is outlined by Snodgrass whose arguments gained wide currency when the modernisation perspective was dominant in economics. His thesis is that the rise of the estates in Ceylon created 'a veritable model of what might be called a dualist export economy' [*Snodgrass*, 1966: 56]: 'There was scarcely any economic interrelationship between the two segments of the economy and hence the "spread effects" of modern sector growth to the traditional sector were negligible ... neither was there much unfavourable "backwash effect". The two sectors were virtually in different countries' [ibid: 5]. Imported capital, imported labour, imported supplies: the Ceylon plantation economy was according to him the model of a foreign 'enclave'.

Leaving aside Boeke's thesis [*Boeke*, 1953], which he disdainfully labels 'sociological dualism', Snodgrass substantiates his quantitative analysis with an array of technical arguments. First, intersectoral flows were minimal: 'Relative to the tremendous size of the trade flow linking the modern sector to the rest of the world, the volume of [intersectoral] flow must have been minuscule' [ibid: 59]. Then, the efficiency of the modern sector depended on its 'closed' working: outside interferences would have affected its smooth exploitative system and its profits. Growth in the modern sector depended on resources 'which could be brought into play without producing more than a minimal disturbing impact on Ceylon's traditional sector'; 'labour ebbed and flowed between the plantation areas and South India according to seasonal, cyclical and secular variations in the estate's needs for workers', capital was easily obtained from outside and through reinvestment, 'the banking system was almost exclusively in British hands and lent almost exclusively to the planters and the import–export trade'; 'land was the only factor of production which could not be brought in from outside' – a limit unless yields could be raised, but a minor factor of disturbance for the peasant sector according to the author [ibid.: 64 sqq].

Snodgrass recognises the existence of an extensive smallholding sector, 'overresponsive to market stimuli' [ibid.: 67], which however had only a marginal impact upon 'the giant mass of the traditional economy': 'Most peasants remained tied to the traditional modes of cultivation, in which paddy farming assumed an almost mystic [*sic*] position at the center of rural life, and the raising of coconuts and a wide variety of secondary crops for use within the village economy played a secondary role'. Hence a situation of general stagnation in the traditional sector.

Snodgrass concludes that of the enormous surpluses generated by the plantation sector a large part was reinvested, another used to maintain a high consumption level among the upper classes, another remitted to Europe: 'The growth of the estate sector then increased dividend levels for foreign shareholders and increased consumption levels within a growing enclave in Ceylon itself, but it did not materially raise levels of expenditure on Ceylon-produced goods and services, nor did it provide a surplus to be invested in other parts of Ceylon's economy' [ibid.: 70].

A Sri Lankan economist, S.B.D. de Silva, developed a strong critique of the modernisation perspective, while keeping a dualistic approach [1982: 210–23]. He rejects Snodgrass's theory on the grounds that the plantation is not a capital enterprise and the 'farm' a labour enterprise: both are labour-based. The plantation is no more 'modern' than the peasant unit, and the traditional rice economy is no less able than the modern capitalist economy to generate surplus: therefore underdevelopment cannot be the result of the absence of spread effects from the plantation to the peasant sector. It is not the lack of links between the two sectors which is responsible for the stagnation of the peasant sector, but the lack of dynamism of the plantation economy itself [ibid.: 34]. However, De Silva maintains that the plantation was of necessity an enclave because of the planters' need to secure labour in a context of scarcity and high intensity of labour (more in tea than in rubber or coconut, where the cost of labour represents a less important proportion of the cost of production) [ibid.: 291].

Bandarage [1982; 1983] and Gunasinghe [1990] are the best marxist critiques of the classical dualistic theory applied to the Sri Lankan case.[1] Bandarage's object is 'to expose the empirical invalidity and theoretical inadequacy of these facile dualistic assumptions' [1983: 8]. She adopts the approach of Gunder Frank, and expands it with the observations of Godelier and others, on the articulation of modes of production within the dominant-dependent system. But she accepts their version of the dualistic thesis according to which the modern sector derives decisive benefits from the preservation of traditional economies and systems of labour relations: in the Ceylon case, the preservation of a peasantry with links to the land and 'semi-feudal' institutions, used as a 'semi-proletariat', is interpreted as a deliberate policy of the colonial State, which Gunasinghe [1990: 44, 137–8, and Ch.VIII], conceptualises as the 're-activation of archaic relations' to serve the new system.

While in Java the colonial strategy was to force plantation production into the village rice economy, in Sri Lanka, as in Sumatra, the colonial state fostered large separate units on highlands and undertook to provide them with every facility by way of legislation, finance, infrastructure, and labour control devices. But highland expropriation adversely affected the peasantry, which reacted through developing plantation crops in smallholdings, rather than by regular employment on estates. The relationship between estate and smallholder production was conflictual and the competition for land between the two was severe, but the latter lacked state support.

Meanwhile paddy production remained a neglected sector, as estates obtained rice from outside, and the state did not reinject into the paddy sector the revenue it derived from it. After rejecting the dual economy thesis, Bandarage retains the opposition of two sectors, or rather modes of production: 'the incompatibility of the interests of capitalist export agriculture and that of peasant subsistence and smallholder production ... resulted in a conflict between opposing modes of economic production, in which the plantations were victorious because they had the support of the colonial state at every turn.' [1982: 20]. The two sectors were linked through several channels: peasant response to market stimuli was quick and extensive indeed, but the author insists on the fragility and lack of autonomy of that kind of development: 'these village agricultural sectors became residual to dominant plantation agriculture' [1983: 331]. Her conclusion is that 'the plantation sector was developed at the expense of village agriculture and peasant subsistence' [1983: 325].

In a recent book, Moore [1985: 66–75] contrasts two different approaches of dualism. The 'macro-economic perspective' (as exemplified by Snodgrass) identifies two quasi-autonomous networks, one commercial, the other subsistence. 'It is admitted that there are "leakages" or interactions between the two sectors. These are, however, least evident in the case of tea production and because tea so dominates the economy in value terms, the kind of exceptions provided for example by the coconut industry can be dismissed as peripheral phenomena' [ibid: 72]. By contrast, the historical-sociological perspective, concerned with longer-term interaction, social relations and mid-country areas, insists on estate–village integration. According to Moore, 'estate-village isolation most nearly approaches descriptive accuracy in the case of tea production ... a crop little grown in surrounding villages', while rubber represents an intermediary situation, and coconut a rather integrated crop. A regional typology of estate–village contact is therefore necessary.

It is significant that dualistic models have usually been adopted by economists in macro-level generalisations, but that their followers ran into difficulties as soon as they tried to apply them in local socio-economic studies. In other words, they were not evolved from the study of the 'peasant sector' and its inner dynamic, but from the study of the colonial impact. Impact-oriented research has its limitations, and we shall try to re-examine the 'classical' Sri Lankan case during the British period in a perspective giving more emphasis to endogenous dynamics, following the model set up by Adas [1974] for Lower Burma. This will be preceded by some observations on dualism and integration in the period before the nineteenth century.

The Pre-British period

Organised commercial agriculture in Ceylon antedates the advent of the British: under Portuguese and, moreover, Dutch domination there was in the island an organised production or collection of cinnamon, tobacco, arecanuts, and surplus production of paddy and other crops was organised by the Kandyan kings in their royal villages. In most cases, specialised agricultural work was assigned to labour imported from South India. Tobacco was fully integrated in the Jaffna economy and its development led to the employment of Pallar and Nalavar agricultural labour castes from Tamilnadu [*Arasaratnam*, 1982]. The colonial powers assigned to South Indian immigrants of recent origin, the Salagama, originally a weaver caste, the delivery of a fixed amount of cinnamon bark; in exchange they were given paddy and other land for their subsistence [*Roberts*, 1980]. These cultivation systems were not basically different from that which evolved under the Sinhalese kings; paddy production in royal, temple and 'seigneurial' villages was generally assigned to low caste labourers of more or less recent South Indian origin, often designated as coolies [*Meyer*, 1990: 167–70].

Labour scarcity and socio-cultural resistance were already major constraints for any attempt at planned production, in a system in which the dominant form of surplus extraction was labour rent [*Gunasinghe*, 1990: 27], and recourse to South Indian agricultural labour castes prepared to migrate to Sri Lanka appeared the only answer [*Kumar*, 1965]. But then migration resulted in integration: caste-based labour service (or *rajakariya*) provided the machinery to control labourers and at the same time to socially integrate them, while economic rationality did not require separation. On the contrary, the logic of service tenure was to link the two 'sectors', allowing subsistence paddy land in exchange for labour or produce. As a result, although cinnamon was a specialised crop, no dualism evolved; but the system had its limits, since the excessive requirements of the Dutch often compelled the Salagama to neglect their subsistence agriculture and sometimes to revolt.

The British coffee plantation system was a radical departure from the Dutch system, in that it was based on direct production under imported management with imported labour. Moreover, as the British were unwilling or unable to use the existing systems of labour control, integrative mechanisms progressively ceased to operate. It is often said that with the development of mid-country estates, economic integration progressed during the British period [*Farmer*, 1963: 43]; but the reverse thesis may be defended. The timing and forms of village–estate relationships are complex: flexibility prevailed during the coffee era with its slow transport, seasonal work on estates and temporary migration of males only; during the tea era with its permanent resident family labour, improved transportation, and professional management, the plantation sector may have severed some of its links with the villages.

Village–Estate Continuum and Intersectoral Flows

My basic argument is that the village–estate continuum was in many cases significant enough to 'blur the border' between the two sectors, that in a micro-perspective, intersectoral flows played a decisive role in the socio-economic life of most villages in the plantations areas, and that economic depressions in the world market resulted in a dramatic deterioration in the so-called 'peasant sector'.

In the case of coffee, coconut, rubber, and to a lesser extent tea, there was a continuum between village gardens, small holdings belonging to villagers or to outsiders, larger estates owned by members of the Ceylonese bourgeoisie, and large plantations, the property of European individuals or companies.

Coffee originally was a peasant crop grown at medium elevation in village gardens, or in *chenas* [*Ameer Ali*, 1970: 66], and up to the 1840s the bulk of exports came from that sector; even after plantation production exceeded peasant production, the latter increased sixfold until its decline during the 1870s and 1880s [ibid: 70]. But the statistics which distinguish peasant coffee from estate coffee by the methods of processing and quality of the product, conceal the fact that most of the growth took place in an intermediary category made up of small-scale plantations opened by local traders and village worthies.

The case of coconut is even more telling. Initial capital expenditure was minimal and returns could not be expected before ten years. It was a safe investment rather than a speculation, cultivation could be started by most according to their means, and the distinction between estate and smallholding is in a sense a statistical artifice. At the turn of the century, the Colombo bourgeoisie invested in the 'coconut belt' the huge profits derived from its participation in the modern economy [*Roberts*, 1982]. With the progress of settlement and public sale by government of small blocks around the villages, a class of 'peasant-planters' also developed (which often transformed itself into a class of 'small traders-planters'), improving the profitability of their enterprise by planting other cash crops while coconuts were maturing [*Meyer* 1983: 35].

In the case of rubber production [*Pieris*, 1965: 184–98] there is no real difference between smallholding and estate: a large proportion of smallholdings belong to non-peasants, many estates are worked by villagers, and as equipment is less expensive than for tea, the distinction between micro-holdings, medium size holdings and large estates is again in large measure a statistical artifice.

Only in the case of tea is there no self-evident continuum, since it requires expensive machinery which imposes a break-even point. But village-level cultivation and smallholdings did exist; in 1938, plots of less than ten acres represented 11 per cent of the tea area. They were linked to the large estates because they had to sell their crop to estate factories. This is a case of interflow rather than of continuity, but on closer scrutiny, it appears that many smallholdings were actually cultivated by enriched workers (mainly *kanganis*) from the neighbouring estates, or by Sinhalese or Muslim artisans and traders working for the local plantations [*Meyer*, 1990a: 183]. Thus even in the case of tea cultivation, the distinction between estate and smallholding is not clear-cut.

Linkages and Leakages

Linkages between 'village' and 'estate' in the production and manufacturing process were not limited to the sale of green leaves to tea factories. For example, a similar and larger interflow linked small coconut growers and owners of oil mills attached to bigger estates. And according to my research, seeds and seedlings of tea, coconut and other plants were freely exchanged between large and small producers; there were even cases of low-caste peasants opening up tea nurseries, making a fortune by selling young plants to larger estates, and reinvesting their profits in the purchase of paddy land: a process of socio-economic differentiation in the village was thus started, in which new categories challenged the established landowners.

What appear now as linkages were considered by the early European planters as leakages. Reports written between 1873 and 1903 are full of 'evidence' provided by planters of theft of tea leaves and young plants, rubber latex, cocoa pods, coconuts and coconut seedlings; it appears that the dualistic prejudice was so ingrained in the planters' minds that they could not believe that villagers were able to succesfully plant the same crops as themselves [*Meyer*, 1990a: 175].

Contrary to dualistic assumptions, estates became a market for village produce, and villages a market for excess 'cooly rice', flows which resulted from the established practice of paying estate workers partly in rice [*Wesumperuma*, 1986: 183]. Ironically enough, a non-negligible amount of rice, the basis of the so-called peasant subsistence sector, was obtained by villagers from the estate sector. On the other hand, village production met the plantation demand for such goods as straw, vegetables, coconuts, arecanuts, poultry, jaggery (palm sugar), toddy (palm 'wine') and illicit arrack [*Roberts*, 1965].[2] There was also demand for *kurakkan* (a millet called *ragi* in India), the traditional meal of the poor in Tamilnadu, which was grown on Kandyan *chenas* (high lands). Estate population increases and the growth of plantations in mid-country areas intensified these flows. In rubber areas such as Kalutara [Administration Reports for Kalutara 1912–13 and 1918] 'the large immigrant cooly population probably takes up all the surplus food produced by the villagers, the sale of it taking place in numerous Sunday or other weekly markets'. In order to put a stop to these 'leakages' many planters set up estate bazaars and licensed toddy and arrack shops.

Land and Labour

In his detailed study of the Kandyan countryside, Gunasinghe [1990: 59, 81, 124–7] maintains that no kulak class emerged because no accumulation by peasants could take place. In the period 1848–1915, which he defines as that of consolidation of the capitalist structure, concentration of ownership benefitted non-residents (temples or aristocratic families for paddy land, low country entrepreneurs for plantation land). With the prevalence of sharecropping in paddy cultivation, and the reactivated power of the aristocracy, upward mobility for peasants was not through agrarian accumulation, but, especially after 1930, through education to modern sector employment. As a result, 'the class structure consisted of a land owning aristocracy, a non-Kandyan petty bourgeoisie, and a basically undifferentiated peasantry which however contributed more and more to the stratum of landless workers' (through population pressure on land) [ibid.: 59]. The relative lack of peasant class differentiation helped sustain dualistic theories. But at this stage, a critical appraisal of the category 'villager' becomes relevant. Following the colonial categorisation, Gunasinghe excludes migrants of low-country origin who settled in the Kandyan planting districts. It is true that many resided in small bazaar towns catering for estate needs, but at the turn of the century practically every village contained some migrants as petty traders, craftsmen, small scale planters leasing village lands, or later as government employees – especially teachers. They became progressively integrated, by marrying village girls, purchasing paddy land, even acting as village spokesmen. They were the people who benefited most from the new economic opportunities, being on the borders between estate and village, and able to employ landless villagers as casual labourers. This emerging rural petty bourgeoisie effectively challenged the influence of the landowning aristocracy.

I have argued elsewhere that the role played by Sinhalese labour in plantation economy has always been underestimated, and that such employment became necessary to the economic processes of both villages and estates, at least during certain periods and in certain areas [*Meyer*, 1979: 459].

The employment of immigrant cooly labour on estates is the major plank of the dualistic models. This issue is tackled by most other contributors to this volume, and I am prepared to agree with them that preferential employment of imported coolies was the basis of the system. In the case of Sri Lanka, why Kandyans refused to become coolies on estates has been hotly debated [*Jayawardena*, 1963; *Bandarage*, 1983; *De Silva*, 1982; *Wesumperuma*, 1986; *Kurian*, 1989]. It is generally admitted that there was a convergence of factors: the absence in Sri Lanka of landless agricultural castes (numerous in Tamilnadu), refusal to reside on the estate itself, wage retention and unfair treatment by planters, control by headmen over dependent villagers, and at the beginning, absence of population pressure.

However, all the initial work of clearing land was done by villagers, often under low country Sinhalese contractors, and Sinhalese plantation labour was not negligible at least during the twentieth century: in the early 1940s, on large estates employing resident workers, 11.5 per cent of the labour force was resident Sinhalese, 17.7 per cent non-resident regularly employed Sinhalese and about five per cent employed on contracts. Whenever resident employment under direct supervision and with wage retention devices was not imposed, Kandyan villagers eagerly sought employment on estates, especially when they confronted the disappearance of highland subsistence cultivation and problems with paddy cultivation due to plantation development, the collapse of their own coffee cultivation and grain tax crises, land hunger increased by population growth, or monetary needs induced by new consumption models. Before the growth of state employment in rural areas in the 1940s, estates were the main source of cash within reach.

The flow of labour between villages and tea estates was regulated by the varying needs of the planters, who employed villagers as extra labour in rush periods or on contract for specific tasks such as weeding, while keeping their regular Tamil labour nucleus.[3] It also depended on village agricultural and festive seasons, and on the ability of larger paddy land owners, in areas such as Sabaragamuwa where they remained powerful, to retain their grip on their labour force, which generated competition for labour, especially during the first decade of the twentieth century, and attempts by the colonial government to induce more villagers to seek employment on estates [*Ceylon*, 1908]. The nature of work in rubber and coconut cultivation made it rather easy for villagers to work part time as tappers, or to pick and transform the nuts, while keeping an eye on their own 'farm'. The high incidence of part-time farming in these areas is noted by Moore [1985, Ch.7]. However it must be remarked that village labour was not the only or even main factor of flexibility in the system: Kurian [1989] shows that Tamil labourers commuted more often than is generally thought between estates and South Indian villages.

Conversely, in time of labour shortage, planters became worried by the 'leakage' of Tamil coolies from the estate to the village: these people were either 'bolters', or coolies still residing on the estate and seeking outside employment. In mid-country areas, they functioned as village casual labour, being employed for short spells in the paddy fields, or by the shop keepers, to redeem drinking or gambling debts. Some were dismissed workers who settled permanently in the villages, but most of them continued to reside in the lines with their families [*Meyer*, 1990a: 173].

On the whole, it appears that dualistic theories opposing village and estate as self-contained labour markets cannot be sustained. Gunasinghe [1990: 91] gives a typical case of integration of estate and village labour markets for the 1970s, which I consider valid for most localities and previous decades as well. An outsider owns both paddy land and a small tea estate; his tenant sharecrops a paddy plot but must supplement his income by casual work (himself and/or family) on the estate. Any attempt to resist landlord pressure as tenant will be met by a refusal of the landlord to employ him as labourer. However, labour flows from village to estate could not create a unified plantation market because the two categories of workers (full-time and part-time estate labourers) were not employed in the same capacity: in a sense, dualism existed inside the estate. No class consciousness across the divide emerged even if village workers sometimes mingled with Tamil coolies in attending trade union meetings.

Economic Depressions and Integration

Historical moments like the coffee crisis of the 1880s, and the general depression of the 1930s, provide a decisive test of the role played by the plantation economy in village life. In the 1880s it appears that peasants not only lost revenue from coffee sales, but that 'now that the number of coolies has decreased, and the salary of the managers been reduced, the villager has to travel a long distance to find a market for his goods, and has great difficulty in acquiring sufficient to meet the taxes' [*Ceylon – Administration Report for Ratnapura 1881*]. As a result, paddy cultivators who had paid their paddy tax with income from selling coffee or vegetables, or from working on estates, roads or in bazaars, fell into arrears and many had their paddy lands sold for default. The village was obviously not shielded from fluctuations in the estate sector, and even falling back on subsistence agriculture was uncertain. This led a small group of members of the Ceylonese elite, British officials, and journalists linked with the Cobden Club, to campaign for abolition of the paddy tax, which occurred in the 1890s.

The crisis of the 1930s revealed the extent of interactions between different sectors of the Ceylon economy. While in many countries peasant producers fell back on subsistence (mainly paddy) agriculture, in Sri Lanka no such tendency was perceptible. Small producers of tea and rubber were saved by government-issued coupons under a quota scheme. As only large estates (and not all of them) continued to produce, wage labourers employed by medium and small estates were thrown out of work. Manipulation of migration flows with India in the early 1930s finally worked against village workers, taken on temporarily to replace Tamil workers as soon as there was a hint of recovery, but not kept on when recovery was deemed steady enough to allow the return of the regular Tamil coolies [*Meyer*, 1980: 147]. Their bargaining power was not strong enough, even with a halt to migration from India in the early 1940s, and growing unionisation of Tamil workers on the estates, to persuade planters that they could dispense with their core of skilled Tamil labour.

Trade unionism was soon to be recognised by planters as an acceptable substitute for the *kangani* system in their dealings with their labour force.

The crisis was compounded in the case of coconut, by the indebtedness of many landowners. As active in Ceylon as in South East Asia, Nattukottai Chettiar moneylenders had granted extensive credit facilities to Ceylonese purchasers of coconut (and to a lesser extent rubber) land, generally upon mortgage. Their practice was to collect heavy interest without asking for repayment of capital. But as early as 1925, European banks which used to extend credit to Chettiars restricted their facilities. Some Chettiar firms collapsed while others foreclosed on their mortgages when it became clear that their debtors were unable even to pay interest. Thus, Chettiar firms which linked European finance with the indigenous plantation sector, channelled the world crisis into the coconut belt.

When in 1933 tea prices improved, the impact on up-country villagers was immediate:

> This is a factor which has far more bearing on the prosperity of the Sinhalese villager than I ever realized before the bad time through which tea has recently been passing. Not only do enormous numbers of villagers own small patches of tea of their own, but they are dependent to a very large extent on work on tea estates, while their market for vegetables in particular is very largely dependent on the prosperity of the estate labourer [*Ceylon – Administration Report of the Registrar of Cooperative Societies 1933*: E3].

Thus recovery also demonstrated integration.

Whether the 'impact' of the estate on village conditions has been beneficial or harmful is not the subject here but it cannot be totally avoided. When plantation economy came under nationalist criticism in the 1920s, several observers insisted on the symbiosis between village and estate: to quote two of them [*Ceylon – Administration Report Ratnapura 1927*: I4]:

> It became evident in the course of mapping out that those villages were most prosperous and the increase of population most marked where planting activity was greatest. From the village point of view estates are desirable.

> Not only do the estates offer remunerative employment to the villagers when they have time to spare from their own agricultural pursuits, but they are also employers of skilled labour ... The inhabitants are better fed, better clothed and better housed. On the other hand, permanence of progress is dependent upon a delicate adjustment of the distribution of land as between the capital and the villager.

The recent study by Kemp confirms the extent to which village/estate contact has shaped village life in a tea area, and what is true for tea areas up-country is a fortiori valid in rubber and coconut areas where a large and influential section of the village population was made up of immigrants from other Kandyan villages and from the low country, attracted by the opportunities created by the plantation. Purchase of paddy fields, a very important feature of social mobility in the village, was made possible by resources drawn from the outside, mainly plantation employment [*Kemp*, 1984: 53; 1982: 177–98]. But the remark of Samaraweera [1981: 128] still holds true: 'The critics have exposed the flaws of the dualistic concept but they have failed to offer a suitable explanation of their own to account for the evolution of the colonial economy of Sri Lanka.' In a general assessment of dualistic theories, Roberts concludes [1973: 164]:

> Duality has been overdrawn, [but] the question remains whether, subject to the use of more qualified language, the main tenets of the dual economy hypothesis cannot be sustained. The contrasts in resource use between the two sectors remained significant ... How close does Ceylon approach the blurred-border zone economy type, between the two ideal types of dual economy and its unnamed opposite?

It cannot be denied that the peasant economy never became fully intergrated in that the great bulk of the peasantry did not take to the cultivation of plantation crops, nor become coolies on the estates, nor radically extend their food production to supply the food requirements of the estates. In other words, neither revolution nor involution occurred in the countryside as a result of estate impact. If any attempt at generalisation has to be made, it may be said that: (1) villagers' ability to subsist in a new economic environment was based on diversifying activities, obtaining resources from the estate sector, adaptability, and *bricolage*. (2) The question is not so much whether these relations blurred the border, as in what measure they were vital to the working of both sectors. The 1930s' depression revealed the extent to which villages depended on estates; estates were less dependent on villages, but for economic and political reasons they could no longer ignore them. The relation may be described as asymmetrical, but not dualistic.

UNDERSTANDING THE ENCLAVE MYTH

Enclave myths are the popular version of dualistic theories; they are especially potent concerning Sri Lanka because they are cast in the mould of previous representations of space, and help reinforce them as well as justify politically powerful interpretations of history. My object here is to show that there are at least two myths mirroring each other, in a dialectical relation, which played a significant historical role. Samaraweera [1981] and more recently Moore [1985 and 1989] have paved the way for understanding economic facts and ideologies in the context of Sinhala ethno-nationalism, without however explicitely linking the enclave image with previous myths which flourished in the island [Meyer, 1990b]. On the planter side, the version of the enclave myth remains to be analysed.

Stereotyped Representations of the Village and the Estate

We have argued that 'estates' and 'villages' were linked in many ways. Now we turn to representations. During the colonial period, 'villages' and 'estates' were conceived as 'total self-contained institutions' belonging to two separate worlds inhabited by two different species, the peasant and the coolie. Whatever links they may have had with each other, peasant and coolie were seen as essentially irreconcilable. Progressively, they came to be loaded with conflicting values, and their images built into myths which reinforced their opposed identities.

The image of 'enclave', which generally implied the existence of relations of domination-dependence, was employed as material for such myths. Those who used it considered that the 'foreign enclave' was dominant and the body which it occupied dependent. This links with the metaphor of the 'enclave' as a cancer, destroying the economic and social fabric from inside, although village was generally described as passively 'hemmed-in' by the estate rather than 'enclaved'. The attributes of each entity were systematically opposed and the reciprocal character of both myths suggests that, in a sense, they were the dual faces of a single myth.

Before exploring the genesis and meaning of these opposed stereotypes, it should be recalled that the categories 'village', 'estate', 'peasant' and 'coolie' are in a large measure modern constructs in the whole of South and South East Asia [*Breman*, 1988; *Moore*, 1989]. Breman [1988] argues that the rural sphere throughout Asia was characterised by a high degree of fluidity. The reification of the category of 'village' and its representation as a self-contained entity in colonial discourse fits in quite well with the dualistic assumptions of the planters, and may have been reinforced by them. The Sinhalese term for village, *gama* (sanskrit *grama*) was originally applied to a family estate, and *govi* (cultivators, rather than peasants) bears caste connotations. It was Europeans (from Maine to Phear) who built up the myth of a self-contained 'aryan' village, followed by the nationalists who developed the image of the 'poor' peasantry (the term poor being applied in a moral sense, to the peasantry as a whole).

The enclave myth is grounded in popular versions of Darwinism prevailing during (and after) the Victorian era, especially in the colonial world, which attributed specific virtues and vices to different races and advocated their separate development. The well-known myth of the 'lazy native' [*Alatas*, 1977] is derived from the same approach. In a sense, Boeke's cultural dualism is its theoretical expression, and 'grand apartheid' in South Africa its extreme outcome. Hence the British were characterised as enterprising, the Sinhalese as lazy and unreliable, and the Tamils as hard-working and obedient.

But two other factors may explain the strength of this myth. One is connected with social psychology, the other with economic rationality. Most British settlers in Ceylon, especially planters, were obsessed by what they perceived as the threat of native society and withdrew into themselves without trying to understand it. For example, investigation of the evidence given by planters in successive reports (1873 to 1903) on the theft of plantation produce [*Rogers*, 1987; *Meyer*, 1983:33, 1990a: 174–5], revealed that their psychosis was not based on any increase in theft. Their fear was to be surrounded by unlawful, uncontrollable villagers, imagined to loaf and spy all day, steal plantation produce at night, waylay and rob estate coolies, block access roads, and direct their cattle to graze on the estate. Surrounding villagers were perceived as invaders in British territory, commented a British civil servant wryly. Later, when barbed wire came into use, some planters undertook to fence their estates to materialise separateness, and a spate of 'border incidents' resulted.[4]

Total independence was the ideal of any entrepreneur in the era of laissez-faire. The tropical estate provided the closest approximation to a capitalist version of an earlier seigneurial ideal: the chief with full power over his workforce and his territory. The enclave myth was consistent with the planter's *raj* system: the colonial planter was not prepared to rule over the whole of the island, nor to be ruled from outside but wanted to govern his own little principalty, evade taxation, keep the labour he brought in, and so on. Relaxing the rules of separateness would mean jeopardising his authority, making control over labour more difficult and reducing his profits, as shown by Breman [1989, *passim*] and by Murray in his contribution to this volume.

There was however a minority of long-established colonial planters, especially in mid-country areas, who were able to speak Sinhala, had regular contacts with village headmen and employed villagers. When planters found it difficult to get enough labour from India, at the beginning of the twentieth century, many attempted to attract village labour. The Report and Proceedings of the Labour Commission of 1908 [*Ceylon*, 1908] are full of planter testimonies that they got on well with villagers and were happy with non-resident village labour commuting from village to estate more or less regularly [*Meyer*, 1979]. Whether there was any change in labour relations is open to doubt, but this representation of the villager was obviously adapted to suit changing conditions. When planters were increasingly criticised by the nationalist elite in the 1920s, they reacted by insisting on the benefits derived by villagers from the estate. This became a leitmotif during the sittings of the Land Commission in 1927–1928, when planters who feared they would be prevented from acquiring new land, defended the thesis of co-development. When the chairman invoked the problem of villages isolated by the development of estates around them, some planters maintained that 'it is the non-development of the areas in the neighbourhood of the village that drives away the villager'. A.M. Clarke, spokesman for planters in the province of Uva where land was still available, claimed that relations of mutual benefit prevailed, or should prevail [Evidence of the Planter's Association of Ceylon before the Land Commission, 31.03.1928, Sri Lanka National Archives]:

> I think the interest of all sections of the people would be best served by development on mutual lines enabling large and small landholders to advance side by side. In Uva, villages exist in fairly large numbers near estates; when I first went to that district, about ¾ of the villagers paddy fields were cultivated once a year. The villagers today not only cultivate all their available paddy lands once a year but in some cases harvest two crops ... The villagers have really advanced side by side with the planters. It is not a case of turning the villager into a coolie. He works for a part of his time on the estate and devotes the rest to the development of his own land.

Those were prosperous days. As soon as the depression set in, most planters again insisted on separate development. While the malaria epidemic of 1934–35 enabled them to argue that estate medical facilities could also serve villagers in case of need, the temporary halt to migration from India which resulted in the first serious Sinhalese/Tamil clashes in estate areas [*Meyer*, 1990a] reinforced the idea that dualism was the best formula for social peace in plantation areas.

The Dilemmas of the Colonial Administration

According to Ward, Governor of Ceylon in the 1850s, the aim of the colonial administration was 'to do justice to both [planting and peasant interests which] in many respects are completely identified' (quoted in Roberts, [1965: 329–31]). But in reality these interests often clashed, and less optimistic civil servants on the spot were faced with dilemmas. According to Bandarage [1983: 64], the intention of the liberals to carry a 'dual mandate' (developing both British capitalist interests and the welfare of the natives) was a piece of wishful thinking, and 'what followed during the nineteenth century was not harmony, but conflict between capitalist and peasant interests'.

Many nineteenth-century civil servants were so involved in the plantation industry that they systematically took the planter's side, to the point of attempting to help him 'coolify' the villager – a hopeless enterprise. Contrary to Jayawardena [1963, *passim*], Bandarage [1983: Ch. 6], tries to substantiate the thesis that there were recurrent attempts to uproot villagers, to force them out of their supposed laziness and equip them to become coolies: the use of corvée labour for road construction in the 1830s, the legislation of 1840 and 1897 expropriating land, the compulsory commutation in money of the paddy tax, and the subsequent forced sale of paddy land, are cited. The cynical assertion by the Fergusons (spokesmen of the plantocracy), that 'the pinching of the stomach is morally good because it will induce the peasants to work on the plantations' reflected a position common among the planters and a section of the officials [*Meyer*, 1979: 464; *Bandarage*, 1983: 179–80].

But the position of the administration was not so clear-cut: whether to ensure separation to 'protect' the villager, to 'coolyfy' him, or to open new employment opportunities, was the dilemma of the British *raj*, best analysed by L. Jayawardena [1963: 145]:

> On the one hand, as exemplified by the attitude of the Government Agents, they wanted the native to remain a peasant: this meant seeing that he had enough land to maintain himself in the manner in which he was accustomed ... On the other hand, they wanted him to improve his lot, to shake himself out of his lethargy: this meant placing the European in his vicinity. He could work voluntarily on the estate if he chose, or learn habits of thrift and enterprise by imitation, and wean himself from the wasteful *chena* [slash-and-burn] cultivation ... The compromise solution for which they were groping, was one where the European was near enough to the native to influence him, but not so near as to dispossess him.[5]

When local officials were faced with the task of settling village high lands, the issue was raised of whether or not to maintain separation between estate and village land. 'In principle, they were supporters of the British investment empire. But it was not uncommon for individual officers to try to protect village reserve lands from the encroachment of expanding plantations' [*Blackton*, 1978: 36]. Some of them tried to establish a kind of *cordon sanitaire* to ensure separation (see, for example, the Assistant Agent of Kegalla [diary 18.03.1885, Sri Lanka National Archives]). When the Land Commission was the scene of a confrontation between nationalists and planters, several district officers, especially in the mid-and low country, again insisted that there was no contradiction in the development side by side of village and estate, as long as balance was kept between both interests.

> When capital is hungry for land, it is apt to overlook the requirements of the village in this direction. Capital must be kept within bounds, but if this can be achieved by the exercise of a reasonable foresight as regards the alienation of Crown land and the provision of adequate safeguards against the improvident alienation by small holders, the villagers have everything to gain by capital enterprise within reasonable limits [*Ceylon – Administration Report for Ratnapura 1927*: I4].

Another major concern of the administration was to preserve communal harmony which was threatened by the proponents of the enclave myth.

Even after the depression and the crisis in communal relations up-country, it was still the view of many enlightened British colonials that the villagers had been able to overcome hard times by getting the best out of the dual structure: 'The villager has two strings to his bow: if there is a drought or his crop fails for some other reason, he can fall back upon the estate. If on the other hand the tea and rubber industries are hit by a depression, he will at least have his paddy or his *chena*' [*Jennings*, 1948: 45].

Feelings and Representations of the Kandyans

The myth of the golden past when village life, undisturbed by the Europeans, was pure and moral, seems to have been common in the Kandyan countryside by the end of the nineteenth century, at a time when the colonial policy of betting on the aristocracy led to a 'reactivation of the feudal order' [*Gunasinghe*, 1990: 200]. The image of the village as the antithesis of the plantation in the writings of Kandyan headmen has to be understood in that context. But whether or not the common villagers felt threatened by the development of surrounding estates, and what image they formed of the plantations, cannot be ascertained on the basis of these writings.

Aluvihare, a chief headman of Matale, in a report, written in 1896 at the request of the Government Agent, on villagers whose lands had been taken over for a plantation, gave free rein to his imagination in drawing a picture of village life before and after estate development, as far as I know the earliest substantial text of its kind. It is worth quoting extensively. After a lenghty and idyllic description of village life and New Year celebrations, Aluvihare depicts the disruption resulting from plantation development:[6]

> Agents and explorers are at work. Large extents of government lands are lying near the village, tracts of virgin forests where the sound of the axe is not heard, except that of a solitary bee-hunter. Soon this is acquired, the contractor's axe is soon clearing it up, and where once was a vast forest, a smiling estate springs up with neat bungalows, lines, engine houses, cattle sheds and other buildings required for the successful growth of coffee, tea or cocoa, and for its manufacture into a marketable form. Soon a cart road skirts the border of the village. Much benefit this gives the villager no doubt. Straw that had been burnt or thrown into the fields for manure after thatching the houses, or had to be carried far to be sold, can now be sold at the *kanata* [threshing floor]. Fowls and eggs are now bought for a better price by the estate workmen than the *thambi* [muslim trader] ever paid for them before. Many an idle day could now be brought to profit by working for the selling or other contracts. Jak fruits that were thrown away are now eagerly bought by Ramasamy. Solid advantages certainly.
>
> But see the other picture:
>
> The buffaloes of the poor villager soon get into the tea or coffee garden or to the grass fields of the estate. They are caught and many a rupee has to be paid for damages. It often follows that this mode of recovering damages cannot protect the fields from cattle, and great is the damage no doubt they do to the plantations and therefore we cannot blame the European owner for taking stricter precautions, as to fence 500 or 800 acres with a fence strong enough to keep out a buffalo or a bull is out of the question. The nearest magistrate is applied to, and a licence or permit to shoot all cattle trespassing is soon obtained. Sad havoc now among the cattle. High caste Sellambaram or Kandesamy and his Meenatchi will not touch a dead cow, though they daub their floors with its dung and wash their faces sometimes with its water. But Paliamuttu or Chakkiliya Karpen think a dead buffalo is a godsend. Soon the poor villager finds he has no cattle to plough his fields, he has to borrow from a more favoured village, but he has no buffalo of his own to return the compliment. So he sends for buffaloes to the neighbouring village, promising to

pay for their use in money or in grain, a great loss indeed. The milk or the *ghee* [clarified butter] which was abundant is no more. Another evil yet, a buffalo or two from the neighbouring village has strayed into the estate and is shot down. No more buffaloes to be had, even for hire, so now the poor man must take his *mammoty* [hoe] and turn the sods. But is this all? No.

From one ravine a constant stream of water flowed down and irrigated his few *pelas*. But the vandal axe of the low-country contractor has cleared all the trees on both sides with the consequence that the stream is dry and the fields useless, at least to grow paddy in. From his sheltered hillside bungalow, the European superintendent or owner sees his ploughed acres sweeping down to the village edge and between that and the cart road lies the village. What if this can be aided? It is a sore point with him to see this miserable stretch between his estate and the road; like Nabboth's field, it is much to be desired. Soon a small holding is acquired at a high price of course. The wedge at the thin end is driven in. Now steps in the negotiator. What persuasion, what honeyed words, and maybe in some instances what threats are used to induce the villager to give up his patrimony and his birth right. Having the estate just behind their dwelling houses and thinking that they would often get into trouble should they live so close to the estate was another cause the villagers had in connection with the inducement referred to above to part with their ancestral property.

Then follows the exodus of the villagers. Lucky the few who had relatives in villages situated in healthy districts. But most go to the unhealthy places where a hundred causes put them into hardships, chiefly the unsalubrious malarial nature of the country, so that often a few of the younger members of the family had died. They were obliged to come back to the site of their villages and labour as coolies. It had further resulted thus that they not only lost the lands that came to them from their ancestors, but the seclusion and privacy of their dwellings is gone. All sense of decency is lost, and they have brought themselves lower than the Tamil coolies without a place of abode which can be called their own, and compelled to work sometimes in the very fields which were once their own as day labourers and not as owners of the soil. For the Tamil coolies have hopes that by saving some rupees they will get to their native villages some day, redeem their pledged fields and gardens, and once more they will be landlords. But the native who has parted with all his lands for a sum of rupees has no village, no home, no household goods and no pets.

Most of the arguments later repeated in successive reports, such as the Reports of the Land Commission [*Ceylon*, 1927–29] or Report of the Kandyan Peasantry Commission [*Ceylon*, 1951] are already present here, before the nationalist discourse was fully articulated: shooting of buffaloes by planters, drying up of streams, and the grabbing of village land by estates, supposedly lead to the abandonment of village homes and dramatic outmigration to the then malarial dry zone. But Aluvihare adds another dimension. As could be expected of a high caste headman from an ancient family, status and moral considerations are given a prominent place. The worst fate for villagers is to become coolies and lose their caste, by associating with beef eaters and shameless drunkards. More than just an enclave, the estate is perceived as a corrupting body, and its superintendent in his bungalow as a malevolent power hanging over the village. Moral hierarchy is inverted: the low and impure are above the high and pure. This negative perception was sufficiently rooted in the Kandyan psyche to lead W.E. Davidson, a revenue officer who wanted to control the development of plantations in Kegalla district, to circulate in the 1890s an appeal to villagers not to sell their highlands, where planters and estate workers were depicted as the incarnation of vice [*Meyer*, 1992: 333].

In a report annexed to the Census of Ceylon for 1901, another chief headman, Mahawalatenne, expressed a similar anxiety. In 1897, a number of villagers went to work on estates:

> such persons in my return are described as coolies or say labourers and not as *goviyo*, cultivators, a name too respectable for a coolie ... I have seen estate coolies and I have seen road coolies, and I would rather have our Sinhalese *goviyo* cultivate chena and get 50 per cent less of produce, than see such great calamity as going on an estate or the road as coolies befalling them ... It is no good for our Sinhalese *goviyo* who never heard of death from starvation to leave hearth and home, kith and kin, and go to the estates, and according to his own crude notion, live a sort of low-caste life'.[7]

The number of immigrant Tamil coolies was considered 'too great', and, significantly, other non-Kandyan communities were similarly treated: 'The relation of the Low-Country Sinhalese to the Kandyan is the same as that of the foreign-born [*sic*] Tamil, Moor or Malay. He comes and goes.'

The unsettled population had always been a problem for headmen, and they feared competition from those they called 'low-country adventurers', as well as the formation of an uncontrollable rural proletariat; more than the intruding plantation as such, it was social mobility and employment prospects increased by its development which created a problem for them, both as police officers and as landowners. When asked by the British Agent at Matale whether the villagers liked having large estates in their midst, the local headmen replied: 'Rogues like it but good men do not' (quoted in Meyer [1979: 466]). For headmen, control over local people was the key to power and wealth, as crucial for them as control over their labour force was for planters. When they felt it endangered by the competition of the estates, they tried to reassert duality, as did the planters, and adopted the enclave myth (see Administration Reports for Ratnapura 1907 and 1916 quoted in Meyer [1979: 465]). Dualist representations could serve the purposes of various classes and ideologies.

The perceptions of common villagers, who could use employment opportunities created by the estates to free themselves from their dependence on headmen, might have been quite different. As far as terminology is concerned, there was no dual representation: in Sinhala, the word used by villagers for their gardens planted with fruit trees or bushes and vegetables, is *watte* (cf. Tamil *thottam*). The same word was generally used for estate plantations, possibly with the adjective *maha* (big). On the other hand, the hostile reaction of Kandyan villagers when forests were extensively cleared for the first coffee estates reveal another image: many forest-covered mountain areas were considered as the gardens of the gods or the abode of demons and there was a belief that no one could fell them without incurring their wrath.[8] Many villagers believed that their buffaloes were attracted into the estates to be killed and eaten by low caste coolies; a dramatic decrease in the number of cattle was attributed to the malevolence of planters and coolies. Buffaloes were needed for paddy cultivation, and their efficiency was proportional to the amount of freedom to graze they enjoyed in the surroundings of the village. On the whole, it seems unlikely that the 'hemmed-in village' myth originated in the village itself, but some of its elements were certainly already there.

The Enclave Myth and the Contradictions of the National Elite

Elaborating national myths is an essential exercise for any national elite, especially in the era of mass education and 'print capitalism' [*Anderson*, 1983], in the process drawing on pre-existing ideas as building material. The enclave image occupies a prominent place in what Moore [1989: 189 sqq] calls 'the Sinhalese nationalist myth'. It is used to validate the view that the original Aryan–Buddhist–Sinhalese village was destroyed by the British planter's introduction in its midst of a Dravidian element which will act as a fifth column and help India once more to invade Sri Lanka.

The 'rediscovery' by low country Sinhalese of the ancient Buddhist chronicle of the island, the *Mahavamsa*, played a decisive role in the elaboration of a Sinhalese–Buddhist ethno-nationalism at the turn of the twentieth century. According to the founding myth of the *Dhammadipa* (island of the Doctrine), developed by Theravada monks to explain the resilience of Buddhism in Sri Lanka and its obliteration in India, Sri Lanka was the land chosen by the Buddha to become the repository of his pure doctrine, and the Sinhalese the people chosen to accomplish his will.[9] The 'hemmed-in' and 'enclave' myths were connected to the *Dhammadipa* myth; foreigners again threatened the living space of the Sinhalese, and Tamils again were involved (now as agents of their European masters). This time the danger was more pressing, because the evil force was inserted in the heartland of the Sinhalese, spying, acting like a fifth column, and corrupting the mores of an hitherto pure and virtuous peasantry. India was considered as permanently famine-striken, and the source of every possible epidemic: coolies were suspected to be germ carriers and therefore doubly untouchable, physical degeneration of the Kandyans was feared. The estate itself degraded the village environment through soil erosion, drying up of springs, and the filth sent down to the village area by the estate drains. In the south Asian cultural context, equating the estate with impurity and the village with purity made the myth highly pervasive.

The writings of *Anagarika* Dharmapala (in *Guruge* [1965 533–6]), the most influential (but not the most radical) propagandist of Sinhalese–Buddhist ethno-nationalism, support this interpretation. After condemning the British land policy and comparing the fate of the Sinhalese peasant to that of the Irish, he observed in an article published in the 1900s:

> The Sinhalese are isolated, they are between the Devil and the deep sea, they have no other land to go to, they have no ethnological relationship with any other existing race or country [*sic*]. Two things only are possible for them, to progress as a nation under the aegis of British protection or to degenerate and live as slaves working for their day's wages under inhuman British planters like the Negro slaves in the Southern States of North America before the days of the American Civil War.[10]

Elsewhere Dharmapala and his followers depicted planters as the incarnation of European vice: drunkards, hunters, beef eaters, debauchees. Their low caste coolies were supposed to connive at their misdoings, while the villagers were deemed proud – witness their refusal to work on estates; but another image of Kandyans popular among the low country westernised elite was that of uneducated peasants with loose sexual behaviour and lazy habits, who had to be rehabilitated. This concern for 'moral upliftment' was expressed in particular by the Ceylon Social Reform Society in the years 1900. The attitude of the emerging low-country elite towards the up-country Kandyans was characterised by ambiguities: myths might have acted as a screen to conceal these contradictions.

Insecurity among the Sinhalese also reflected some uncertainty about their identity (see Spencer [1990a]). A recurrent fear was that Kandyan villagers mixing with Tamil coolies would themselves become Tamil [*Meyer*, 1979:468], which confusedly expressed the then unspeakable thought that their ancestors might have been of Tamil origin. The vision of a total victory of the plantation enclave haunted the Sinhalese. If the Government sold grass lands to planters, wrote a journalist in the 1920s, 'the day [was] not distant when cooly lines [would] be seen on the ruins of Kandyan village homes' (*Ceylon Daily News*, 6.01.1927). 'The Sri Lankan villager, as was asserted time and again, always rejected estate work to preserve his integrity and identity: the transformation of the peasant into an "estate cooly" was nothing short of a "national calamity"' (Samaraweera [1981:135] quoting R. Aluvihare, a Kandyan politician and a descendant of the headman who wrote the above quoted report).

The socio-political implications of these myths are analysed by Samaraweera [1981], and Moore [1985; 1989]. Members of the elite were 'the most unlikely champions of the cause of the peasants' [*Samaraweera*, 1981: 130] since they were themselves heavily involved in plantation economy and grabbing peasant land [*Jayawardena*, 1963; *Meyer*, 1982a, 1992]. Another contradiction was that the supposedly strong paddy peasantry had actually been converted into half-cooly in the very areas dominated by the elite, the mid-and low-country rubber and coconut belts. 'In delineating a battle line between the peasant and the plantations, they were drawing attention to an area which clearly impinged on their own interests' [*Samaraweera*, 1981:136]. What were the means employed to overcome the contradiction? As soon as self-government became a possibility, prominent members of the low-country elite such as D.S. Senanayake found that they needed the support of the Kandyans to fulfil their own political ambitions, and that the enclave myth was the most convenient tool to win them over and block attempts by the colonial administration to foster Kandyan regionalism. Land hunger was then being felt in several Kandyan districts, and among a section of the civil service, led by C.V. Brayne, there was a wave of sympathy for the 'poor peasant' and a desire to stop indiscriminate estate development [*Moore*, 1989: 194; *Meyer*, 1992]. Even before the Donoughmore reforms in 1931 instituted a measure of self-government in Ceylon, the recommendations of the Land Commission of 1927–28 led to the adoption of a new land policy along these lines: this was the first defeat of the planter lobby, and the first success of Senanayake who (after 1931) became Minister of Agriculture and Lands and the leading nationalist figure.

At this juncture, the prevailing attitude of the Sinhalese elite was paternalist: the 'poor peasant' had to be shielded from the influence of plantations, and 'rehabilitated' (that is, weaned from his improvident habits) to make him strong enough to resist. The malaria epidemic of 1934–1935 (especially severe in the rubber and coconut areas) came at the right moment to reinforce the conviction that villagers badly needed help, and created the opportunity for young social workers from Colombo, some politicians and the press, to 'discover the village'. As most plantations escaped the worst effects of the epidemic, the idea that coolies were better off than villagers became widely accepted, and the government was requested to treat 'its peasants' as well or better than the planters treated 'their coolies'. This was the starting point of the 'welfare state system', and of socio-economic studies of the peasantry [*Meyer*, 1982b].

The myth of a 'disintegrating village' and of a peasantry in need of 'moral rehabilitation' had its roots in the idealised image of the pre-plantation village, but numerous enquiries and reports on rural society after 1935 gave it a scientific backing. After the official report of the Kandyan Peasantry Commission of 1951 [*Ceylon*, 1951], the academic enquiry by S.J. Tambiah and N.K. Sarkar gave wide currency to the thesis of the disintegration under the impact of plantation growth of an hitherto 'well integrated and self sufficient social and economic system, well adapted to certain ecological factors' [*Sarkar and Tambiah*, 1957: XI].[11]

The principal response of the elite was to focus on the large areas in the dry zone where colonisation schemes could be started and a strong peasantry rebuilt, unhampered by the proximity of any plantation. These areas were the heart of the ancient irrigation-based Rajarata Kingdoms which flourished well before the Kandyan Kingdom and over a much longer period, and whose reconquest became the major project of Senanayake and his successors up to the present day (the Mahaveli development scheme being its crowning achievement). Another way out, notably in times of depression, was to lay the blame on outsiders, in imitation of European dictators: in addition to British planters, Indians (Tamil coolies and *kanganis*, Chettiar moneylenders and Borah traders) provided convenient scapegoats [*Samaraweera*, 1981: 141]. Legislation was introduced to exclude most Indian immigrants from landownership, from certain categories of employment, and finally from citizenship [*Samaraweera*, 1977; 1981], measures which led to a very serious deterioration in Sinhalese–Tamil relations up-country [*Meyer*, 1990a].

Dualistic theories and enclave myths were in the nature of self-fulfilling prophecies.[12]

Dualistic Models and 'Back to the Land' Utopias

I follow Breman's [1988] suggestion that the image of a stable and autonomous village community in Asia is a colonial invention. Ironically, while rural mobility was a regular feature of the *ancien régime*, villagers themselves subscribed to an image of stability because it was in their interest to pass as original villagers in the eyes of the colonial administration. Breman also rightly points out [ibid: 38] how the nationalists assumed the myth of the village: 'It is ironic that the village, once declared the cornerstone of colonial policy, is now proclaimed a symbol of resistance by those who profess nationalist sentiments'. Hence utopias like 'village rehabilitation schemes'.

Dualistic theory should be perceived as an historical phenomenon, a dimension of both colonial and nationalist thinking. Accordingly, presenting the economy in dualistic terms and pretending that the peasant sector was safe from fluctuations generated in the modern sector, while the crisis provided full evidence to the contrary, was a piece of politico-ideological displacement. Colonials were anxious to shift their responsibility and make others believe that they had left the traditional structure unscathed. Nationalists adopted the same course, but gave it a pessimistic twist: according to their myth, the colonial plantation system affected the village negatively and it was their duty to come to the rescue of the 'poor peasant', to help him reconstruct a mythical golden age. Such a view could reconcile traditionalists with the emerging elite, and contribute to offset the contribution of the latter to the spoliation of the peasantry.

These are common reactions in times of depression, which produce recession in minds as well as in production. In Europe the same attitude was widespread, for example, among the French bourgeoisie under Petain. Those intellectuals and technocrats subscribed to the nostalgia for the *ancien régime*, and to the 'back to the land' myth as a panacea for the ills of a supposedly disintegrating French society, in which the moral values of 'the peasantry' were assaulted by the perverting influence of 'cosmopolitan capitalism'.

NOTES

1. As compared with the shallow analyses of Ponnambalam [1981:8] according to whom the two sectors of the economy 'co-existed in splendid isolation, establishing no contact whatsoever between them'; and of Shanmugaratnam [1981: 70–73] speaking of 'virtual

encapsulation of villages by estates': 'the plantation and peasant sectors constituted a disarticulated bimodality and the so called modern sector did not serve as development pole ...'

2. Manufacture and distribution of arrack (palm and especially coconut alcohol) was a state monopoly farmed out to wealthy low-country Sinhalese entrepreneurs: information on illicit trade in drinks is abundant.
3. In his monograph on Spring Valley, Kemp [1982:145] shows that 'whilst most of the productive work of tea cultivation was indeed left to the regular Tamil workforce, the work required by maintenance and improvements, rush periods or fluctuations in the Tamil population was given to local Sinhalese. In the Uva highlands, the physical juxtaposition of village and estate was more than a case of mutual avoidance.' For the period 1908–1950, the estate employed regularly between 180 and 320 village workers, under 12 to 18 *kanganis*; in the 1970s, there were about 575 villagers on the books, against 883 adult Tamils and 665 young Tamils.
4. Planter psychosis leading to violence is analysed by Breman [1989; esp. p.170 sqq.], Stoler [1985], and by Kelly in the present volume, as a basic feature of Sumatran and Fiji plantation life.
5. Jayawardena draws heavily on the writings of one of the ablest civil servants of the late nineteenth century, W.E. Davidson, assistant agent at Kegalla and later settlement officer. In his contribution to this volume, Kelly shows that in Fiji the policy of Gordon, the first governor, was to keep indigenous Fijians separate and prevent them from being employed by the planters; Gordon when he became governor of Ceylon followed that same policy.
6. Aluvihare Ratemahatmaya to Assistant Government Agent Burrows, 9.11.1896, encl. in Burrows to Govemt Agent Central Province 11.11.1896, Sri Lanka National Archives, Pending File 178, Sale of Lands by Villagers.
7. Report of the Chief Headman in charge of Kadawata and Meda Korale in the Ratnapura District, [*Ceylon, 1902*, Vol.1, app. F: 113–27].
8. The interdict could be overcome if properly recognised, and I am inclined to think that many conflicts were 'settled' at the local level. Witness the story collected in 1939 by a settlement officer in the Kurunegala district [Diary of the Assistant Settlement Officer Wijekoon, January 1939, Sri Lanka National Archives]:

   > The opening up of Delwita estate in the early days had been greatly hampered by the high rate of mortality among the coolies and the mysterious collapsing of any building that was constructed. The neighbouring villagers went in deputation to the superintendent, and suggested that an annual offering of 10 rupees and a bushel of rice be made to the *devale* [an old temple right in the middle of the estate]. This was done, and a complete cessation of all acts of destruction by the powers above was the immediate result. This offering in money and kind has thereafter been made every year without a break. The *kapurala* [priest] however whispered to me that the present superintendent has reduced the tribute from 10 to 5 rupees [a result of the depression!]: he seems rather anxious and fearful of the effect this reduction may have on the powers above.

9. Among the favourite episodes of modern readers is the story of the pretender Dutugemunu who succeeded in expelling from the throne the Tamil king Elara. When a child, Dutugemunu, according to the chronicle, answered his mother who asked him why he laid crouched on his bed, that he could not stretch out his limbs because he was hemmed-in between the sea and the power of the Tamil king Elara in Anuradhapura. This is the first symbolic representation of space to be found in the Buddhist chronicle, and its constant use in modern 'geopolitical' discourse is highly significant. What may be called the Dutugemunu complex (the obsession of helpless islanders fearing invasion from the mainland and their living space shrinking away, with no place to migrate to), became a crucial component in the modern Sinhalese consciousness. The modern use of history and legend is analysed in Spencer [1990a] – see especially Gunawardena's essay entitled 'The People of the Lion'.

10. Dharmapala's position appears contradictory, as was that of the British-educated elite as a whole, whose interests were linked with those of the Raj: for them, British rule had two faces, and the planter epitomized the ugly one.
11. Another volume was published two decades later with the same title (*The Disintegrating Village*), but with a question mark added in the introduction; however the deconstruction of the myth was limited to that of the idyllic image of the pre-plantation village and to the absolute dualistic model, and the idea of a 'traditional paddy-centered communal village', disintegrating under modern impact, was retained [*Morisson et al.*, 1979: 7, 15–17].
12. Marxist intellectuals who founded the Lanka Sama Samaja Party in 1935 rejected the dualistic view and insisted on the class solidarity of estate and village workers. With a view to undermining their growing influence in the rubber belt (two of them were elected in 1936), a programme of 'rural reconstruction' was initiated by Senanayake. In the most congested districts, a systematic policy of estate acquisition for redistribution to peasants was undertaken in the 1940s, which led in a few cases to the expulsion of Tamil coolies [*Samaraweera*, 1981: 156–7; *Meyer*, 1990a: 186–7]. The politicians who undertook in the 1940s to retransfer estate land to villagers, and the more radical SLFP politicians who nationalised plantations in 1972 and 1975 under Mrs. Bandaranaïke's leadership, may have unwittingly primed a time bomb. Though such a programme was never clearly defined in their manifestos, it was and possibly is still lingering in the minds of the young JVP rebels who played a leading part in the 1971 and 1989 revolts against the established order. They converted the Maoist stategy of conquest of the town from the countryside to the conquest of the estate enclave from the villages, with its possible outcome the destruction of the conquered entity, and a return to subsistence agriculture [*Samaranayake*, 1987; *Bastian*, 1990: 330]. The disastrous results of the Pol Pot experience in Cambodia had obviously not been analysed by them.

REFERENCES

Adas, M., 1974, *The Burma Delta: Economic Development and Social Change on an Asian Rice Frontier*, Madison, WI: University of Wisconsin Press

Alatas, S.H., 1977, *The Myth of the Lazy Native*, London: Cass.

Ameer Ali, A.C.L., 1970, 'Peasant Agriculture in Ceylon 1833–1893', M. Phil. thesis, University of London.

Anderson, B., 1983, *Imagined Communities*, London: Verso.

Arasaratnam, S., 1982, *The Historical Foundations of the Economy of the Tamils of North Sri Lanka*, Jaffna: Thantai Chelva Memorial Trust.

Bandarage, A., 1982, 'The Establishment and Consolidation of the Plantation Economy in Sri Lanka', *Bulletin of Concerned Asian Scholars* Vol.14, No.3, pp.2–22.

Bandarage, A., 1983, *Colonialism in Sri Lanka: Political Economy of the Kandyan Highlands 1833–1886*, Berlin: Mouton.

Bastian, S., 1990, 'Indian Tamils: Emerging Issues', *Lanka* (Uppsala) Vol.5, pp.329–58.

Blackton, C.S., 1978, 'The Europeans of the Ceylon Civil Service: The View from the Kachcheri', *Ceylon Journal of Historical and Social Studies*, Vol.VIII, No.1, pp.32–46.

Boeke, J.H., 1953, *Economics and Economic Policy of Dual Societies*, New York: Institute of Pacific Relations.

Breman, J., 1988, *The Shattered Image: Construction and Deconstruction of the Village in Colonial Asia*, Amsterdam (*Comparative Asian Studies* No.2).

Breman, J., 1989, *Taming the Coolie Beast: Plantation Society and the Colonial Order in Southeast Asia*, Delhi: Oxford University Press.

Ceylon – *Administration Reports*, Colombo: Government Press, Annual series (1867–).

Ceylon, 1902 – *Census of Ceylon 1901*, Colombo: Government Press, 1902, 4 vols.

Ceylon, 1908 – *Report and Proceedings of the Ceylon Labour Commission*, Colombo: Government Press, 1908.

Ceylon, 1927–29 – *Interim Reports and Final Report of the Land Commission, Sessionnal Papers Nos. 18–32 of 1927 and 18 of 1929*. Colombo: Government Press.

Ceylon, 1951 – *Kandyan Peasantry Commission Report, Sessionnal Paper No.18 of 1951*, Colombo: Government Press.

De Silva, S.B.D., 1982, *The Political Economy of Underdevelopment*, London: Routledge & Kegan Paul.

Farmer, B.H., 1963, *Ceylon, a Divided Nation*, London: Oxford University Press.

Gunasinghe, N., 1990, *Changing Socio-economic Relations in the Kandyan Countryside*, Colombo: Social Scientists Association.

Guruge, A. (ed.), 1965, *Return to Righteousness: A Collection of Speeches, Essays and Letters of the Anagarika Dharmapala*, Colombo: Government Press.

Jayawardena, L.R.U., 1963, 'The Supply of Sinhalese Labour to Ceylon Plantations, 1830–1930; A Study of Imperial Policy in a peasant society', Ph. D. thesis, Cambridge.

Jennings, Sir I., 1948, *The Economy of Ceylon*, London: Oxford University Press.

Kemp, Ch., 1982, 'Spring Valley: a social, anthropological and historical inquiry into the impact of the tea estates upon a Sinhalese village in the Uva Highlands of Sri Lanka', Ph. D. thesis, University of Sussex.

Kemp, Ch., 1984, 'Politics and Class in Spring Valley, Sri Lanka: An "Anti-Structuralist" Interpretation', *Journal of Peasant Studies*, Vol.12, No.1, pp.41–64.

Kumar, D., 1965, *Land and Caste in South India*, Cambridge: Cambridge University Press.

Kurian, R., 1989, 'State, Capital and Labour in the Plantation Industry in Sri Lanka', doctoral thesis, University of Amsterdam.

Lewis, R.E., 1848, 'The Rural Economy of the Sinhalese, with some Account of their Superstitions', *Journal of the Ceylon Branch of the Royal Asiatic Society*, Vol.2, No.4 pp.31–52.

Meyer, E., 1979, 'Between Village and Plantation: Sinhalese Estate Labour in British Ceylon', in *Asie du Sud, Traditions et Changements* (eds. M. Gaborieau and A. Thorner), Paris: Editions du CNRS (Colloques Internationaux du CNRS No.582).

Meyer, E., 1980, 'Dépression et Malaria à Sri Lanka, 1925–1939: l'impact de la crise économique des années 1930 sur une société rurale dépendante', Paris, doctoral thesis, Ecole des Hautes Etudes en Sciences Sociales, 2 Vols.

Meyer, E., 1982a, 'Bourgeoisie et société rurale à Sri Lanka (1880–1940)', *Purushartha*, Vol.6 (*Caste et classe en Asie du Sud*), pp.223–50. English translation, *South Asia Research*, Vol. 11, No. 1, May 1991.

Meyer, E., 1982b, 'L'épidémie de malaria de 1934–1935 à Sri Lanka: fluctuations économiques et fluctuations climatiques', *Cultures et Développement*, Vol.XIV, No.2–3, pp.183–226, and No.4, pp.589–638.

Meyer, E., 1983, 'The Plantation System and Village Structure in British Ceylon: Involution or Evolution?' in *Rural South Asia: Linkages, Change and Development* (ed. P. Robb), London: Curzon Press.

Meyer, E., 1990a, 'Aspects of the Sinhalese–Tamil Relations in the Plantation Areas of Sri Lanka under the British Raj', *Indian Economic and Social History Review*, Vol.27, No.2, pp.165–88.

Meyer, E., 1990b, 'Representations of the Past and of Space as Obstacles to Peace in Sri Lanka,' *Lanka* (Uppsala), Vol.5, pp.297–320.

Meyer, E., 1992, 'From Landgrabbing to Landhunger: High Land Appropriation in the Plantation Areas of Sri Lanka during the British Period', *Modern Asian Studies*, Vol. 26, No. 2, pp.321–61.

Moore, M., 1985, *The State and Peasant Politics in Sri Lanka*, Cambridge: Cambridge University Press.

Moore, M., 1989, 'The Ideological History of the Sri Lanka "Peasantry"', *Modern Asian Studies*, Vol.23, No.1, pp.179–207.

Morisson, B. *et al.*, 1979, *The Disintegrationg Village. Social Change in Rural Sri Lanka*, Colombo: Lake House.

Pieris G., 1965, 'The Economic Geography of Rubber Production in Ceylon', Ph. D., Cambridge.

Ponnambalam, S., 1981, *Dependent Capitalism in Crisis: The Sri Lankan Economy, 1948–1980*, London: Zed Press.

Roberts, M., 1965, 'Some Aspects of Economic and Social Policy in Ceylon', Oxford, D. Phil. thesis, 1965.

Roberts, M., 1973, 'Aspects of Ceylon's Agrarian Economy in the XIXth Century', in *History of Ceylon Vol.III* (ed. K.M. De Silva), Peradeniya. University of Ceylon Press Board.

Roberts, M., 1980, 'From Southern India to Sri Lanka: The Traffic in Commodities, Bodies and Myths from the XIIIth Century Onwards', *South Asia*, New Series, Vol.III, No.1, pp.36–47.

Roberts, M., 1982, *Caste Conflict and Elite Formation: The Rise of a Karawa Elite in Sri Lanka, 1500–1931*, Cambridge: Cambridge University Press.

Rogers, J., 1987, *Crime, Justice and Society in Colonial Sri Lanka*, London: Curzon Press.

Samaranayake, 1987, 'The Changing Attitude Towards the Tamil Problem within the J.V.P.', in *Facets of Ethnicity in Sri Lanka* (eds. C. Abeysekera and N. Gunasinghe), Colombo: Social Scientists Association.

Samaraweera, V., 1977, 'Land as "Patrimony": Nationalist Response to Immigrant Labour Demands for Land in the Early XXth Century Sri Lanka', *Indian Economic and Social History Review*, Vol.XIV, No.3, pp.341–62.

Samaraweera, V., 1981, 'Land, Labor, Capital and Sectional Interests in the National Politics of Sri Lanka', *Modern Asian Studies*, Vol.15, No.1, pp.127–62.

Sarkar, N.K. and S.J. Tambiah, 1957, *The Disintegrating Village*, Colombo: University of Ceylon Press Board.

Scott, J.C., 1976, *The Moral Economy of the Peasant*, New Haven, CT: Yale University Press.

Shanmugaratnam, N., 1981, 'Impact of Plantation Economy and Colonial Policy on Sri Lanka Peasantry', *Economic and Political Weekly*, Vol.XVI, No.3, pp.69–80.

Snodgrass, D.R., 1966, *Ceylon: An Export Economy in Transition*, Homewood, IL: Richard Irwin.

Spencer, J. (ed.), 1990a, *Sri Lanka: History and Roots of the Conflict*, London: Routledge.

Spencer, J., 1990b, 'Writing Within. Anthropology, Nationalism and Culture in Sri Lanka', *Current Anthropology*, Vol.31, No.3, pp.283–300.

Stoler, A.L., 1985, *Capitalism and Confrontation in Sumatra's Plantation Belt, 1870–1979*, New Haven, CT: Yale University Press.

Wesumperuma, D., 1986, *Indian Immigrant Plantation Workers in Sri Lanka: A Historical Perspective, 1880–1910*, Colombo: Vidyalankara Press.

Strategies of Labour Mobilisation in Colonial India: The Recruitment of Indentured Workers for Mauritius

MARINA CARTER

INTRODUCTION

Indenture has generally been defined as an economically-induced migration combined with coercive contractual obligations. In order to explain levels of out-migration, correlations between years of famine and volume of labour exported have frequently been cited, whilst recruitment has been characterised as highly structured, and involving the use of paid personnel to deceive or coerce marginalised individuals into emigration depots. Changes in the terms and conditions of migration are largely ascribed to reforms imposed from above through legislation [*Tinker*, 1974; *Mangru*, 1987].

The indenture system has therefore been distinguished in the literature from *kangani* recruitment to Malaysia and Ceylon, described as a more informal, kin-based labour mobilisation.[1] The view of indentured recruitment as founded on duplicity but kept in check by a benign Indian government, has been challenged by other scholars who see migration as a rational and conscious decision for individuals faced with economic adversity or political and cultural oppression [*Lal*, 1980: 66; *Emmer*, 1985; *Sain*, 1983: 179; *Ali*, 1980: 5]. The assertion that recruits were already mobile and seeking work is an important feature of this literature.[2]

A central flaw of both theories is their failure to analyse crucial changes in recruiting strategy which helped to maintain the dynamism of overseas labour mobilisation. The purpose of this article is to demonstrate the increasing use and role of returnees in sustaining migration to Mauritius – the principal importer of indentured labour from India in the nineteenth century.[3] This analysis of the shift from commercial contracting to the use of ex-indentured labourers as recruiters provides a means of correcting

Marina Carter, Department of History, Royal Holloway & Bedford New College, University of London, Egham, Surrey, TW20 OEX. The author is grateful to the editors for their useful comments on earlier drafts.

the weakness of the deception argument which does not take into account migrants' increasing familiarity with the conditions of overseas labour, without surrendering to the revisionist claim that indenture necessarily or generally benefited migrants.

The use of returnees (sent back to India to recruit labourers and bring them to Mauritius) began informally within the first decade of indentured migration. The strategy of returnee recruiting was adopted because it was cost-effective, and avoided many of the problems associated with directly coercive methods of labour recruiting. It was also a means of maintaining a migrant stream to the colony in the face of rising competition between labour exporters. The study of returnee recruiting helps to explain the manner in which strategies for reproduction of the work force evolved over the indenture period (far from remaining static as previous studies of indenture have implied). Such recruiters, who often received sirdarships when they returned to Mauritius with a band of new immigrants, played an important role both in the management of labour on sugar estates, and as pioneers in land acquisition and settlement off estates. By integrating new arrivals into Indian social and economic networks in Mauritius which were increasingly independent of plantations, returnees fulfilled the function of attracting immigrants to the colony in spite of poor prospects for such workers.

Immigration levels to Mauritius peaked in the 1850s and thereafter successively declined as a further shift in labour strategy was adopted, replacing the emphasis on immigration with the utilisation of existing labour through job contracting and the rise of small cane farming. This article therefore charts the pattern of labour recruiting for Mauritius over the three decades (1830s to 1860s) which saw the mass immigration of Indian labour.

I. THE ADOPTION OF AN IMMIGRANT LABOUR STRATEGY

That indenture was conceived and set up as a cheap labour option in the face of threats to the servile labour supply is widely accepted [*Rodney*, 1981: 332–3; *Adamson*, 1984: 47; *North-Coombes*, 1984: 84–5]. If planters initially sought to use immigration as a means of imposing lower wages on the existing labour force, it quickly became apparent in Mauritius that continued recourse to Indian labour would enable estate-owners to dispense entirely with a more costly creole labour force [*Carter*, 1991b]. This fact would prove crucial to the later adoption of policies designed to encourage the settlement of Indian families in Mauritius.

Even before the abolition of slavery in 1834, the expanding plantation economy of Mauritius had begun the search for an alternative labour supply. India was a natural choice for an island with a long tradition of recourse to the subcontinent for goods, skills and manpower [*Carter*, 1991b]. Having identified India as an appropriate source of labour, planters and their commercial partners in India targeted potential recruits by tapping existing economically-induced seasonal rural-urban or rural-rural migrations. Local collaborators or intermediaries were then employed to divert such labour to the new destination.

The number of labourers exported to a particular territory was, however, contingent upon the capital available to engage and transport them, legislative controls in India, and competing inland or overseas recruiting. A capital squeeze or restrictive legislation in Mauritius could equally affect the labour flow. Thus, the incidence and level of migration was controlled by economic and political factors both at the site of mobilisation and the destination [*Mookherji*, 1959: 368; *Miege*, 1982: 24]. Various means of acquiring an effective immigrant workforce from this labour stream whilst keeping importation costs down were tested during the three decades following the initiation of this experiment in 1834.

II. THE RATIONALE OF RETURNEE RECRUITING

Prior to 1842, and the imposition of government controls, the collection and transportation of indentured labour was undertaken by agency-houses located at India's principal port towns and aided by local recruiters known as *duffadars, arkatis* and *maistries*, remunerated on a payment by head basis. The coercive character of this merchant controlled 'coolie trade' and the public antipathy it excited led to the suspension of indentured emigration in 1839.

In order to deflect such criticism, whilst ensuring that the immigrant labour strategy remained economically more viable than the coercion of newly freed slaves, planters searched for a labour importation strategy which was cheaper than wage bargaining in Mauritius and at the same time acceptable to legislators still susceptible to abolitionist opinion. Merchant exporters were accordingly replaced with state regulated emigration agencies in the 1840s, but the difficulty of securing an appropriate and regular immigration from recruiters motivated solely by pecuniary interest was exacerbated by the multiplication of competing labour-seekers for both internal and overseas employment from the mid nineteenth century onwards.

With capitalists bidding against each other for possession of groups of migrants arriving at the port towns, it became more difficult to prevent local *maistries* and *duffadars* from collecting for one colony and selling recruits off to another. The abuses associated with efforts to obtain labourers for a minimal outlay whilst preventing rival recruiters from offering better terms to those collected, prompted repeated government interventions [*Tinker*, 1974: 63–80; *Carter*, 1987: 50–118]. The employment of returnee recruiters, often actual or would-be foremen or *sirdars* who were despatched to India by their employers to collect recruits and to accompany groups of emigrants from the depot to the plantation, was seen as a partial solution to such problems.

III. THE OPERATION OF RETURNEE RECRUITING

The principle of appointing a 'headman' or *sirdar* to take charge of bands of immigrants on the passage to Mauritius had been adopted during the 1830s, and in many cases such individuals were the *de facto* recruiters of the group they led. South Indians reported having emigrated to Mauritius along with their *maistry* recruiters in 1836; Ramanath and Roghoonath, both *dhangas*, recruited amongst members of their own tribal community in Calcutta and in Chota Nagpur and travelled with them as *sirdars* to Mauritius [*Bates and Carter*, 1987]. In this respect, indentured migration had already acquired characteristics more commonly associated with later *kangani* migration. Planters reported that such recruits, often migrating collectively from one locality and organised under the appointed *sirdar*, provided more satisfactory labour than migrants collected singly by local recruiters from Calcutta, Madras or Bombay.

In an effort to secure a labour force which more closely approximated to the former, some estate-owners initiated the strategy of sending back *sirdars* or other trusted workers to recruit from their own districts. The first ships bringing indentured labourers back to India from Mauritius in 1839–41 on the expiration of the first five year indentures thus also brought returnee recruiters such as Dhibby Deen, who carried a letter authorising the payment of his return passage and that of 'any able-bodied men who may wish to accompany him, not exceeding 50 in number'.[4] Other returnees acted independently in the collection of family members. As early as 1839, a movement was underway to collect spouses and return with them to Mauritius. Even those who went back to India after several years' work simply to deposit savings, or to take advantage of the free return passage, were more likely to be given employment as commanders once back in Mauritius.[5] Others combined family collection with paid labour recruitment: Mohun, of Nagpur, having left his brother in Mauritius when he returned to India in 1839, intended to collect his sister from their home village: 'I will take her with me, and rejoin my brother; if my sister is married I will take her husband also. There is plenty of good work. I have orders to get 25 men, if I can, to take with me'.[6]

The use of returnees was officially sanctioned following the visit of an envoy of the Mauritian government to India in 1843, with a brief to report on the working of the emigration establishments. He recommended that 'a decided and fatal blow' should be dealt to the *duffadar* system and that the onus in recruiting be placed on returnees such as those whose return to India with their savings had in his view already created 'un puissant stimulant pour une emigration croissante'.[7]

Local recruiters in India now also began to be drawn from the ranks of returnees and were given money to go into the provinces to recruit labourers 'of the same caste as themselves'.[8] The proportion picking up recruits on the way to Calcutta rather than collecting them from natal villages was believed to be high, however.[9] Closer inspection of returnee operations in this period reveals a wide range of activities. Some, like Mauderbuccuss and Chuttoo who brought 43 men from their home district of Azamgarh in 1848, were indeed mobilising their fellow villagers; others brought smaller bands – often consisting simply of their spouses and close relatives; still others arrived at the depot with bands of unrelated individuals, collected en route to or at the ports.[10]

The relative importance of returnees in Mauritian labour mobilisation by mid century is shown by the number of individuals arriving at the depot without having been recruited by crimps, and by preferential treatment and pecuniary benefits offered to these recruiters by planters.[11] One of W. West's returnees was shipped to India on the *Fyzel Currim* 'in order that he may have a cabin and other comforts which he could not possibly have, if embarked with other immigrants'.[12] The Emigration Agent of British Guiana at Madras, visiting the Mauritius depot in 1850, counted 45 returnees with 215 new immigrants on one ship alone. This method of recruiting was more 'attractive' to the native of India, he believed, and added 'while it continues open I apprehend the natives will give a preference to it over all other emigration'.[13]

The Mauritian government initiated a series of measures designed to benefit such emissaries and to encourage family immigration: in 1852 returning immigrants bringing legitimate married couples, or the wives of labourers already in the colony, were given a bonus of £1 for each such woman. By 1852, local recruiters in Calcutta were being paid only for bringing families or particularly sought-after groups such as tribals.[14] To extend this cost-effective recruitment strategy to Bombay, the government sent 50 immigrants from that presidency back there as recruiters.[15]

Nevertheless, both at Bombay[16] and Madras[17] local contractors continued to exercise overall control of the labour supply because of the lack of a large depot establishment as at Calcutta, using *maistries* and *mukadams* to collect recruits and maintain them until embarkation, occasioning conflict between these local employees and the returnees. The agent at Madras defended the retention of contracting, claiming that competition from French exporters who paid their recruiters higher wages, had attracted both *maistries* and returnees away from his service: 'the greater part of the men who have been sent up from Mauritius to collect coolies for their master are now taking coolies to Pondicherry ... they have all proved false as they cannot withstand the temptation of the great inducements held out'.[18] Returnee recruiters certainly could not operate as effectively in a situation of competition for labour, when rival capitalists were recruiting in the proximity and offering financial inducements to attract indentured labour to other overseas destinations.

The difficulty of defining the part played by the 'unofficial' returnees in recruitment led to claims that migrants for Mauritius now came into the Presidency depots largely 'unassisted'.[19] Such views lent currency to a growing belief that returnees were merely using their position as planter emissaries to bribe or persuade individuals already en route for Mauritius to join their bands. They were shown to have been embarking ostensibly as new immigrants, and once on board seeking to win recruits for their employer by distributing small gifts. Prior to 1858 the law prevented engagements from being signed in India, and allowed new arrivals 24 hours to 'select' an employer. In this way an employer who had invested heavily in the collection of labour in India could lose recruits to another who had contributed nothing to those costs. Returnees were accordingly blamed for exacerbating competition between Mauritian planters at times of perceived labour shortage and denounced as 'cunning rascals' practising 'a truly eastern despotism'.[20]

The combined evidence of the counter-productive absorption of Mauritian returnees into rival recruiting operations in India and of skirmishes at the immigration depot in Mauritius led the colonial government to adopt, in 1858, a system whereby engagements could be signed in India. This was designed to effect a link between migrant and employer which the will of the recruiter could not influence, whilst legally defining the status and role of 'special agents' sent from Mauritius to recruit for individual employers.[21] Unless such returnees or special agents had been provided with funds to ship bands in their own names, new immigrants could be engaged to serve only on those plantations which had sent a requisition for labour to India.

Thus, returnees who had been concealing their status to avoid problems with local depot officials in India, found their position further threatened. Nagoo, who wished to rejoin relatives and friends on one estate, was obliged to engage her band to another planter because her former employer had not formally applied for labour.[22] A *sirdar*, Matadoo, explained the difficulties he had experienced at Madras:

> I recruited 17 men for Mr Raffray. I could not enter the depot with them otherwise than by applying to the contactor Soobrayen. He received payment from the Agent for having recruited my men, whereas I was their real recruiter. I received no remuneration whatever. All the coolies who wish to come to Mauritius have to apply to Perria Maistries [Great Recruiters]. They divide and engage the men as they like. Six of my men were put into other gangs against my will. I consented because I was told that I would otherwise not be allowed to embark.[23]

Others complained of forced re-indenture along with their bands, or of employers in Mauritius failing to pay them for their recruiting services.[24]

In 1861, a further enquiry instituted by the Mauritian government again supported the role of returnees claiming that recourse to Mauritius had become an established custom in certain areas, encouraged by them. Chummun's statement that he and 19 others 'were advised to go to Mauritius by a relative of mine who had just returned from that colony' was reported as evidence. Beera, a Mauritian 'old immigrant' had recruited 27 men in his district during a ten month return trip to India, intending to take them back to his previous employer, ostensibly 'because I had been kindly treated by him'. Continuing recruiting difficulties were therefore blamed on rival labour exporters: the bands of both Beera and Chummun had reportedly been intercepted at Raniganj, dissuaded from going to Mauritius, and were en route to the Bourbon depot when employees of the Mauritian agent found them in Calcutta.[25] The consolidating Indian Act XIII of 1864 incorporated measures designed to minimise the likelihood of such recruiting frauds by requiring recruits to obtain passes from local magistrates.[26]

These measures had little effect upon the rivalry between local recruiting officials and returnees.[27] At Madras, depot clerks and contractors were accused by several Mauritian returnees of sending their bands to Natal. As one said: 'I recruited 8 men at Trichinopoly for Mr Piat. I took them to Madras to Vencata Soobyah, from whom I had been obliged to take a pass. He sent me back to Trichinopoly to recruit more men. When I came back I found that he had sent to Natal the 8 men I had left in his charge'.[28] The emigration agent again defended the contractor, claiming that he was the only supplier who had not left the Mauritius service to recruit for the French or the tea companies.[29] The evidence that returnees were the best proponents of family migration was by now incontestable, however: 'old emigrants are constantly in the habit of returning to their former employers with whole families of women and children', wrote the Calcutta agent in 1871. Yet, operating alongside local contractor-led labour recruitment and despatch, and competing with rival colonial and internal demand, returnees had been effectively blocked from exercising complete control over their recruits.[30]

IV. THE ROLE OF RETURNEES IN INDENTURED MIGRATION

Was recruitment by returnees in any sense less coercive or exploitative than the contractor/*duffadar* mobilisation? Returnees were certainly recognised to be an efficient, low-cost means of maintaining the migrant stream to Mauritius, even when conditions there were less favourable than elsewhere, as the benefits offered to them demonstrate. It can hardly be argued, however, that they were more likely to deceive or profit from migrants than local recruiters, who had an equivalent pecuniary interest in labour collection. Returnees, moreover, could disseminate information about the colonies in India and were important mobilisers of family groups. The continuance of commercial recruiting practices and in particular, competition between exporters, limited returnees' independence of action. Nonetheless the activities of returnees marked a significant break with other forms of contract labour recruitment.

Popular perceptions of emigration were undoubtedly shaped by returning labourers. Time-expired migrants who returned to India penniless and disappointed could effectively dissuade recruits from indenturing overseas: many rumours which circulated in the recruiting districts such as those which foretold of forced conversion or commensality, conceivably originated from comments of returned migrants. From the late 1840s, evidence is available of bands collected by returnees absconding en route as a result of 'the misrepresentations' of other, 'dissatisfied' returned labourers! As the pattern of migration to a particular territory became established ex-indentured labourers began to provide more accurate information to new recruits about changing conditions in the colony. Events of 1867 and 1868 in Mauritius when a malaria epidemic swept the island, occasioning heavy mortality, followed by the passing of a stringent vagrancy law, prompted what the emigration agent called 'a growing dislike towards the island, from malicious reports circulated by disaffected return emigrants'.[31]

Such individuals may have been amongst the considerable numbers of

Indians ejected from the colonies at regular intervals as 'incorrigible vagrants' because they were considered unsuitable to the rigours of plantation production or discipline [*Emmer*, 1986: 197–9]. Mauritius was not prepared to bear the maintenance costs of those deemed 'unfit' or 'insubordinate'. The colony's denunciation of such individuls as 'mischief making ringleaders' is itself evidence of the effectiveness of the counter-propaganda which some returnees sought to disseminate amongst their compatriots.

On the other hand, the favourable or exaggerated reports of returnees acting for planters could be used to create attractive images of colonial destinations, drawing new recruits into the depots. Returnees' familiarity with the colonial situation and hence credibility as recruiters was undoubtedly used to serve the interests of labour importers. Those acting for employers could significantly enhance their own status and position in Mauritius if successful, obtaining appointments as *sirdars*, higher wages or cash payments per recruit. Grierson's comment therefore remains ambiguous: 'in districts where there are returned emigrants, emigration is popular. In districts where there are none, it is the reverse'.[32] Caste-based fears, he reported, were countered with talk of the piety of Hindus in Mauritius, and the numbers of priests there.[33]

At the same time, the success which prompted planters to rely increasingly on such intermediaries could be turned against them. Old hands from Mauritius were frequently sought out by rival labour exporters to recruit for them. Some worked for more than one agency at a time, creating confusion amongst recruits, for example by calling the West India depot the 'New Mauritius' depot. In so doing, they crossed over into the ranks of the *duffadar*, or professional local recruiter, thereby contributing a not insignificant part to the mobilisation of labour in India for the overseas enterprises. Other returnees like Chengleray Naidoo, of Chittur, who was already a veteran of migrations to Mauritius, Bourbon and Guadeloupe, travelled with the bands they took to the various agencies, indicating the increasingly international experience of these individuals.[34]

Of more significance than the inducements offered to join rival agencies, however, was the potential for independent recruiting. One returnee, Rama, wrote to his brother Lutchmun in Mauritius: 'if you and Hittoo bring money we shall be able to recruit 1,000 people in one month'.[35] The independence of action exhibited by such individuals and the confidence vested in them by migrants, made returnees a threat to employers' authority. Attempts to counter their position, in one case refusing returnees free passages, met with a sharp down-turn in applicants at the depot. They were also at odds with the established practices of emigration officials: returnees who brought predominantly male bands were criticised for failing to comply with the required sex ratios; others were accused of bringing 'rejects' to be registered. The old and dependent who would have been turned away by the agent found an ally in the returnee: for fear of losing whole bands the agent was obliged to pass them. The high wage rates or advances promised by some returnees to their recruits were honoured for the same reason.[36]

In some cases such conflicts simply reflected existing tensions between the plantocracy – under whose orders many returnees acted – and colonial administrators. Nevertheless, a clear distinction can be made between such planter emissaries and individuals returning independently to Mauritius with family and kin, or acting as go-betweens for labourers abroad and their relatives in India. The case of Ramasamy is illustrative. His father, Mecken, had migrated to Mauritius in 1843. Three times he gave money to returning labourers to pass on to his son together with his wish that the latter join him. In 1861 Ramasamy embarked from Karaikal on the *Jacques Coeur*, but he had been tricked: it was bound for Martinique. He jumped overboard when the ship docked at Mauritius en route, and was eventually reunited with his father after a separation of two decades.[37] This case serves both to demonstrate the role of returning emigrants in maintaining links with India and the persisting evidence of fraud in indentured recruitment. The deceptions now practised, however, tended to be specific rather than general, related to destination rather than to conditions of labour, and were often a result of competition between exporters.

The active part played by returnees such as those who visited Ramasamy gave to indentured migration a momentum which could exist independently of government control. Thus, when falling profits put a temporary stop to organised recruitment, returnees and their bands sought alternative means to travel to Mauritius. Madoo, with a band from Satara, waited eight months in Bombay for an available ship, then journeyed to Cochin and finally to Madras, where his remaining recruits were embarked on the *Phoebe Dunbar*. Cassiram Juggernauth travelled from Bombay to Calcutta to find a ship available to return him to Mauritius.[38] The colonial government used such incidents to demonstrate the 'attractiveness' of Mauritius to immigrants. Of course, it was not the ordinary recruits but the returnees who were most motivated to reach Mauritius. They were often driven by more than the promise of pecuniary reward: ties of property and family in Mauritius acted as powerful incentives to return. Cassiram owned an acre of land, and had left his family on the island when he was sent back after a residence of eleven years to recruit for his employer. Some returnees closely resembled the *kanganis* who recruited in their own locality and travelled with their recruits to estates, where they acted as headmen or sirdars.

Other returnees acted as more informal recruiters, or merely re-migrated with their families. Contemporary officials liked to emphasise the contribution of these individuals, who were 'not sent by anybody'; they often paid the passage costs of relatives to avoid indenturing them.[39] Women who migrated under the indenture system, were not generally formally contracted as indentured workers. Family migration to Mauritius was not designed to secure cheap female labour as such; it acted rather to ensure the reproduction of the work force in a colony which had come to rely totally on immigrants for the production of its principal crop [*Carter*, 1991a]. Neither were women treated as passengers however, because bounties were paid for their introduction and they were routed through the depot. Returnees could claim female relatives from the depot if they refunded the costs of their passage. Well-established migrants did send their families to and from Mauritius as 'passengers' and thereby operated outside the indentured system altogether.

Was the average returnee a small scale mobiliser of family labour or a planter emissary recruiting large bands to work on estates? Table 1 below presents data from ships arriving at Mauritius between 1849 and 1853, when on average, 10 to 20 returnees were enumerated per ship.

The predominance of bands numbering fewer than 20 individuals, often bringing only family, kin or local recruits, generally all from one district, indicates that the average returnee was not a professional employee of the agent, or, like local recruiters, acting as a collector of disparate individuals. Those at the head of larger bands, which comprised a miscellaneous collection of migrants recruited from various districts or 'on the road', were generally designated special agents with a brief to recruit. Some returnees came alone and amongst them were recorded the victims of robberies and others who reindentured because they had no funds to return to their native villages. But in general their presence on arriving ships was less a sign of rootlessness than of entrepreneurship.[40]

V. THE IMPACT OF RETURNEES UPON INDIAN SETTLEMENT IN MAURITIUS AND WIDER IMPLICATIONS OF THE STUDY

By the 1860s sustained immigration to Mauritius had created an over-supply of labour and depressed wage rates, in theory enabling employers to control and discipline the work-force 'with less resort to extra-economic devices' [*Adamson*, 1984:48]. Immigration as a labour strategy had largely outrun its usefulness and was henceforth on the decline. Labour policy was now increasingly directed at shifting the costs of labour production and reproduction away from the estates, in the guise of job contracting. Returnees and *sirdars* progressed from recruitment and supervision to becoming employers of labour. This coincided with a fall in world sugar prices which forced the fragmentation or *morcellement* of large plantations, as a consequence of which significant tracts of land under cane were acquired by such intermediaries who ploughed the profits of job contracting back into land purchase [*Allen*, 1983: 246–52].

TABLE 1

BANDS COLLECTED BY RETURNEES (1849–1853)

No. Returnees Per Ship	No. Bringing Family or Alone	No. Bringing Family &/or Recruits	Size of Bands (persons)			
			0-5	6-10	11-20	>20
17	3	14	10	2	1	1
13	2	11	10		1	
17	12	5	5			
15	6	9	7	1	1	
29	18	11	5	3	2	1
15	5	10	3		3	4
15	6	9	7	1	1	
11	5	6	5		1	
7	3	4	3		1	
29	17	12	6	3	2	1
8	4	4		2	2	
19	3	16	7	5	1	3
6	1	5	5			
7	4	3	2		1	
15	9	6	5			1
9	5	4	3	1		
13	5	8	8			
25	13	8	6		2	
5	2	3	3			
19	9	10	8	1		1
26	11	15	8	1	5	1

Source: CO 167/304 to 332 Reports on Ships, 1849–53.

In 1867 planters launched the first of several attempts to check the growing independence of time-expired Indians or those who had not completed a ten-year 'industrial residence' and who lived by cultivating rented or freehold plots, or hawking produce, by reviving sweeping vagrancy legislation [*North-Coombes*, 1984: 109–13]. The criminalisation of the non-indentured Indian was both an attempt to deproletarianise workers now selling their labour-power to the non-estate sector and to inhibit the growth of a smallholder class.

The activities of returnees and the recourse to them on the part of the colonial authorities had launched a semi-independent settlement process. This had served the planters' interest in shifting the costs of labour reproduction away from the estates and was followed by the adoption of labour contracting, but ultimately presaged the break up of large chunks of the plantation sector in the late nineteenth century. Access to land had acted as a powerful incentive to permanent settlement and the establishment of Indian villages in Mauritius permitted a reaffirmation of cultural traditions during the indenture period the like of which would have been impossible under slavery. Returnees who became *sirdars* and job contractors thus formed a 'new' elite which had emerged out of indentured immigration to join the 'old' Indian mercantile and landowning elite in Mauritius.

Was the pattern of indentured settlement outlined in this paper peculiar to Mauritius? Certainly the distance between the source of labour and its destination made it easier for some Indians in Mauritius to afford their own passage costs and that of family members, thus creating greater independence of movement. The relatively high ratio of returnees to new immigrants inevitably influenced the latter's knowledge of local conditions, and family settlement was actively encouraged by the colonial state. The impact and timing of immigration in Mauritius, making Indians the largest population group within a few decades also differed from the experience of other indenture-importing countries. The island had received the bulk of its total immigrant intake by 1870; which helps to explain, for example, the greater proportion of Indians who had moved off estates relative to Malaysia by the beginning of the twentieth century [*Jain*, 1985: 183–4]. The rapidity and scale of the demographic revolution wrought by Indian immigration in Mauritius was without parallel amongst British colonies in the post-emancipation period. The success of the Mauritian strategy of labour mobilisation is attested to by the volume of immigration: when the indenture system was abolished in the early twentieth century Mauritius had received an estimated 453,000 Indian labourers, compared with 238,909 recruited for British Guiana, 152,184 for Natal and 60,965 for Fiji [*Tinker*, 1974: 39–113]. Analysis of such intermediary groups as the returnees in Mauritius, or the *kanganis* of Malaysia and Ceylon, in the more distant West Indian and Pacific Ocean markets may nevertheless demonstrate that the Mauritian experience outlined here was shared, albeit to a lesser degree, by other areas of Indian settlement.

CONCLUSION

For immigrants under indenture in Mauritius, the inequality between contracting parties was as evident, and working conditions on plantations as harsh, as elsewhere. In assessing the character of indentured recruiting, however, structure and process should be differentiated. Whilst the structure was highly exploitative, it could occasionally be subverted; returnees were uniquely placed to effect this. As a result, labour mobilisation could be experienced very differently by individuals recruited at the behest of or by family members. The implications of this must be tied into theoretical attempts to place indenture as a category on a slave/free continuum.

Direct coercion was a less efficient form of labour recruitment than the use of returnees. By creating an intermediary class of recruiters-cum-overseers, and promoting a family-based migration, planters obtained cheap labour whilst avoiding the difficulties associated with commercial recruiting, but fuelled a semi-autonomous migration they could not subsequently control. Returnee-led recruiting fostered a level of independence from the formal structures of labour selectivity and allotment, and a disjunction from the pattern of labour requirements, which prompted planters and the colonial state to curb the actions and influence of returnees with restrictive legislation. At the same time the cost-effectiveness of returnee-led immigration and the impetus created for family and kin migration, limited both the government's will and capacity to forestall returnee influence.

By using returnees to provide a stable, family-oriented work-force, planters hoped to refine the exploitative potential of indenture, sustaining an immigrant stream when terms on offer elsewhere were demonstrably more advantageous. The focus on family migration evoked a settlement rather than a short-term labour contract and migrants appraised indenture not simply in terms of wages for labour but as a prelude to permanent settlement. For some recruits, the presence of relatives or kinsmen, even the assurance of an established Indian community in Mauritius, outweighed the purely material consideration of relative wage rates. In this and other respects, indenture may be seen as closer to white indentured servitude than to the system of slavery which had immediately preceded it [*Marks and Richardson*, 1984: 8–11]. Several important qualifications remain: the conditions of labour were such as to make completion of contracts difficult and Indians were generally prevented from competing for access to resources on equal terms with white colonists. This impediment was partially alleviated in Mauritius by the prior existence of an Indian mercantile and landowning class who loaned money and rented land to ex-indentured workers. Equally significantly, many migrants were in practice prevented by competition between labour exporters in India from reaching the colony where for family or other reasons they aspired to work or settle.

NOTES

1. For a discussion of the characteristics of *kangani* migration see Jain [1987: 12], and Miege]1982: 19–20].
2. Lal has emphasised that colonial recruits were already part of a 'great process of spatial mobility' and had thus made the 'initial break' from home *before* encountering the overseas recruiter [1980: 59–60]; see also Emmer [1986: 187]. Contemporary observers estimated that such internal migrants formed a sizeable proportion of colonial recruits: Bengal Emigration Proceedings [*BEP*] 15/77 Protector to Secy. Govt. Bengal, 30 April 1862; Mauritius Archives RA Series [*RA*] 1176 Protector's Report, 17 March 1852.
3. The term 'returnee' is used in this article to refer to migrants who returned to India and subsequently re-migrated to Mauritius, usually bringing new recruits. Indentured labourers who returned to India but did not re-migrate to Mauritius as recruiters are identified as 'time-expired' or 'returned labour'.
4. *Parl. Papers* 1841 (427) Chapman Barclay to Colville, Gilmore & Co, Calcutta, 23 Nov. 1840; Returnee recruiters were also active in Madras as the case of Ramaswamy from Madura showed: Mauritius Archives PL Series [*PL*] 57 Emigration Agent [EA] Madras to Secy. Govt. Madras, 24 Feb. 1843.
5. *Parl. Papers* 1840 (58) Mauritius Committee of Enquiry, Lucas and Lesieur; Bengal Public Proceedings 13/53 Anderson to Col. Secy, 10 May 1845.
6. *Parl. Papers* 1841 (43) Examinations of Five Coolies, 1841.
7. 'A powerful stimulus to a growing emigration', *Parl. Papers 1844* xxxv (356) Anderson to Dick, 11 Nov. 1843.
8. Public Record Office, Colonial Office Series [CO] 167/254 Gomm to Stanley, 17 July 1844 (202); *RA* 777 EA Calcutta to Col. Secy 5 Nov. 1844; *RA* 1026 Caird to Col. Secy, 18 Jan. 1849.
9. *RA* 1081 EA Calcutta to Col. Secy, 15 Jan. 1850, 23 May 1850.
10. *Parl. Papers 1849* xxxvii (280) Caird to Col. Secy, 1 June 1848, list of return emigrants on the *Julia. Parl. Papers 1846* xxxviii (691–I) Finniss' report on *Sophia*, 1 Dec. 1845 enc. in Gomm to Stanley, 12 Jan. 1846.
11. *CO* 167/319 (44) Report on the *Champion*, Feb. 1850.
12. Mauritius Archives, PA Series [*PA*] 2 West to Col. Secy, 10 March 1845.
13. Colonial Land & Emigration Commissioners' Report [*CLEC*] *Report*, 1852, Appendix 42, White to Governor of British Guiana, 8 Nov. 1850.
14. EA Calcutta to Col. Secy, 17 Feb. 1852; EA Madras to Col. Secy 10 April 1852; *RA* 1176 EA Calcutta to Col. Secy, 28 March 1853; *RA* 1314 EA Calcutta to Col. Secy, 20 Oct. 1855.
15. *PL* 33 Col. Secy to Protector, 18 Nov. 1851; a circular encouraging immigrants to bring their wives and families was translated into vernacular scripts and circulated in the colony in 1852: Mauritius Archives PB Series [*PB*] 5 Protector to Col. Secy, 4 March 1852. Family immigration was particularly important to Mauritius because of the failure to retain an ex-apprentice workforce on estates.
16. *RA* 1361 Minute of Pratt, 23 May 1856.
17. *RA* 1081 EA Madras to Col. Secy 2 Sep. 1850, encl. 'Contract between Henry Wood Gabb (EA) and the contractor P.A. Pillay, March 1850'.
18. *RA* 1129 EA Madras to Col. Secy 11 and 15 April 1851, *CO* 167/329 EA Madras to Col. Secy 12 May 1851.
19. RA 1081 Report of Hugon, Protector 11 Oct. 1850 and 17 March 1852. He suggested that migration could be carried on 'simply and naturally' as in the migration from Madras to the Tenasserim coast.
20. *Parl. Papers 1847–8* xliv (61) Pamplemousses planters to Gomm 30 June 1847. Similar epithets were not used to decry those planters who, having employed returnees to supply new immigrants, reneged on the agreed payments, *PB* 5 Protector to Col. Secy, 6 May 1852.
21. The special agents were to be licensed and under the jurisdiction of the emigration

agent, but planters were responsible for their conduct. The signing of contracts in India, effectively bound the recruit to his or her future employer through the recruiter-emissary. Indian Parliamentary Papers [*IPP*] 188/58 Col. Secy to Secy. Govt. India, 24 Sept. 1858; *Parl. Papers 1859* xxi (31–I) Appendix, Ordinance 30 of 1858.

22. In 1862 Vengadoo [or Vincata] seems to have been involved in French recruiting also, having concluded an agreement with a commercial house in Yanaon to bring *coolies* to their depot on commission: Weber, Vol.4, p.1083.
23. *PB* 11 Protector to Col. Secy 4 Oct. 1864, encls. See also RA 1587 EA Madras to Col. Secy 14 Feb. 1860.
24. *PB* 6 Protector's Report 13 Nov. 1855; *PB* 9 Protector to SM Plaines Wilhems, 25 Jan. 1862.
25. *IPP* 188/64 Col. Secy to Beyts, Jan. 1861; *Beyts Report*, Beyts to Seton Karr, 13 April 1861 encl. 5, Statements of Chummun and Beera, 23 March 1861. Bourbon was an Indian Ocean island under French control (present day Reunion).
26. *BEP* 15/77 Caird to Protector, 3 April 1862; *IPP* 188/70 Secy Govt of Bengal to Secy Govt India 28 Dec. 1864; *CLEC Report*, 1865, p.18; Royal Commissioners' Report p.242; *CO* 167/454 Barkly to Duke of Newcastle, 3 Dec. 1863.
27. Previous signs of trouble between returnees and the depot employees to whom they were obliged to present their recruits had been noted from 1846 when the protector at Mauritius discovered that returnees were attempting to conceal their identity as such. Subordinate staff of the Calcutta Agency were found to have extorted money from them, and particularly from those sent to recruit by planters: *PL* 32 Protector to Col. Secy 12 Nov. 1847; *PB* 11 Protector to Col. Secy, 16 Jan. 1865.
28. Statement of Carpanen, engaged to Mr. Piat. *PB* 11 Protector to Col. Secy 17 Jan. 1865; *RC Report*, p. 244.
29. *RA* 1788 EA Madras to Col. Secy 11 April 1865; *PB* 11 Protector to Col. Secy, 17 Sep. 1866.
30. *BEP* Vol.168, Eales to Protector 2 Jan. 1871; *PB* 20 Protector to EA Madras 4 March 1874; *RA* 2075 EA Madras to Col. Secy, 10 June 1871.
31. *Parl. Papers 1847–8* xliv (61) Gomm to Grey 23 Aug. 1847, encls. He also blamed other measures for the backlash against Mauritius by returnees, such as the abolition of sirdarships for those bringing bands. *RA* 1945 EA, Madras to Col.. Secy, 5, Sep. 1868; *RA* 2075 EA Madras to Beyts 8 Feb. 1871; CLEC Report, 29th General Report, 1868.
32. *Grierson Report*, p.18; Cumpston, p.157.
33. Ibid, p.19, and *Diary*, interview with a woman returnee from Mauritius named Sukhiya, p.9.
34. *BEP* 15/76 Governor minute to Immigration Council, Jan 1861; *Parl. Papers* 1849 xviii (691–I), Neave Report, 12 July 1845; *BEP* 15/77 Deposition of Kisto duffadar, 6 Dec. 1860. One such returnee subsequently worked for Assam: *BEP* 432/17 Biden to Secy, Govt. of Bengal, 30 April 1866.
35. *PB* 5 The letter, translated from the Nagri script, was sent by the Protector to the Col. Secy on 18 March 1852.
36. *PB* 6 Protector to Col. Secy 26 June 1855, Protector's Report 28 April 1856, Petition of 'Telingas'. A rebellion broke out on the *Futtay Mubarak* after recruits failed to be given a wage advance promised by 'old *sirdars*' *PA* 3 EA Bombay to Protector 21 April 1856.
37. *PB* 10 Protector to Col. Secy, 19 March 1861, 17 April 1861.
38. *Beyts* Report, 'Notice to Recruiters of Emigrants Mauritius', p.54; *PA* 4 Protector to Col.onial Secretary 7 May 1851; *RA* 1361 Superintendent of Police to Bombay Committee of Enquiry 13 April 1856; CO 167/370 Hart to Col. Secy 20 July 1855.
39. Migrants whose passages were paid in this way were dispensed from indenturing: *PA* 3 EA Calcutta to Protector, 28 Jan. 1856; *PB* 6 Memorandum of Hugon, 30 Oct. 1856.
40. The reports on ships in *CO* 167/304–332 also provide some information on individuals' reasons for returning to Mauritius. Details of larger bands collected by 'special agents' are available in selected immigration registers (MA: *PE* Series).

REFERENCES

Ali, A., 1980, *Plantation to Politics, Studies on Fiji Indians*, Suva, Fiji.

Adamson, A., 1984, 'The Impact of Indentured Immigration on the Political Economy of British Guiana', in K. Saunders (ed.), *Indentured Labour in the British Empire, 1834–1920*, London: Croom Helm.

Allen, R.B., 1983, 'Creoles, Indian Immigrants and the Restructuring of Society and Economy in Mauritius, 1767–1885', Ph.D., Illinois.

Bates, C. and Carter, M., 1987, 'Tribal Migration in India and Beyond', conference paper, Princeton, and in G. Prakash (ed.), *The World of Rural Labourers*, Oxford University Press Delhi: (forthcoming).

Carter, M., 1987, 'Indian Labour Migration and the Indenture Experience in Mauritius', Oxford, D. Phil.

Carter, M., 1991a 'The Family under Indenture: A Mauritian Case Study', *Journal of Mauritian Studies*, Vol.4, No.1.

Carter, M., 1991b, 'The Effects of Immigrant Labour on a Local Workforce: Mauritian Ex-Apprentices, 1834–1861', conference paper, University of London.

Emmer, P.C., 1985, 'The Great Escape: The Migration of Female Indentured Servants from British India to Surinam 1873–1916', in D. Richardson, *Abolition and its Aftermath: The Historical Context, 1790–1916*, London: Frank Cass.

Emmer, P.L., 1986, 'The Meek Hindu: The Recruitment of Indian Labourers for Service Overseas, 1870–1916', in Emmer (ed.), *Colonialism and Migration: Indentured Labour Before and After Slavery*, Dordrecht: Martinus Nijhoff.

Jain, R.K., 1985, 'Multi-Ethnicity and Overseas Migration: The Case of Tamils in Malaysia', A. Pant and S. Gupta (eds.), *Multi-Ethnicity and National Integration*, Allahabad.

Jain, R.K., 1987, 'Overseas Indians in Malaysia and the Caribbean: Comparative Notes', conference paper, Oxford.

Lal, B.V. 1980, 'Approaches to the Study of Indian Indentured Emigration with Special Reference to Fiji', *Journal of Pacific History*, Vol.15, pp.52–70.

Mangru, B. 1987, *Benevolent Neutrality: Indian Government Policy and Labour Migration to British Guiana 1854–1884*, Hertford, Hansib.

Marks, S. and P. Richardson, 1984, *International Labour Migration: Historical Perspectives*, London: Temple Smith.

Miege, J.L., 1982, 'L'Indenture Labour dans L'Ocean Indien et le Cas Particulier de l'Ile Maurice', Conference paper Leiden.

Mookherji, S.B., 1959, 'Indians in Mauritius', *India Quarterly*, 15, New Delhi.

North-Coombes, M.D., 1984, 'From Slavery to Indenture: Forced Labour in the Political Economy of Mauritius, 1834–1867', in K. Saunders (ed.), op. cit.

Rodney, W. 1981 'Guyana: The Making of the Labour Force', *Race and Class*, Vol.XXII, No.4.

Sain P.B., 1983, 'The Problems of Indian Immigrants to Mauritius, 1834–1878', *Journal of Indian History*, Vol.LXI.

Tinker, H., 1974, *A New System of Slavery*, Oxford: Oxford University Press.

Weber, J. 1987, 'Les Etablissements Français en Inde au XIXe Siècle, 1816–1914', doctorat d'etat, Aix.

'Coolie' as a Labour Commodity: Race, Sex, and European Dignity in Colonial Fiji

JOHN D. KELLY

APPROACHING COLONIAL CAPITALISM

I see this study first and foremost as a contribution to efforts to understand colonial capitalism. Let us begin with questions about capitalism raised in the important, if polemical, recent exchange between Michael Taussig, Eric Wolf and Sidney Mintz. According to Taussig [1989], Mintz and Wolf neglect the fetishised character of capitalist commodities and thereby naturalise commodity metaphysics in their recent studies of world capitalism. In reply, Mintz and Wolf [1989] criticised Taussig's interpretive style, and defended their efforts to study 'the growth of a world in which the relationships between producers and consumers come to be analysable through commodities' [1989: 28] [cf. *Taussig*, 1980; 1987; *Wolf*, 1982, *Mintz*, 1985]. I would like to take both sides seriously. Taussig is right that 'critique cast in conventional terms sustains convention' [p.15]; Mintz and Wolf are surely wrong that their type of study can be separated from questions about commodity fetishism [*Mintz and Wolf*, 1989: 28]. But Wolf and Mintz are right to seek an understanding of large systems of relationships and their histories, and are persuasive in their defense of the search for real causes and consequences. In study of capitalist plantations in colonial Asia I think we can meet the double burden these scholars impose, and reach a clearer understanding of the logic and history of a major system of commodity production, without losing sight of the culturally-constituted character of its forms like 'commodity', 'labour', 'coolie' and 'race'.

John D. Kelly, Department of Anthropology, Princeton University, Princeton, NJ 08544–1011, USA. This article is based on field and archival research conducted in the Fiji Islands in 1984–85. The author thanks the Fiji government for permission to conduct his research and the Fiji National Archives (Setariki T. Tuinaceva, archivist) for permission to consult relevant documents. His understanding of colonial Fiji has benefited greatly from discussions of history with many Fiji Indians, whom, though he does not name them, he sincerely thanks. For their particular insights into colonial European society in Fiji he especially thanks Margaret Patel and Fergus Clunie. This study also benefited from discussion at the workshop on 'Capitalist Plantations in Colonial Asia' at CASA in Amsterdam, 26–29 September 1990. Thanks are owed to all participants, and to the editors of this collection. Special thanks are also owed to Martha Kaplan.

What I do not want to do is use the categories of political-economic discourse to constitute the terrain in which we find colonial capitalism. After all, European colonial capitalism existed long before political-economic discourse did. If one version of the problem is whether every society is organised by markets and/or modes of production and consumption, the problem can also be posed in Weberian terms, as a question about 'rationality'. Weber was deeply ambivalent about capitalism, about whether it represented *the* rationalisation of *this* world, or whether it was one among plural systems rational in relation to their own particular ends. I will follow the line in Weberian thought which connects the idea of rationality to the commitments and entailments of particular social positions.

In this approach the question of whether something is rational or irrational can never be a question about things in themselves, and can be replaced by the search for the commitments that *make* something rational for a particular, informed, and interested agent. The discovery of apparent irrationality can then be the device for launching such a search, and unearthing presuppositions about forms like 'commodity' or 'coolie', and the historical relationships that enact such conceptions and commit agents to versions of their meaning. Let us interrogate the organising categories of capitalist plantations in colonial Asia. We could try to ask whether categories like 'commodity', 'labour', 'coolie' and 'race' are rational or natural, just as an historian of medieval Europe can ask whether leeches are rational or natural as medical means. But we might learn much more if we ask about the vantage point that made the categories and practices seem rational and natural, and about the causes and consequences of their use.

In this article, I will use apparent irrationalities to open and reopen discussion, irrationalities arising from the premise that indenture was an economic system organised for maximum efficiency in commodity production. From Fiji's extraordinary penal system will come discussion of indenture as a labour form, the 'coolie' in colonial imagination, and contradictions in law. From creative use of shit will follow discussion of commitments in British colonial culture to organising categories of 'race' and 'prestige'. What shaped the history of these plantations? What commitments led to what contradictions?

This way of shaping our inquiry runs the risk of logocentrism and reification of a delineated colonial culture, but it has the advantage of avoiding prior commitment, on our part, to a political-economic vision of the terrain. The goal is to track not simply points of view but also their histories in efforts to do things. The search for the logic in a fragment of colonial capitalism will lead not to neat logical systems but to extremely messy ones, to contradictions between interests, and even to anomalies, blind spots and crisis points from contradictions *within* particular interests.

THE CONSTITUTING OF COLONIAL FIJI

Fiji is an island group in the South Pacific. 'We have the honour of being', wrote J.B. Thurston, 'the most outer or remote British settlement in the world' (quoted in Young [1984: 25]). Thurston was one of the most influential European settlers, and later a Governor of Fiji. Fiji's colonial government, a late-comer to Empire, was like many others created primarily to govern and protect European settlers, markets, and interests. European settlers first came to the Fiji Islands in significant numbers during the 1860s in a cotton boom provoked by the American civil war. In 1874, a gathering of indigenous Fijian leaders ceded sovereignty to the British crown, by signing their names and marks to a Deed Of Cession. But the problem remained how to govern the indigenous Fijians; also, there remained the other typical colonial problem, how to pay for the colonial government itself without wasting any metropolitan resources.

Scholars [*Legge*, 1958; *Gillion*, 1962; *France*, 1969; *Moynagh*, 1981; *Scarr*, 1984] debate the particular motives behind colonial Fiji's solutions: an extreme form of 'indirect rule', massive land reserves and paternalist protectionism for the indigenous Fijians, and the introduction of indentured labour from South Asia as the labour for a plantation economy. Sir Arthur Gordon, the first Governor, chose sugar as the principal export crop for the new economy, enticed the Australian-based Colonial Sugar Refining Company to establish large-scale milling and refining facilities (wiping out many local refining entrepreneurs), and began the immigration of Indians in 1879. The Indians called 'coolies' did the work on the sugar plantations, and lived in 'coolie lines' far from the indigenous Fijian villages. Thus the identities called 'racial' in colonial Fiji were constituted – ruling 'Europeans', savage, vulnerable 'Fijians', and labouring, coolie 'Indians' – making a world wherein, to this post-colonial day, indigenous Fijians own the land inalienably, Fiji Indians predominate in the sugar business and much of the rest of the economy, and race still vies with class as an organising principle.

Fiji did not invent the idea of Indian 'coolies' as a natural and perpetual working class. The system for recruiting wage-seeking rural Indians had long been in operation by the time the Fiji colonists began what they often called the 'importation' of indentured labour. But Fiji, developing its plantations so late in the history of indenture, was a major coolie destination in the last years of the system, and became perhaps its most notorious locale, as indenture critics and the indentured labourers themselves pondered the 'dirty trick' of the indenture contract. Marx suggests in *The German Ideology* that all workers in capitalism are confronted, sooner or later, with the dirty trick inherent in the transformation of living labour into growing capital, 'more and more enslaved under a power alien to them' that is, 'in the last instance', the world market [*Marx*, 1978: 163]. But in the first instance, at least, differences in forms of labour commoditisation matter. The disillusionment of wage-seekers finding themselves indentured overseas would suggest that some capitalist commoditisations of labour are both dirtier and less mystifying than others.

LABOUR COMMODITIES AND CITIZENS OF EMPIRE

Let us enter the world of the Fiji plantations first by way of the courtroom. Large numbers of complaints about direct violence against the indentured Indians found their way to the Fiji government's Inspectors of Immigration, and to the press in India; in overwhelming numbers, they can be found in Fiji's court records. But the labour discipline of indenture did not depend primarily on the frequent slaps and kicks. Indenture relied on penal sanctions. In one early year, 1886, a total of 8,853 charges of labour law violations were filed against a total Fiji indentured population of 5,237: the number of prosecutions was 169 per cent of the number of labourers under indenture. Generally speaking, from 15 per cent to 40 per cent of the indentured labourers were convicted of at least one violation of labour laws each year: for absence from the plantations without permission, for absence from work, for refusal to work, for failure to complete tasks [*Tinker*, 1974: 194–5]. Only in the last, reform conscious years of the system did the rates drop appreciably, to under ten per cent. The penalties for 'labour law violations' included fines, jail sentences, and extensions of the term of the indenture contract. Until the late reforms of the system, an Indian charged and convicted of failing to complete a task could be, simultaneously, paid nothing for what he did do, jailed, and forced to accept an extension covering the time in jail.

In 1892, when 40 per cent of the 'coolies' were convicted of labour law violations, only one employer charged by an Indian was convicted [*Tinker*, 1974: 194]. Even in 1919, the last year of indenture, when five per cent of the labourers were tried and convicted, only three employers were convicted of violations of the rights of the labourers (Annual Report of Immigration Department, 1919). Clearly, these courts worked against the Indians. But now we come to the puzzle, an anomaly. On the one hand, as Tinker [1974: 223] sums it up, 'To the Indians, the magistrates and the Protector [in Fiji, the Agent-General of Immigration] stood as guardians against the unbridled power of the planters'. On the other hand, the European establishment was not always happy about Justice 'as she was extended to the coolie' (G.L. Barrow's phrase, discussed below). On 15 October 1912, Fiji's principal newspaper the *Fiji Times* condemned use of the death penalty to punish Indians for murdering other Indians, 'simply and solely on the ground of the insane wastefulness of it all ... nor can we conceive of any human being', the *Times* argued, who would kill a vicious horse, or destroy a bicycle arguing 'an eye for an eye'.

> Why? Because a bicycle that can go is worth money, and a heap of twisted metal is not! But if a man – a useful worker, worth let us say 365 pounds in the year – in a fit of passion kills another man, his fellow men ... solemnly murder him to vindicate that very incompletely defined diathesis called Justice.

Superficially, the *Fiji Times* was speaking of all people. But in Fiji some people were more quantified in value than others: each coolie ship delivered 'labour units' contracted for at a specified price. This very capitalist form of labour ownership was also very peculiar, suspended uneasily between slavery and wage-labour, and it led to great tensions in the operation of law.

For the years of indenture, the 'master' owned the labour of the labourer as if he or she was a slave – controlling his or her time and movements – but could not as completely control the penal system. Ironically this labour, owned more totally than other forms of contracted labour, entailed a new sort of risk, a threat to profit. The 'coolies' owned (temporarily) and invested in (significantly) by the plantation owners could be lost (even permanently) following their criminal convictions, in procedures which treated them as people, not property. In 1913 the *Fiji Times* commiserated with a plantation owner whose plantation had been the scene of a series of violent crimes, including murder and rape. 'Mr. Harper will lose the services of no less than eleven units of labour for several weeks. He will be in a still worse position if the accused are convicted.' Indian accusations in court of Harper's own participation in the crimes were simply ignored.

Different tensions arose in other areas of law. After the expiration of the indenture contract an Indian was called 'free', a telling term. The first regulations concerning 'free' urban Indians were designed to control them and to restrict their freedoms – a curfew, a prohibition on selling

Indians alcohol, a restriction of municipal voting franchise to English-speaking and literate voters. These laws are interesting not simply in the banal sense that they reflect an interest to control the Indians, but because special legislation was needed for control. Before the specific modifications, the 'free' Indians had rights legally equivalent to those of the Europeans. This legal situation was the basis of what the Fiji Europeans called the 'Indian problem', a threat of Eastern pollution of the young and vulnerable body politic of Fiji. Let us reconsider how this formal equality for the Indians came to be, how the Indians came to be free, and unfree.

Fiji's version of indirect rule put significant restrictions on the indigenous Fijians. Indigenous chiefs worked with colonial officials to codify 'customary laws' that required obedience to indigenous chiefs and labour on behalf of chiefs, villages and kin groups. To the British, elaboration of chiefly authority was part of their civilising project, and the restrictions on Fijian 'commoners' (as they came to be called) were entirely for their own good.[1] In contrast the Indians were not socially bracketed by any similar form of 'indirect rule' or 'customary law'. The 'coolies' were thought to be without leaders or social organisation of their own. Apart from a few married couples and families, they were recruited as individuals, and not in their home village or within kin and caste networks of the recruiters, but generally in bazaar towns and pilgrimage centers. This individual recruiting was not a universal feature of indentured labour emigration from India in the nineteenth century, but it was basic to Indian social identity in Fiji.[2] The Fiji Indians were taken to be individual, mobile and socially unattached, in the modes of recruitment, in the legal terms of the contract, in the definition of 'labour units', and in the arrangement of authority in the lines. In India the indenture contract was secured when the recruits appeared one by one before a magistrate who verified that they had contracted for indenture and understood its terms. Indian critics of indenture emphasised the inadequacy of these interviews. Sanadhya and Chaturvedi[3] wrote: '... recruiters fool them and take them to the magistrate. The magistrate asks, "Are you agreed to go to Fiji?" As soon as the word "yes" leaves the lips, the registry is done. What would be the registry? From saying only "yes", there was five years of black water!' As lines of new 'coolies' were processed by the magistrate (165 in twenty minutes, Sanadhya and Chaturvedi claim), a social transformation was wrought in a fleeting moment. However, as a ritual reconstructing the Indians' social and political identity, the formality of this interview and the theory of contract it reified were far from inconsequential. It established the Fiji Indians, in the course of binding them temporarily to indenture, as British subjects otherwise adult, competent and 'free', capable of making contracts. It committed British courts to finding their rights and responsibilities in law thereafter. The Indians called the contract *girmit*, from the English word 'agreement', and called themselves *girmitiyas*, the *girmit* people. They understood that the contract had fundamentally changed them.

What then had these 'free' individuals contracted for in indenture, and how was it to be enforced? They were to be 'labour units', and were to receive wages, but were not to be 'wage labourers' in the classical political-economic sense. They were not paid by the hour but rather by the task. Six times a week, an overseer or *sardar* (work boss) defined for them a specific work allotment. If they were to be paid for that day's work, they had to complete it. The system located considerable power in the authority to define the task, power that could be abused both to reward favorites, such as mistresses, and punish 'troublemakers'. By most accounts [e.g., *Sanadhya*, 1914; *Gill*, 1970; *Burton*, 1910; 1912] both types of abuse were common. But let us not lose sight of the labour form itself.

E.P. Thompson [1967] argued that the introduction of work-time, clocks in factories, and wage-labour geared to work hours, served to discipline labour. Following Marx on the advantage of buying labour power rather than its product, he argued that such temporal reorganisation of labour was important to the rise of 'industry' and new levels of productivity. Why then was 'coolie' labour not organised by work-time, rather than by task, a hold-over from slave-labour systems? There was also an objection to tasking and slave-labour in classical economics: it was 'not economic in its exploitation of labour', even when working slaves to death, because it gave its workers no incentive to raise production, and set tasks below the capacities of the best workers. Tinker [1974: 8] quotes the famous view of Adam Smith: 'The work done by slaves, though it appears to cost only their maintenance, is in the end the dearest of any. A person who can acquire no property can have no other interest than to rest as much, and to labour as little as possible.'

As Thompson emphasised, the issue for the capitalists was work discipline: in England itself it was 'not until the second half of the eighteenth century that "normal" capitalist wage incentives begin to become widely effective' [*Thompson*, 1967: 81]. For Fiji, in the late nineteenth century, these 'normal capitalist wage incentives' were not the chosen form. In the Immigration Ordinance regulating importation of indentured labourers to Fiji there was provision for either time work (based on a nine-hour day) or task work (in theory, the amount 'an able-bodied male adult immigrant could do in six hours' steady work', with the women's tasks three fourths of men's [*Gillion*, 1962: 82]). For the first month of indentured Indian labour in Fiji, the Colonial Sugar Refining Company managed its labourers on a time-work basis, but then shifted to tasking. In response to a query from India regarding the 'tasking' system and low wages resulting, Governor Thurston explained in 1884 that when the Fiji coolies were paid for time-reckoned work days, they 'would do comparatively and sometimes literally no work' (quoted in Tinker [1974: 184]; see also Gillion [1962: 82]).

One might speculate that the Fiji 'coolies' did not perform well as wage labourers because they had not yet learned that 'time is money', that they still had what E.P. Thompson would call a more 'natural' sense of time, better suited to task than time definitions of labour. I suggest, however, that the selection of tasking over time wages in Fiji's indenture system was not the consequence of the historical stage reached by Fiji's indentured Indians as labourers, but rather an outcome of the enacted vision of history in colonial capitalist practice. Rather than inscribing capitalist history in a political-economic terrain, in which labour has a history of transformations from natural rhythms into pure capitalist forms, let us consider colonial capitalism as a capitalism in which political-economic definitions of labour efficiency were considered and rejected. In Fiji, as in most of the indentured labour systems in Asia, the 'coolie lines' were built from scratch. In the late nineteenth century, well into the industrial revolution and use of time wages at home in Europe, and more than fifty years after the abolition of slave trade, colonial Europeans all across Asia, in locales wherein slave plantations had never defined the genre, put 'coolies' into lines, denied them control of their persons, made them servants to planter masters. The attempt of a new colonial government in Fiji to make the colony pay for itself was framed within a system of well-established colonial agencies, the management by white masters of racial inferiors.

Most of the Indians recruited for indenture were looking for work, for wages, when recruited.[4] Nevertheless, the 'coolie' in the colonial imagination was different from a 'labourer' in political-economic theory. A 'coolie' was a different type of human being than a 'European'. The 'coolie' was suited to a particular type of labour and life by race: by physical, mental and emotional characteristics, characteristics determined (some thought) by the action of the tropical environment over vast time. Thus his identity and substance as an individual were not shaped by his capitalist transactions – he did not relate to his employers primarily as a seller of labour, relating to a buyer of labour. Rather, his suitability for tropical labour was determined by his racial nature.

In quite specific ways the indenture contract, while a contract, was informed by this colonial imagination. The flat wage rate did not try to make increased wages a work incentive. The 'coolies' were not seen as people working for wages, but instead as workers by nature, but workers requiring discipline. Thus, even when experiments were tried with time wages, it was in a context of plantation labour organisation. The operating premise was that the key to labour discipline was external constraints and enforcements, not incentives for internalisation. The European settlers and CSR, the dominant refining company, were always against 'coolie' education.

We return then to the penal discipline system, its logic and its contradictions. On the surface, it operated with great efficiency to meet the needs of the employers. The rules of evidence for labour ordinance cases in magistrate's courts were extremely lax. Routinely, the assertions of the overseers were sufficient evidence for conviction, and conviction rates in typical years were 90 per cent and higher. Not only was this a focus of protest for the Indian critics led by Sanadhya and Chaturvedi, but it was lampooned, in public, by Europeans themselves. The following is an excerpt from a satire of official callousness in Fiji published in Australia by a long-time Fiji resident [*Barrow*, 1921: 4–5] in a call for reforms to benefit indigenous Fijians:

> Scene: Stipendiary Magistrate's House. Time: 6 A.M. S.M. in pyjamas, about to enter bath room.
> 'Good morning, Mr. Sweetener.' ...
> Mr. S.: 'I have a coolie fellow, Damnum Singh. He said something to me yesterday. I don't know what it was, but it must have been something bad.'
> S.M.: 'Impudent fellow! Where is he?'
> Mr. S.: 'I have brought him handcuffed to the Station. By the way, allow me to thank you for those handcuffs. Our overseers find them awfully useful.'
> S.M.: 'Don't mention it. Constable! Bring in Damnum Singh.'
> S.M. to Damnum Singh: 'What the — do you mean by speaking to Mr. S. yesterday?'
> Dam Singh: '*Ham bola* (I said) ——'
> S.M.: 'You would speak to me, would you, impudent fellow? Constable. Six months. Oh, no; he might have to go to Suva; three months.' ...
> Occasionally this gentleman sat upon the Bench.
> A coolie enters with policeman.
> 'What, you again, Rumpus Persad? Six months.'
> Constable: 'Beg pardon, Sir, he's not a prisoner; he's a witness for the police.'

> S.M.: 'Oh, very well, three months.'
> Policeman: 'He's a witness, Sir.'
> S.M.: 'I don't care if he is a jury. Planting season is almost over, and I want another man very badly.'
> Constable: 'Yes Sir. What shall I charge him with, Sir?'
> S.M.: 'Desertion.'
> Constable: 'But he's a free man, Sir.'
> S.M.: 'Oh, well – contempt of Court.'
> Constable: 'Yes, Sir.'
> S.M.: 'Ah! I see Nastyman Das, Lippit Singh, and Nuisance Maharaj. I give you each six months.'
> Inspector: 'But, Sir, they are charged with not completing their task. You cannot send them to gaol for a first offence.'
> S.M.: 'Who says I can't? I have sentenced them to six months.'
> Inspector: 'Yes, Sir, but what about the evidence?'
> S.M.: 'I had all the evidence by 'phone early this morning. Mr. Phatdyner, the manager of the company, told me all about it.'
> Inspector: 'But these are not the men about whom Mr. Phatdyner spoke, Sir.'
> S.M.: 'Never mind. Three months each.'
> Inspector: 'What for, Sir?'
> S.M.: 'D— it, Sir. Am I to do all the work? They are guilty of something; take them away.'
> It sounds ridiculous, but that was justice as she was extended to the coolie by one S.M. The cases we instance are by no means exaggerated.

Even in satire, colonial premises are telling. The issue for Barrow was 'justice as she was extended to the coolie', properly and improperly, by European authorities. To understand the colonial social formation, I think we should examine this phrasing very closely.

We have already noted the costs to the planters in the way justice 'was extended to the coolie', for example the money lost to an employer when an indentured murderer was hanged. The *Fiji Times* was wrong to depict this as justice 'incompletely defined'; it was rather justice too completely defined to function smoothly for profits. But there were also areas where justice was incompletely defined in the form 'extended to the coolie', especially when it came to the rights of Indians as plaintiffs. On the one hand, the indenture system restricted quite ruthlessly the possibilities for social and political organisation among the Indians. Strikes by indentured labourers led to increases in labour ordinance penalties, and prohibition of complaints made by groups of more than five [*Gillion*, 1962: 83]. But if this channelled the Indians into the role of 'petitioners: and humble petitioners too' [*Tinker*, 1974: 227], the courts were not always comfortable when called upon to formally and publicly 'extend justice' to Indian petitioners, or even Indian victims. When one of the few European overseers charged with physical abuse of labour was found guilty for the sixth time, the Sub-Inspector of Police, who was prosecuting, said in his summation that 'he was compelled to ask His Worship to inflict a severe sentence', 'though it was very painful to be obliged to do so' (quoted in *Fiji Times*, 28 Sept. 1912; the overseer got two weeks with hard labour). In other cases, and especially when the crime was rape, evidence against overseers led more quietly to their transfer, in extreme cases to their transfer out of Fiji. J.W. Burton, a leading European critic of indenture, defends this practice and laments its necessity. First I quote one of his examples, then his depiction of the real problem:

> Din Muhammad was a man of good caste and education. He had a history in India, but it was more foolish than vicious. He was placed in charge of the women as *sardar* (head man) on a certain plantation under an European overseer. Din Muhammad was no saint; but he had spasms of ethics – and more than spasms of obstinacy. When his *sahib* made improper proposals to him regarding some of the better-looking women under his care, he became suddenly virtuous, and virtue mounted itself upon a mule-like intractability. He even preached a little on the subject, quoting from the Quran – of all books! The overseer was exasperated, and beat Din Muhammad so savagely that he had to be sent to the hospital. The coolie inspector heard of the case, and made some inquiries. The overseer sniffed trouble, and so laid a charge against the *sardar* of attacking him, paying four coolies two rupees each to swear that he himself struck only in self-defense. The four coolies told a consistent story – for two rupees is good money to pay in a country where a witness for any purpose can be obtained for a single rupee. Din Muhammad was sentenced to six months' imprisonment in the Suva jail. The coolie inspector, however, was not satisfied. He 'Sherlock Holmesed' the case, and found out that Din Muhammad had been ill-used, and that he had really sought to protect the women under him. The facts were laid before the Governor: Din Muhammad was released; and the overseer had to leave the country.
>
> [Again, in second example,] European prestige among these people suffered another loss.
>
> It is really only by prestige that the people can be kept in check; and when that goes, there goes all along with it all control. The

Indian can scarcely be frightened. When his temper is fully roused, no power on earth can hold him back.

To the European critics, grave risk lay in the mistreatment of Indians by Europeans, whether Europeans 'extending justice to the coolie' or just managing them day to day, *not* because Indians like Europeans would protest violation of their natural rights but because Indians, *unlike* Europeans, were violent, emotional, dangerous and animal-like.

In the late twentieth century, political-economic discourse has been so widely employed, making real a capitalism based on apparently 'free market' transactions, that the political-economic vision of human nature is hard even to question. Daily life, especially in elite circles, provides overwhelming evidence of human agency as intrinsically calculating, utilitarian and transactional; all seems reducible to Adam Smith's rubric of trucking and bartering, or what Bourdieu calls economics 'in the widest sense' [1977: 58, 183].

Marx was a major critic of this tendency in social theory, insisting that reflection began *post festum*, after the feast, and that we realise the historical specificity of social forms and logics first imagined to be natural, especially the commodity form itself [1976: 168]. But he himself fashioned a new political-economic vision, of a human nature made historically and socially through relations of production. My point here is that the colonial Europeans, in Fiji at least, did not imagine the Indians to be made or remade by market transactions or other political-economic processes. Their vision of Indians, on the basis of which they organised indenture, and 'extended justice to the coolies', was of a people racially and historically different, and inferior. The colonial self was not thought to share a human nature, an economic rationality, with the colonised, but was rather seen as the civilised intruder into wilder lands; justice was not the product of natural rights and a general will, but something to be extended in a civilising process, something extended to 'natives' and 'coolies' by their civilised superiors, their superiority marked by such designations as 'His Worship'.

Corrigan and Sayer [1985] detail the degree to which the English national past, and even the British present, are organised by a logic quite unlike that of the American Declaration of Independence or the French Declaration of the Rights of Man. Rather than the creation, from the bottom up, of a people for whom it is not the crown, but rather the market that comes before all, the British national polity is imagined as descending both in history and in structure. Its real crown was English civilisation, its 'society' and 'nation' at first designations of narrow elites, but then grown more inclusive in a series of reconsolidations. Thus the colonial imagination of rights and justice, depicting them as extended from an authority, rather than arising from a human nature, follows the lines of a British political imagination that considerably antedates the discourses on political economy. If political-economic revolutions at home could present the colonial world with a series of shocks, such as the abolition of slavery, the colonial world itself, like the British state, could compromise and reconsolidate. It could devise a form of contract, a form of market-place transaction, that could bind a dangerous, uncivilised natural labourer, into useful if temporary subservience. It could try to extend justice in ways that enforced the contract and controlled the danger.

The constant, through all the shifting situations, confrontations and contradictions that were labelled 'the Indian problem' in Fiji's colonial history, was the struggle of the Europeans to maintain their social height and contain the Indians' social place, against the implications of their commitment in the indenture contract to the Indians' formal and legal 'equality' and 'freedom'. But let us keep our eyes, a bit longer, on the plantations and indentured labour. We turn now to a particular, highly physical part of the struggle to maintain and abridge European dignity.

GENDER, SEX, VIOLENCE, AND EUROPEAN DIGNITY

Consider the following story, told by a surviving Fiji Indian indentured labourer in the 1970s (from Ali [1979: 55]):

> There was a European who came here with a son and had no wife. He played around with Indian women. There was an Indian who used to assist him and I scolded this man. He told the *sahib* about my remarks and I became a marked man. I was given a job, digging drains. I worked for five days finishing at 5 p.m. *Sahibs*, *sardars*, and the water-carriers all colluded against me. Besides the *sardar's* wife was the woman involved with the European overseer. At the end of the week I received only 3d in wages.... That evening my shipmates and I got hold of the *sardar* responsible and beat him up. Only my shipmates and I were involved, the other inhabitants of the line kept aloof. But a Muslim chap stealthily stole away and reported to the overseer that *sardar* Aiyappa was liable to be killed. The overseer loaded a gun and came down. The women of the lines, whom I called mother or sister and who treated me well, took up their hoes. He retreated, pleading to the women not to hit him, moving backwards he landed in a sewer pit. The women then threw shit on him.
>
> The overseer ran away. Afterwards we also released Armogam, another *sardar*, whom we had locked in his room so he would not

get involved. Later the manager asked Armogam what had happened, he stated that the European overseer had fired a shot and it had grazed the lips of one Murgan. The manager informed Farquhar [of CSR; cf. 'Phatdyner' in Barrows' court satire] and the overseer concerned was sent back to Sydney. This overseer, apart from his involvement with Indian women, was a good man. All this happened in 1916.

Fiji overseer Walter Gill [1970: 34–7] described work-gangs of women 'coolies' disciplining their European overseers by capturing them, beating them, immobilising them and urinating on them. He described an attempt to do this to him when he was overseer for a work-gang of women. At Holi, in particular, European overseers were at risk of polluting bombardment; Gill claimed that favored overseers were sometimes doused with perfume, instead. Finally, a fragment of another story was told to me by a descendent of Fiji 'European' settlers, who heard it from his grandmother: An overseer's horse rode up to an afternoon tea on an estate house veranda, the overseer tied naked to its back.

Gill wrote that 'Coolies had earthy ways, especially the women' [1970: 38]. I raise this set of examples to open a new dimension to explore in plantation social relations – relations not simply gendered but sexual, despite an extraordinary European inability to address the sexual aspects openly. I don't think we can understand plantation relationships of control, hierarchy, 'race' and discipline if we look only at the efforts to make money profits by extraction of surplus value in commodity production. The sexual dimension is also there. We will discuss sexual violence generally, and then come back to this effective use of shit to smear European reputations.

The rate of violent crime on Fiji's plantations was extraordinary. Even when the Indians were still a small minority of the total population of Fiji, cases of Indian murder, wounding with intent to cause grievous bodily harm, and attempted suicide dominated the docket of Fiji's Supreme Court. But efforts to explain the violence as a direct outcome of labour-relation antagonisms have been frustrated by the character of the violence. Studies of the violence in indenture Fiji have puzzled over the fact that most of the acts of violence were not conducted on lines of labour relations, but instead concerned what the Europeans called 'sexual jealousy'. Even when Indian *sardars* and European overseers were attacked, the motives were frequently related to sexual relationships; most commonly, the attacks were by one indentured labourer on another, violence directed by indentured men on indentured women, on male rivals, or on themselves.

I have written elsewhere [*Kelly*, 1989] about this violence, the contradictions engendering it in the social life of the coolie lines, and its role in the abolition of indenture. To summarise, we would obviously deceive ourselves if we accepted the colonial explanation, that South Asian men 'naturally' become violent when there is a 'scarcity' of women, as in the Fiji coolie lines where the gender ratio was five men to two women. Can we see the sexual violence of indenture as a function of the relations of commodity production? A connection can be made between the labour system and the sexual exploitation of women in Fiji, at least in a negative sense. The connection lies in the way this particular capitalist project reproduced its labour force. The European operators of indenture in Fiji had no real need for the 'free' Indians, except as extra hands on occasion. Nor, to 'reproduce' coolie labour, did they need an Indian marriage or family system. 'Coolie' labour lasted for five years and was replaced by further recruiting.

In the indentured-labour systems of colonial capitalist production, unlike even slavery systems, the social reproduction of labour had no relation to the social careers or sexual reproduction of the labourers. What happened to the 'coolies' before and after indenture, and on their own time when they were not working, mattered for the profits of indenture only in so far as it interfered in the productivity of the labour units. And this fact of economic efficiency should not lead us to presume that a commitment to seeking maximum money profits directed all the behaviour of the adventurous young white colonial men, mostly from Australia, who controlled Fiji's coolie lines, some of whom, wrote J.W. Burton [1910: 290], 'imagine that because a woman is brown she has, therefore, no rights of person; and ... profess incredulity if either a woman or her husband are above selling virtue'.

Outrage over the sexual abuse of Indian women in Fiji became the trump card of the anti-indenture campaign in India in the 1910s. But the Fiji Europeans literally did not see that it was happening. The rules for marriage and for sexual relations in the lines varied from planter to planter, and sometimes from overseer to overseer. Most respected the marriages recorded in indenture records, and had others officially recorded; some approved divorces and remarriages, sometimes led by the advice of the *sardars*. If the memoirs of Walter Gill are to be believed, many took one or more indentured Indian women as concubines (see Stoler [1989] on concubinage elsewhere in the colonial world). But despite European domination of the situation and its rules, the European observers of indenture in Fiji put the blame for the problems elsewhere: on the 'degraded' women themselves.

Even the leading European critics of indenture attributed the problem ultimately to the shameless immorality of Asian sirens. 'Morally, the Indian in Fiji is outside the decencies of description' wrote missionary Burton [1912: 116]. C.F. Andrews, Gandhi's associate, said in a speech to Fiji planters [1915: 7–8], 'The women of India are very chaste; but these women, well, you know how they are, and how can it be different, situated as they are, living the lives they do, brought up in this atmosphere of vice and degradation?' Andrews did not want to offend his audience; Burton wanted to portray the heathen as immoral; Gill wanted his memoirs to be spicy and entertaining. But their rhetorical figures are too consistent to be attributed all to such aspects of intention, genre or context. Whether the European author was an apologist or critic of indenture, church person, government official or overseer, the lines were depicted as the place of 'vice', the women first as degraded or immoral, then as undescribable, and finally, despair was expressed about the influence on children. Consider another slice of Andrews [1915: 7–8]: 'You know the state of immorality in the coolie "lines". I have seen it with my own eyes. I have watched their faces and have seen the children; I simply can't bear to think of children growing up in that atmosphere.'

The rhetoric mixes insistence on direct, sensory evidence ('I have watched their faces') with refusal to be specific about any sexual practice or relationship. Of course, it shares this characteristic with western discourse about sexuality generally, as Michel Foucault [1980] has so convincingly demonstrated. But in context, a system of sadistic sexual exploitation was hidden, even from the consciousness of the perpetrators.[5] Apart from Burton's reference above (Sanadhya and Chaturvedi praised Burton's courage) Fiji Europeans rarely discussed their role in the 'immorality'. The occasions on which European men discussed their participation were specially situated: gossip, Gill's tell-all memoirs fifty years after the events, and moments of crisis, as when the constable investigating an attack on Gill presumed that Gill must have taken the attacking man's wife or daughter. Beyond this, sexual exploitation could not be seriously discussed, despite the increasing anxiety in the late years of indenture to specify the facts of plantation life with increasing exactitude. When the issue was labour conditions, required reporting led to ridiculously specific official labour statistics, totals and averages per plantation always calculated to two decimal places, even for the smallest estates and their three or five workers. But when press accounts in India provoked public protest and official concern about abusive sexual relations, official inquiries from the colonial government of India were turned aside by official Fiji, even though the fate of indenture was then being debated.

In short, the Fiji Europeans never really registered the fact that there was a lot of rape going on, as well as other forms of sexual coercion for which we do not have a serious vocabulary. 'Hers was a wide-eyed acceptance of the reason for her being there', Gill wrote of the Indian woman he called from the lines to his own room. 'I needed her body, so she was content' [1970: 80]. The way Fiji's form of capitalism reproduced labour was undoubtedly a necessary condition for this sexual exploitation, but I doubt that any economic rationality can explain the dynamics of the discourses on sexuality, or the sexual abuses themselves. In the hands of the Gandhian nationalists, the exploitation of women under indenture and the refusal of officials to admit to it became powerful examples of the corruption of the colonials generally. Gandhi called the anti-indenture campaign the first national *satyagraha*, insistence on the truth campaign. It was perhaps the easiest truth that the nationalists ever insisted upon, that indenture was not simply a labour system operating for the mutual benefit of workers and employers.

It is important that the Fiji Indians did respond to the conditions of the coolie lines with their own creative social commitments. Most found ways to establish households, new kin networks began, and a community emerged, committed to religious devotionalism, insistent on egalitarian social relations, and uncompromising in defense of women's virtue. But I want to return to the plantation and its courtrooms, in search of the practical commitments shaping this colonial capitalism. The courts were the locus of tension between the plantation realities and the colonial imagination, and the point of contact between European civil society and the world of the 'others' it created, exploited, and tried to keep separate. Court records are the one documentary source in Fiji's archives in which the intrusion of white overseers in the sexual and social lives of the indentured labourers is repeatedly and provocatively discussed – almost always by the Indian defendants and their witnesses. If we can judge by transcripts and judges' official notebooks, the court and its officers only occasionally and perfunctorily inquired into defendant's allegations, which did not absolve the crimes of violence in any case. News accounts rarely even mentioned defendant's allegations. But no doubt, to borrow from Burton above, 'European prestige among these people suffered another loss' each time the accusations were aired, and ignored, and no doubt this (in addition to the 'insane wastefulness of it all') was a principal reason for the Fiji European disquiet over the 'coolies' in court. Surely it was better for 'European prestige' to ship off problematic overseers than try them in public. But what was really at stake? I'd like to come to grips with Burton's racist claim, strangely powerful and peculiar, that 'It is really only by prestige that the people can be kept in check; and when that goes, there goes all along with it all control'.

In 'My Twenty-One Years in the Fiji Islands', a polemic published in India in 1914, Sanadhya and Chaturvedi describe a time when Sanadhya fought back after an overseer hit him. The way Sanadhya tells it, the event changed his relationship with the overseer permanently for the better. Sandhya wrestled the man down, out of sight from any other overseers or indentured labourers. The overseer was left with the choice of struggling on at risk of extending his defeat, perhaps being discovered by others in defeat, and taking a radically new tack. He gave up. Once Sanadhya let him off the ground, he told Sanadhya that Sanadhya had taught him a valuable lesson, and that they were both lucky. Of course, Sanadhya's victorious counter-attack was never discussed in a court of law. Similarly, as far as I know no indentured labourer in Fiji was ever charged in court with attack by urine or faeces. But for the victims such attacks seem generally to have forced transfer, if not exit from the colony. Why was shit, in particular, the weapon against rape? Precisely because the dignity of the overseer, his masculine, sexual, controlling persona, could not survive it.

To Burton 'European prestige' was the essential means to hold the Indians in check because the Indians feared nothing, and could not be reasoned with when their tempers were aroused. In the colonial imagination, intimidation was the only effective vehicle, because the coolies were creatures of their passions, violent, jealous, and vicious. The same excesses in temper and passion which made the men dangerous made the women exciting. Both were in the nature of the 'coolie', and doom awaited the man who went too far into their world – or better, who entered into it, and into his own passionate nature, without taking command of both. As Ashis Nandy [1983] and others have emphasised, the European, especially British coloniser was obsessively masculine. The prestige of the colonial European and his power lay precisely in his civilisation, his achieved control over the heart of darkness both in himself and the world.

We are sometimes charmed by Kipling-esque colonial accounts of calm self-possession in the face of danger. But let us think again about the terms of colonial accounts of their fear of coolie violence, their concern for their own prestige, their calm self-possession in the face of danger. If we accept it as the natural attitude of the person in the jungle to fear the heart of darkness, we might not ask what he or she is doing in the jungle in the first place. It is the colonial project to seek to conquer, possess and control, again and again, inside and outside, against the awesome powers of climate and terrain, and of gender and 'race'. Colonial capitalism was no accident, and it was not guided by an urban industrial prototype, or a political-economic vision of markets.

QUESTIONS: RACE, CLASS, POLITICAL ECONOMY AND COLONIAL CAPITALISM

Most analyses of the history of European capitalism regard the colonial world with the social historical condescension that suffused colonial history itself. The colonial world was often depicted as secondary and dependent, an extension of a capitalism developed first in Europe, then imposed upon the rest of the world. Recent attempts to write better histories of world capitalism have found space for colonial capitalism within a political-economic world history. For example Wolf [1982] begins from the premises that all societies are part of 'a manifold, a totality of interconnected processes' [p.3], and that to describe this manifold we need 'a new, historically oriented political economy' [p.ix]. Must we universalise 'political economy' in order to build an understanding of world capitalism? If we consider the possibility that 'political economy' is not only a reality, but also a discourse [*Tribe*, 1978], then aren't there dangers in extending over the world the categories of a 'political economy' that developed in and for industrialising Europe? As Mintz [1985: 58] and others point out, a colonial form of industry, slave-based plantations, preceded the industrial revolution – and its political-economic discourse – in Europe. How then should we approach the position of race in the organisation of colonial, and especially plantation capitalism? Must we read race, always, as the surface appearance for an underlying class struggle? Or should we listen to Frantz Fanon's insistence that race preceded class in the organisation of the colonial world, that to understand colonial realities 'Everything ... so well explained by Marx, must here be thought out again' [1963: 40]?

Wolf [1982] has demonstrated that world markets can integrate several simultaneous modes of production. But we don't need to follow his map of the modes of production. The Europe of the nineteenth and twentieth century was transformed by the revolutionary development of a new industrial division of labour, based on the factory and the time-wage-labour form, and understood through the new science, 'political economy'. But for much of its materials it depended upon its colonial world, still organised by 'race', and forms of labour commodity and labour discipline that did less to hide the social inequalities being remade. The colonial world did not simply embrace and deploy political-economic discourse. It did not simply create replicas of the labour forms, relations and markets of Europe. At the abolition of slavery, colonial Europeans devised indentured labour systems to sustain forms of plantation order in the Americas, and to create vast new regimes of plantation production across Asia. Indenture was a compromise, a colonial race relation reaffirmed by an economic contract; the plantation industrial labourer continued to be a racially-defined commodity, and only fleetingly appeared as a marketplace exchanger of labour power. For a century, this labour commodity made possible a new version of the colonial, plantation production system with the booming, industrialising Europe as a ready market for products. The thing collapsed – when, where, and to the extent that it has – because the exploitation intrinsic to its labour form was not so well disguised, and the 'others' figured out what was going on.

NOTES

1. On indigenous Fijian chiefship and its transformations in the colonial period see France [1969], Clammer [1973], but also Hocart [1969], Sahlins [1985], and most recently Kaplan [1989a; 1989b; 1990]. I have written at greater length about official Fiji's resistance to codifying a customary law for Indians in Fiji; see Kelly [1989], and Kelly [1991].
2. For general studies of Indian indentured labour emigration see Kondapi [1951], Tinker [1974] and Breman [1990]. For comparison of recruiting systems see also the articles in this collection by Carter, Behal and Mohapatra, and Das Gupta. Fiji's resistance to using or allowing existing forms of social authority lay at one extreme of the indentured-labour plantation systems; the other extreme was the *kangani* systems. Labourers for Ceylon, for example, were recruited by *kangani* or 'jobbers', the men, often kin or caste-mates, who became their *sardars* (work bosses) in Ceylon. The 'jobber' was responsible for his recruits, and could use the threat of separation from this labour group, especially forced return to India, as a means of labour discipline. In contrast, the indenture recruiters called *arkati* dumped their recruits for bounties in socially-mixed urban depots, from where the new coolies were shipped to Fiji (or the Caribbean etc.). On arrival they were assigned in lots of any size as 'labour units' to the planters and others contracting for that number. According to indenture critics Sanadhya and Chaturvedi, overseers on the Fiji plantations went out of their way to deny Brahmans and other high caste 'coolies' special standing, and broke up groups that were organised. In Fiji, the *sardars* (work bosses) were often Muslims set to control Hindus, or other forms of outsider. The *sardars* were chosen, claimed an overseer [*Gill*, 1970] for their ability to understand and follow overseer orders, and for their 'reliability', the likelihood of their loyalty to overseers and not to coolies in a crisis. There was no disciplinary use of a threat of return to India. For that the Fiji labourers had to work years. Even after working the five years of their indenture contract the Fiji labourers had to pay their own return passage unless they remained in the colony for another five years. In the colonial imagination, they were to remain as productive workers employed by Europeans. At the end of indenture, their new 'free' status was signified by issuance of a 'Certificate of Industrial Residence'.
3. Totaram Sanadhya was a Brahman indentured labourer to Fiji who returned to India in the 1910s, and met Benarsidas Chaturvedi shortly thereafter. Chaturvedi turned Sanadhya's life story and materials from other sources into a widely-read Hindi polemic, *Fiji Men Mere Ikkis Varsh* [My Twenty-One Years in the Fiji Islands] 1914], that was translated into several Indian languages and widely used in the anti-indenture movement. Other influential criticisms of indenture in Fiji were in English, especially the reports of C.F. Andrews, friend of Gandhi and Tagore sent to investigate indenture in Fiji, and *The Fiji Of To-Day* [1910] by John Wear Burton, Methodist missionary to the Fiji Indians.
4. Many but by no means all. Judging by the available narrative evidence the women

recruits, in particular, seem often to have been fleeing something else – life as young widows, or cruel husband's households.

5. No one was conscious of a *systematic* sexual exploitation of Asian women by European men. Plantation folk were clearly aware of sexual relationships between overseers and coolie women, but this awareness did not permeate the urban 'civil society', judging by its public discussions of the dynamics of indenture violence. The image of these sexual relations also varied according to the class, occupation and especially gender of the Europeans. When Gill first approached 'his' Indian woman, she suggested that he leave her alone. 'I nearly agreed', Gill wrote [1970: 79], 'but then it seemed to me that if I did, I would be allowing the white women to make an unwarranted intrusion, interfering in a situation beyond their understanding'. Colonial white women are sometimes blamed for the institution of color bars. Stoler [1989] describes how their very arrival was concomitant with shifts in colonial policy from favor for native women concubinage to concern for racial purity. In Fiji at least what the increasing presence of 'the white women' constrained was nothing for us to mourn.

REFERENCES

Ali, Ahmed, 1979, *Girmit: The Indenture Experience in Fiji*, Suva: The Fiji Museum.

Ali, Ahmed, 1980, *Plantation to Politics: Studies on Fiji Indians*, Suva: University of the South Pacific and the Fiji Times and Herald Ltd.

Anderson, A.G., 1974, *Indo-Fijian Smallfarming: Profiles of a Peasantry*, Auckland: Auckland University Press.

Andrews, C.F., 1915, 'Report of Mr. C.F. Andrews' Speech to the Planters' Association Executive Committee, Fiji, Dec. 7, 1915', copy in Tippett Reference Collection, St. Mark's Library, Canberra, Australia.

Barrow, G.L., 1921, *Fiji for the Fijians: A Protest and a Plea*, Korolevu.

Bourdieu, Pierre, 1977, *Outline of a Theory of Practice*, Cambridge: University of Cambridge Press.

Breman, Jan, 1990, *Labour Migration and Rural Transformation in Colonial Asia*, Amsterdam: Free University Press/Center for Asian Studies.

Burton, John Wear, 1910, *The Fiji of To-Day*, London: Charles H. Kelly.

Burton, John Wear, 1912, *The Call of the Pacific*, London: Charles H. Kelly.

Clammer, John, 1973, 'Colonialism and the Perception of Tradition in Fiji', in *Anthropology and the Colonial Encounter* (ed. Talal Asad), London: Ithaca Press.

Corrigan, Philip, and Derek Sayer, 1985, *The Great Arch: English State Formation as Cultural Revolution*, Oxford: Basil Blackwell.

Fanon, Frantz, 1963, *The Wretched of the Earth*, New York: Grove Press.

Foucault, Michel, 1980, *The History of Sexuality, Vol.1: An Introduction*, New York: Vintage Books.

France, Peter, 1969, *The Charter of the Land*, Melbourne: Oxford University Press.

Gill, Walter, 1970, *Turn North-East at the Tombstone*, Adelaide: Rigby Ltd.

Gillion, K.L., 1962, *Fiji's Indian Migrants*, Melbourne: Oxford University Press.

Gillion, K.L., 1977, *The Fiji Indians: Challenge to European Dominance 1920–1946*, Canberra: Australian National University Press.

Hocart, Arthur M., 1969, *Kingship*, Oxford: Oxford University Press.

Kanwal, J.S., 1980, *A Hundred Years of Hindi in Fiji 1879–1979*, Suva: The Fiji Teachers' Union.

Kaplan, Martha, 1989a, '*Luve ni wai* as the British Saw It: Constructions of Custom and Disorder in Colonial Fiji', *Ethnohistory*, Vol.36, pp.349–71.

Kaplan, Martha, 1989b, 'The "Dangerous and Disaffected Native" in Fiji: British Colonial Constructions of the Tuka Movement', *Social Analysis*, Vol. 26, pp.22–45.

Kaplan, Martha, 1990, 'Meaning, Agency, and Colonial History: Navosavakadua and the *Tuka* Movement in Fiji' *American Ethnologist*, Vol.17, pp.1–20.

Kelly, John D., 1989, 'Fear of Culture: British Regulation of Indian Marriage in Post-Indenture Fiji', *Ethnohistory*, Vol.36, pp.372–391.

Kelly, John D., 1990, 'Discourse About Sexuality and the End of Indenture in Fiji: The Making of Counter-Hegemonic Discourse', *History and Anthropology*, Vol.51, pp.19–51.

Kelly, John D., 1991, *A Politics of Virtue: Hinduism, Sexuality, and Counter-Colonial Discourse in Fiji*, Chicago, IL: University of Chicago Press.

Knapman, Bruce, 1987, *Fiji's Economic History, 1874–1939*, Pacific Research Monograph 15. Canberra: Australian National University.

Kondapi, C., 1951, *Indians Overseas 1838–1949*, New Delhi: Indian Council of World Affairs and Oxford University Press.

Lal, Brij V., 1983, *Girmitiyas: The Origins of the Fiji Indians*, Canberra: The Journal of Pacific History.

Legge, J.D., 1958, *Britain in Fiji 1858–1880*, London: Macmillan.

Marx, Karl, 1976 [1867], *Capital*, Vol.1, New York: Vintage Books.

Marx, Karl, 1978 [1846], *The German Ideology*, in *The Marx-Engels Reader* (ed. Robert C. Tucker), New York: W.W. Norton.

Mayer, Adrian, 1973, *Peasants in the Pacific*, Berkeley, CA: University of California Press.

Mintz, Sidney W., 1985, *Sweetness and Power*, New York: Viking.

Mintz, Sidney W. and Eric R. Wolf, 1989, 'Reply to Taussig', *Critique of Anthropology* Vol.9, pp.25–31.

Moynagh, Michael, 1981, *Brown or White? A History of the Fiji Sugar Industry, 1873–1973*, Canberra: Australian National University.

Naidu, Vijay, 1980, *The Violence of Indenture in Fiji*, Suva: University of the South Pacific.

Nandy, Ashis, 1983, *The Intimate Enemy: Loss and Recovery of Self under Colonialism*, Delhi: Oxford University Press.

Prasad, Shiu, 1974, *Indian Indentured Workers in Fiji*, Suva.

Sahlins, Marshall, 1985, *Islands of History*, Chicago, IL: University of Chicago Press.

Sanadhya, Totaram, in press [1914], *My Twenty-One Years In The Fiji Islands*, (trans. Uttra Singh and John D. Kelly), Suva: Fiji Museum.

Scarr, Deryck, 1984, *Fiji: A Short History*, Sydney: George Allen & Unwin.

Sombart, Werner, 1915, *The Quintessence of Capitalism*, New York: E.P. Dutton & Co.

Stoler, Ann L., 1989, 'Making Empire Respectable: The Politics of Race and Sexual Morality in 20th-Century Colonial Cultures', *American Ethnologist*, Vol.16, pp.634–60.

Taussig, Michael T., 1980, *The Devil and Commodity Fetishism in South America*, Chapel Hill, IL: The University of North Carolina Press.

Taussig, Michael T., 1987, *Shamanism, Colonialism, and the Wild Man: A Study in Terror and Healing*, Chicago, IL: University of Chicago Press.

Taussig, Michael T., 1989, 'History as Commodity in Some Recent American (Anthropological) Literature', *Critique of Anthropology* Vol.9, No.7–23.

Tinker, Hugh, 1974, *A New System Of Slavery*, London: Oxford University Press.

Thompson, E.P., 1967, 'Time, Work Discipline, and Industrial Capitalism', *Past and Present*, No. 38, pp.56–97.

Tribe, Keith, 1978, *Land, Labour, and Economic Discourse*, London: Routledge & Kegan Paul.

Trouillot, Michel-Rolph, 1989, 'Discourses of Rule and the Acknowledgement of the Peasantry in Dominica, W.I. 1838–1928', *American Ethnologist*, Vol.16, pp.704–18.

Weber, Max, 1958, *The Protestant Ethic and the Spirit of Capitalism*, New York: Charles Scribner's Sons.

Williams, Eric, 1980, *Capitalism and Slavery*, New York: G.P. Putnam's Sons.

Wolf, Eric R., 1982, *Europe and the People Without History*, Berkeley, CA.: University of California Press.

Young, John, 1984, *Adventurous Spirits*, St. Lucia, Queensland: University of Queensland Press.

Conclusion: The Making of a Coolie

JAN BREMAN and E. VALENTINE DANIEL

INTRODUCTION

The identity of a labourer as a coolie, like the identity of anything or anybody, is complexly constituted and, however exhaustively described, will remain only partially defined. Coolie-identity is as much the product of self-perception as it is the construction of a category by those who did not belong to it. It is these constructions that historically constituted a coolie in the matrix of power relations which this essay seeks to partially comprehend. Mighty little is known, or rather has been written, about so-called coolies' perception of themselves. Unfortunately neither this essay nor the others in this volume, though pointing towards this lacuna, have managed to fill it. Our focus here is on the colonial literature and subsequent post-colonial writings which have helped us better understand how the category, 'coolie' was, and to a large extent still is constructed.

Constructed identities are neither unitary nor consistent over time and space. Even when it is possible to locate identities specifically in time and place, they are fraught with contradictions and ambiguities. If this be true in general, it is especially true of coolies. In a word, a coolie's identity is 'multiplex.' The task we have set for ourselves then, is not to arrive at a unitary – or even a composite – identity of those labourers who came to be known as coolies. Rather we wish to cast a selective spotlight on the historical landscape of Asia at those moments when capitalist and colonial mentalities contributed to the formation of such multiplex identities; moments and identities which have had a lasting and pervasive effect on the formation of the coolie as a category in Asia; moments and identities that reveal contradictions inherent in capitalism in Asia.

According to the *Oxford English Dictionary* [1971], the English word 'coolie' has two different sources: one Tamil, the other Gujarati.[1] In

---

The authors are respectively at the Centre for Asian Studies, Oude Hoogstraat 24, 1012 CE Amsterdam, Netherlands, and Department of Anthropology, University of Michigan, Ann Arbor, Michigan 48109–1382, USA. They would like to thank Henry Bernstein and Tom Brass for their detailed comments and useful suggestions on earlier drafts; responsibility for the final version of the essay printed here remains, of course, the authors'. Thanks are also due to the World Institute of Development Economics Research of the United Nations University, Helsinki, for subsidising part of the Amsterdam conference costs.

Tamil, *kūli* was, and still is, the term for a type of payment for menial work. Being a menial alone did not qualify a person to be called a 'coolie'; s/he had to be a menial who did not have any customary rights. A *paṇṇaiyāl*, for instance, was a menial – often an untouchable – for whose survival his or her patron bore responsibility. If 'responsibility' was the insignia on one side of this relationship, 'unfreedom' or even 'bondage, marked the other'.[2] But regardless of the physical form and substance of the payment a *paṇṇaiyāl* received (until recently, in kind) it was never called *kūli*. There was no one who accepted any responsibility for the survival of a menial who received *kūli*, and *kūli* never served as a token of such responsibility. *Kūli*, by its sheer paltriness, came to index the fact that the recipient belonged to 'the lowest level of the industrial labour market' [*Tinker*, 1974: 43].

But the English 'coolie' signifies more. Its ancestry on the Gujarati side can be traced to the name *Kuli*, which refers to a *person*, to one who belongs to the Kuli tribe. Its members were described as 'thieves, robbers and plunderers, as degenerate and inferior,' in short, 'as a "villainous race"' [*Potts*, 1990: 63]. Indeed these characteristics were attributed to all coolies at one time or another, even though there were hardly any Kulis among them. The semantic and pragmatic point we wish to abstract from this etymological excursus is that, regardless of his or her purported degradation, the Kuli, by virtue of his or her membership in a tribe, is seen as a person, a human being. The English word 'coolie' combines the 'personhood' of a Kuli, with many of its connotations, and the payment of *kūli*, with all of its connotations, to create an essentially new entity: a 'coolie'. In Tamil, it would be ungrammatical to call a person a *kūli*. That requires the addition of the suffix *-al*, making him or her a *kūliāl*, that is, a '*kūli*-person'. In the transformation of *kūli* to 'coolie', the distinct humanity of the individual was, in a single move, appropriated and eliminated; the person collapsed into the payment. The relations of production in which the coolie was deployed were in no way antagonistic to the formation and sustaining of such an identity. In fact they reinforced it.

The first recorded mention of coolies is to be found in a letter written by a Fr. Luis Frois from Japan in 1581, 'regarding a General "arriving from Ishikawa accompanied by 10,000 men and 10,000 coolies"' [*Tinker*, 1974: 42]. What was the intended distinction in this quote? Was it 'men' versus their officers, or 'men' versus coolies (who were less than men, beasts of burden trailing and servicing the soldiery)? If the categorisation of the coolies as *Untermensch* is ambiguous here, it becomes less so in the subsequent treatment of coolies by their employers (see also Breman [1987] and Kelly [1991]).

When capitalism penetrated into the world of Asia, the coolie-to-be was not so much a proletarian who stood ready to be transported to the rural and urban enclaves created by the new economy, but one who was available and vulnerable for recruitment by virtue of being cast adrift. The advent of *kūli* in the South Indian cultural and historical context formalised the severance of a bond on the one hand (regardless of how such a bond is understood), and on the other, it initiated the person who had been thereby cast adrift into a category of proto-proletarian individuals devoid of their personhood.

Being cast adrift like so much 'flotsam and jetsam', – an apt image indeed – implied two further things about the coolie in the growing capitalist economy and the images it generated. The coolie was perceived at one and the same time as both immobile and unstable.[3] These contradictory conceptual nodes served as the two points of the compass, as it were, that drew the contours of the identity of the coolie as an historical category in the literature of the nineteenth and twentieth centuries. At times these points coincided; at others, they were widely separated. The axis of articulation was servile-like submission. These same two nodes, and a critical evaluation thereof, provide the parameters of this essay.

I. THE COOLIE AS *HOMO RUSTICUS*

The discursive background against which colonial officials and European employers alike constructed 'their' coolie was already in place. Indeed, throughout much of the nineteenth century the Indian village was even taken as typical of both [medieval Western] Europe and Asia.[4] The Indian, and by extension the Asian, hinterland was imagined to be a domain populated by an undifferentiated mass of self-employed cultivators engaged in a subsistence economy. According to this notion the relations of production were village based and the social horizon of its inhabitants more or less coincided with the boundaries of the locality. The members of this endlessly self-reproducing timeless village community had, from generation to generation, neither reason nor occasion to move outside their domestic habitat. These villages were self-sustained units and its inhabitants inwardly turned peasants. To the early colonialists, peasants referred to self-employed cultivators working their family farm which produced their subsistence needs. The ascribed rootedness of these peasants was expressed in a proverbial attachment to the native soil, for example:

> *British India*: Centuries of isolated living in the traditional and familiar environment of virtually one and the same village seem to have endowed the Indian peasant with more than the usual reluctance to move associated with his kindred [*Sandhu*, 1969: 57];

> *French Indo-China*: It is the very structure and basic meaning of Annamite society which keeps the native at home and stifles the spirit of adventure. In general this is true of all the people of the Far East [*Robequain*, 1944: 62];
>
> *Netherlands-Indies*: ... migration on the whole is a capitalistic phenomenon and does not fit into the character or need of the precapitalistic agricultural societies. The traditional ties that bind the villager to his community and to his soil holds him back – if not altogether from migrating at least from permanent settlement elsewhere [*Boeke*, 1948: 63].

Such statements sought the cause of inertia in the mentality of an Asian peasantry entrapped in a closed agrarian setting organised along corporate lines. The modernisation process which would open up the countryside and widen the sphere of social interaction had yet to start. Until then migration was to remain a rare phenomenon [*Breman*, 1988]. Even those authors, critical of the cherished colonial image of a uniform and homogeneous peasant community have, for other reasons, adhered to the idea that an essential immobility characterised the inhabitants of the Asian countryside. An example of this school of thought is Standing's view of British India as a feudal-type society in which moneylenders lured peasants into debt bondage. Relationships of dependency not only acted as a structural barrier to migration but also gave rise to a mentality which discouraged people from seeking a better life elsewhere. '[I]n circumstances of poverty and indebtedness', Standing [1979: 15] writes, 'a deeply ingrained sense of fatalism was likely to preserve a set of attitudes conducive to resignation and immobility'.

The idea that men and women of rural Asia were basically unable and/or unwilling to move to new work places made it necessary for colonial capitalism to detach them from their home milieu by buying them out. That is, they could be made mobile only by tying them to a formal, state-mediated contract (with built-in penal sanctions) to their destined employer or, more informally, to a jobber (*arkati, kangani, mukadam, sardar, tandil, mandur,* etc.). The task was to mobilise an essentially immobile work force only to re-immobilise it by tying it to the enclaves of capitalist production. Indenture came to be one of the principal means, not only of coolie recruitment itself, but also of such re-immobilisation.

Colonial employers complained that workers who were bound by monetary advances failed to honour their contracts' stipulations. They further argued that given the lack of a contractual code of conduct, severe sanctions were necessary to compel the fulfillment of written agreements.

The coolies, on their part, had powerful stories to tell on how they were duped into a new kind of bondage by a 'piece of paper' which tore them loose from their native place [*Breman*, 1990: 35–47; *Kelly*, 1991: 27–9]. The essence of the recruitment procedure was to rely on traditional means of exerting pressure in combination with recourse to newer forms of legal coercion. In plantation regimes where workers had access to cultivable land on the estates' peripheries, many of them resorted to living the 'liminal' or 'double' lives of peasant-estate labourers, as 'peasants-manqué' [*Wolf*, 1959: 143; *Mintz*, 1959: 43; *Geertz*, 1968: 57, 58; *Mintz and Price*, 1973; *Taussig*, 1980: 92, 103, 113; *Cooper*, 1980: 80; *Stoler*, 1986]. In these contexts, the procedures of recruitment themselves may be broadly characterised as as a mixture of old and new practices. However, in those plantation regimes which approached the ideal of enclavism (an immigrant labour force with negligible or no access to land on which subsistence crops could be cultivated), the interpretation of these very same methods in terms of a mere 'mixture' of the old and new does insufficient justice to the social dynamics at work. If we consider the characteristics of these methods in terms of their effects rather than their origins, the organised control over labour instituted towards the end of the nineteenth century should be understood in terms of rupture rather than continuity. The rupture in question divided the proto-proletarian from the genuine proletarian. But as we shall see, the discursive forces unleashed made proletarian status itself a precarious one, relentlessly attempting to push its possessors back to a pseudo-proletarian state.

Over time, because of land pressure resulting from population growth (the standard argument of colonial historians) or due to progressive impoverishment in rural Asia (so well documented in the nationalist literature) the need to compel the rural masses into mobility gradually disappeared. Migrants from an agrarian background were reported to arrive at the gates of the mills, mines and plantations in increasing numbers from the beginning of this century onwards 'out of their own free will', that is, spontaneously and voluntarily, because they had lost their land.[5] In the wake of such a changing tide in the relationship between demand and supply of labour, and the consequent freeing of a bonded coolie force, its transformation into a true proletariat should have been the logical outcome. Is this indeed what happened? The answer has to wait until we have further scrutinised the coolie background.

II. THE COOLIE AS A PROTO-PROLETARIAN

The image in colonial literature of the coolie as a peasant is baffling, not because there were not any peasants among coolies, but because the sheer insistence of the equation of coolies as erstwhile peasants throws into clear relief the denial of the non-peasant origins of many coolies. To be sure, there were contingents of peasants among migrant labourers, in different proportions in different streams of migrants and at different times. For instance, nearly all who went to Bengal from Orissa in the early part of this century were young men of peasant castes, while those who went to Assam belonged to the landless castes and took their families with them [*Chaudhuri*, 1985: 201]. In Ceylon and Malaysia a higher number of peasants were to be found in later migrant streams than in earlier ones. But here as elsewhere, it would be a mistake to slap the general label 'of peasant origin' on these plantation workers. This can be seen in the following complaints, typical of employers or officials who inspected and registered the latest arrivals:

> *Mauritius*: ... out of 50 Indians employed (on a plantation), not more than 20 or 23 had even previously handled the implements of agriculture and this party was composed of a motley group of cooks, 'mussulches', sweepers, syces, etc. and every variety of trade except that for which they were particularly engaged (quoted in Saha [1970: 32]).
>
> *British Guiana*: Of the folk they interviewed only thirteen were agricultural labourers, who with one lime-burner, one cowherd, three peons and a sweeper made up the list of those accustomed to outdoor labour; the remaining fourteen were priests, weavers, scribes, shoemakers, beggars, and so forth.
>
> *Surinam*: The majority [of immigrants] formerly having been house servants, soldiers, policemen, barbers, shopkeepers, hawkers, etc. A few years later the Surinam arrivals included 'A batch of dancing girls and women of a similar description, with their male attendants. These people laughed at the idea of becoming agriculturalists' (quoted in Tinker [1974: 51–2]; see also Emmer [1984: 96]).

Nationalist writers in particular pointed out that a large part of the coolie mass consisted of de-classed artisans who had fallen victim to the economic policies of colonial regimes. The same held true for a wide variety of service workers who likewise became unemployed. Nevertheless, these writers felt constrained to persist in describing the coolie as a peasant. Thus they not only observed that unemployed service workers who were driven back to their villages (in so far as they did not live there already) were declassed artisans but added that they went back to live out their lives as peasants. Roy's comment is typical of this genre. Writing of the Indian city coolie in the beginning of the twentieth century he says: 'He did not come out of the ranks of the expropriated artisans; he came mostly from the peasantry. After having lost its (sic) trade, the artisan was pushed back to the land, where he had come to stay two or three generations before the call of modern industry brought him again to the city' [1922: 113]. There is no evidence that displaced workers, hit by de-industrialisation, were first repeasantised before being recruited as coolies. In fact, they were barred from acquiring land for many reasons. Furthermore, they were more likely to join the ranks of an underclass of proto-proletarians which was already in existence in 'peasant societies' of the *ancien régime* long before colonial policies had made their impact felt – a class from which coolies were recruited with relative ease. What should be remembered is that many migrants came from a non-agrarian background.

We agree with Chakravarty [1978: 268] who observed that 'a major component (of South Asian labourers) was not the uprooted peasantry'. As far back as the early and middle of the nineteenth century, there existed floating, 'untied' labour, vulnerable to agricultural slack periods and not merely passively predisposed to move but poised to do so.[6] Already in the early decades of the nineteenth century tribal migrants had left Chotanagpur in the winter months to seek employment in the rice harvest in Bengal, on public works in Calcutta, and in indigo processing plants in Bihar [*Mohapatra*, 1985: 248]. Java studies at the beginning of the nineteenth century called attention to the presence of 'vagrants' and 'tramps' as a significant social element. The Javanese nobility, when mobilising its forces at the start of the major revolt against colonial authority which became known as the Java War [1825–30], could count on the support of sundry 'bad men', a riff-raff without a fixed abode. These were mostly landless people not encumbered by families who were accustomed to wandering around the countryside:

> There is a certain class which is neither tied to the *desa* nor skilled in any craft: the so-called vagrants (*orang melantjong*) or, as they might be called Javanese gypsies, vagabonds who live from hand to mouth, the least esteemed class in Javanese society. The established *desa* inhabitants refused to give them a home; they shun regular employment and only necessity forced them to work for a few days; gambling, seductive *rongeng*, and venomous opium provide their greatest happiness. It is mainly to those classes that the European entrepreneur must turn for the labour he needs. Surely everyone must realise how irregular and uncertain their performance is (Millard [1868], quoted in Breman [1983: 20]; see also Carey [1986: 83]).

Such were those who were the first to be employed as coolies in the sugar mills which European entrepreneurs began to set up in the 1830s. It was from among these 'dregs' of the rural milieu that, as long ago as the mid-eighteenth century, seasonal workers were recruited in other parts of Java for the sugar mills near Batavia [*Levert*, 1934: 57–8].

In the course of the nineteenth century this proletariat *avant la lettre* underwent expansion, particularly in various regions of Southern Asia, as the victims of colonial de-industrialisation policies were forced to seek other means of employment. Coolies intended for distant locations were frequently recruited outside the villages: along the main roads, by river crossings, at markets, in places of pilgrimage, and in other places where people gathered for religious or agricultural festivals. Some of these vagrants drifted toward the harbours where they found themselves in the coolie depots, at points of departure for transport overseas. This fluid and heterogeneous nature of the labour reserve from which agents of capitalist entrepreneurs drew at will is best captured by Tinker:

> They often took the people who were already available in the ports of embarkation, Calcutta, Madras and Bombay. These were the flotsam of humanity, drawn to the big city by the prospect of employment which had vanished and left them stranded. Some were domestic servants, who had accompanied their European master to the city, and had been discharged when he was transferred elsewhere or went home to Europe on leave or retirement. They included cooks, footmen, washermen, grooms and coachmen. Then there were the entertainers who had fallen on bad times – dancers, musicians, prostitutes. The great majority, however, were simple country folk who had been attracted to the big city in search of casual work as burden carriers-coolies in the basic sense [*Tinker*, 1974: 51].

Reading against the grain of colonialist as well as nationalist writings we arrive at, what we see as a more complex, and we hope a more realistic, picture of the origins of these early coolies. We do not need the ideologically charged and pejorative epithets such as 'vagrants', 'dregs', and 'vagabonds', and inappropriate terms such as 'gypsies', to be able to see that the labour in question was not drawn from a 'settled pool' but from a population that was part of a turbulently fluid social order, and plausibly actively adjusting to such an order. As such, they were far from being passive or will-less peasants (quite apart from the fact that such a peasant in general is a fiction) but by virtue of living in an 'urgent world,' were accustomed to making urgent choices, even if they were hard, and at times, Hobson's.

III. BACK TO 'ROOTS' AND THE NOSTALGIA THESIS

In section II we questioned the widely held view that coolies were displaced peasants, and by implication also the concomitants of this thesis: namely that because of their imputed peasantness and poverty, they were predisposed to be politically conservative, culturally traditional, economically and geographically immobile, and fundamentally will-less. By and large this view of coolies was held by non-coolies: nationalists, employers, colonial officials and others, in each case, at various times, serving a particular agenda of interests. In this section we turn to a view of the coolie which is in a sense far more pernicious because of its hegemonic sway over coolies themselves. This may be characterised as the coolie-as-nostalgic-peasant thesis. This view, in its most popular formulation, holds that soon as the conditions which bound the migrants to their work-place – indebtedness, contract – were removed, they wanted to return to their place of origin, where they would be re-absorbed without difficulty. Thus Sandhu [1969: 98] writes, 'almost all of them [men] appear to have been leaving India not in any permanent sense, at least then, but with the aim of returning to it as soon as possible.'

Intentions are psychologically and historically complex and almost impossible to unravel. But we do know that what underlies this view is the much larger theory favoured by the orientalists, to which we have already alluded, namely that the village formed a closed moral community. To this the coolie would return, once his or her labours were done or, more correctly, not needed [*Breman*, 1988: 1–9; *Inden*, 1990: 131–42]. In Ceylon, for instance, where immigrant labour was to pose serious domestic problems in the new nation state with its own growing national consciousness and ethnic exclusivism, this thesis mentioned was actively kept alive by both the national government and the plantation workers themselves, whose ancestors had first left India as immigrants (see the essay by Meyer in this volume). It was only after many of these workers 'returned' to their nostalgically imagined 'homeland' and 'villages' under various repatriation schemes to which the governments of India and Ceylon (later Sri Lanka) agreed that they realised that they had returned not to community and roots, but to the original state of being 'cast adrift' [*Vamadevan*, 1989]. This should have been clear to anyone who examined the origins of most of these workers as (1) cultivators with little or no land, (2) artisans and de-classed artisans, (3) service workers and (4) '*Lumpens*' [*Misra*, 1975: 216; *Das Gupta*, 1976: 315]. Only in the first category was one likely to find anyone resembling a 'peasant', and then too, one for whom the idyllic village was only a figment of his or her imagination.

The figment in question, however, was so overdetermined that it has been nurtured to assume hegemonic proportions. In the case of Sri Lanka, British colonisers, the post-colonial government, the Sinhalese, Sri Lanka's indigenous Tamils, and even the trade unionists who were supposed to have been fighting in the interest of labour, took it to be true. Among the Indian Tamil immigrant labour population itself this view was sustained by folk songs, folktales, oral histories, rituals, political tracts, and manifestos put forth by trade unions. In the coolie-as-nostalgic-peasant thesis, more than in any other, we find a hegemonic convergence of colonial – even orientalist – stereotypes with the self-identities and aspirations of the laborers. The creation and maintenance of the 'homeland' and 'village' in the imagination of plantation labourers had its most significant consequence as a psychological impediment to the creation of a genuine proletarian class. In fact, such ideas implanted in labourers' mentality a constantly deproletarianising devise which prevented their identities as 'coolies' from being ruptured by their becoming genuine workers. As an extension to the nostalgia thesis, colonial authors suggested that coolies went back to their place of origin also because their households remained dependent on their labour power for cultivating the family farm. The choice was between being a coolie or a peasant. This is why the peasant coolie, constrained by this logic, could never rise above a pseudo-proletarian:

> *Indo-China*: When large-scale industries and large plantations run on Western lines were introduced in Indo-China as the result of French occupation, the Annamite people who supplied the labour were almost entirely a race of peasants. It required a long period of adaptation and hesitating efforts before the farmer could become a wage earner, and it may be said that at the present time the life of the worker in Indo-China, in view of his character, his social ties and his economic position, is still essentially that of a peasant [*Goudal*, 1938: 273].

> *Netherlands-Indies*: This periodic emigration made it possible for part of the surplus population to continue to work in agriculture, even in regions where agrarian industry did not exist. The formation of a proletariat was thus slowed down. The population did not become segmented into two sharply divided groups: a landless proletariat and a class of small farmers, but continued to consist of a homogenous mass of individuals who lived in both the capitalist and the peasant economy [*Meyer Ranneft*, 1916: 72].

As far as colonial capitalism was concerned, defining coolies as peasants only temporarily removed from their native village denied them the dignity of confronting their employers as proletarians who had no place else to go but to stand and confront the forces of capitalism. Yet most of those who 'returned' to their 'community' and 'homeland' discovered that they were not 'peasants' but only coolies who had given up their chance of becoming peasants again. They remained what they were: coolies. 'Sucked oranges' was the epithet given to the returned migrants from the Malaysian plantations in the villages of Tamil Nadu. Of the repatriates in Southern India among whom Daniel carried out field research, there were indeed exceptions. These were mainly members of upper castes and the progeny of labour recruiters who had, over several generations, acquired wealth and property in Malaysia and Sri Lanka. And even of these individuals, the derogatory aspersion 'coolie' is whispered by those who had never migrated. But most of those who returned to their imagined villages returned to earning their living as coolies.

As we have observed, the coolie was seen as one whose origins lay in poverty and peasantry, thereby assuring a politically conservative, culturally traditional, and economically and geographically immobile human being, and s/he was also deemed lacking in volition. The laws of supply and demand of labour which regulated capitalist production was credited as the moving force, the mobiliser. We also saw that in the early years of capitalist production, through devices such as indenture and other forms of bondage the migrant coolie was re-immobilised. Why? And here an obvious contradiction was admitted: because the coolie was capable of willfully violating a contract. More than that, the capacity denied was that of willing good, the capacity granted that of willing evil.

What of nostalgia? As a longing for something lost, it may be considered a sentiment both dependent on and independent of conscious volition. Granting coolies the capacity for nostalgia credits them perhaps with a universal human sentiment, but towards what ends and with what consequences? In nostalgia, the will in question turns out to be less wilful and more dreamlike, and as we have remarked, nostalgia's hegemony is incontrovertible and therefore quite pernicious. We don't think it would be unfair to read backwards into history the vision of a contemporary Sri Lankan Tamil trade union leader whose ancestors came to the island from South India as coolies. Observing the plight of unsettledness among tea estate workers, he told of us: 'Our people, from generation to generation have been like dogs that go round and round and round and settle down only when they find the right spot and angle. The right spot for our people is village India. That is where we belong. Until then we will keep going round and round and round. Never settle anywhere, on anything."[7] In this preordained discursive logic to which the labourer too subscribes, the evolution is not from coolie to industrial worker, but from coolie back to being a peasant, despite the fact that most labourers who return to their 'homeland' find themselves beginning the cycle all over again, not as they imagined but as their ancestors had begun, as imminent coolies.

Nostalgia here is about origins, but in a different way from the coolie-as-peasant thesis discussed in section II. There, concerns with origins justified the recruitment of coolies in the first instance. Here, nostalgia invokes origins to tacitly serve the structures and conditions of plantation and other labour regimes. We find in nostalgia then an ironic convergence of the coolie as essentially immobile with a contrary thesis: the coolie as unsettled. In fact each thesis is shot through with its contradictory counterpart. This will become clearer as we work through the view of the coolie as unsettled.

IV. THE UNSETTLING COOLIE

Coolies are unsettling; they are unsettled themselves and unsettle others. In using the gerund we intend the double meaning it – and by extension each of its cognates and synonyms – carries. In considering the view of the coolie as unsettling, we argue that those who benefitted most from the growth of capitalism in Asia, and the growth of plantations and industrial enterprises more specifically, found that the perception of the unstable coolie served their own interests quite well, but also gave them cause to be anxious. Capitalists, in small and large ways, sought to realise this thesis in setting up the work regimes they required.

If a coolie is one whose link to land and place is severed, he or she is also cut adrift from the traditional skill or work-expertise wherein his or her identity may have been found. This claim may seem extravagant given that traditional work expertise seems to have played a role in the recruitment of migrant labourers. It may be argued, for instance, that a significant number of weavers worked in urban textile mills [e.g., *Joshi*, 1981: 1825], iron workers in railway workshops in East India, and in various parts of South Asia unclean work was assigned to castes traditionally associated with it [*Yang*, 1979: 46; *Masselos*, 1982: 105]. Even within such continuities, however, lay radical discontinuities. The independent craftsman was degraded to a wage-dependent factory coolie, within a production process characterised by a new division of labour. A labourer's skill was less something that he or she used or sold, than something that was exploited only if desired by the employer. Kerr [1983] shows that the choice of castes of potters and stonemasons as the source of labour for railway construction in Southern India from the middle of the nineteenth century onwards was neither completely arbitrary nor the result of an explicit choice on the part of the labourers in question. But the 'radical discontinuity' between a labourer and his or her traditional skill that we write of is made manifest by the fact that whenever the skill in question was no longer needed and another required, the labourer was forced to abandon the first for the second.

To be sure, many labourers on plantations were used to working in the fields. As a matter of fact, this was one important reason why employers were so insistent on recruiting peasants' sons. But even when field work *per se* was not new, the new site of discursive practices in which the skill of field work was deployed made it new. Lal [1985: 68] captures this partly when he says that 'fieldwork was not a novel experience. What was unique about plantation labour in Fiji was the absence of any respite from the relentless pace of work, and harsh discipline for even trivial breaches of the labour contract'.[8]

In essence, the capitalist found in the 'coolie' a repository of the contradictions he desired: a person cast adrift from place and skill and yet called upon to be bound to place of origin (even if only in nostalgic imagination), ready to acquire new skills as if they were traditional and throw away traditional ones as if they were as dispensable as an old slipper. The report of the Royal Commission of Labour on conditions in the Punjab during the late twenties of this century could have been written of almost any coolie in colonial Asia: 'He would work as an agricultural labourer in the Canal Colonies for some time, and then would seek work in a flour mill and after some time you will find him working as a tonga driver in Lahore or Amritsar, and then perhaps in some factory again' (quoted in Bhattacharya [1985: 604]).

While this departs from the view of the coolie as immobile peasant, it creates a space for understanding him or her not merely as unsettled but also shifty, that is, deceitful and evasive. On Java a similar category was known as *koeli roepa roepa*, that is, day labourers who circulated throughout a region and were accustomed to tackling any sort of work that would provide them with a livelihood [*Volkstelling Nederlandsch Indië*, 1930: I, 1933: 83, 88; *Breman*, 1983: 96]. Here, as in the Punjab, such mobility was viewed by employers as reflecting the unwillingness of workers to commit themselves to a permanent job, rather than workers' circumstances. In other words, the only will granted to the coolie, other than the will to break contracts, was the unwillingness to settle into something permanent. Employers, however, were not always keen to build up a stable and skilled work force. Planters in colonial Sumatra had a taste for new-comers, regarding old-timers as 'troublesome'. 'Contract coolies who had signed on for a further term were more likely to show a tendency for disturbing the peace and to resist their betters than were workers who had less experience with life and work on a plantation. This explains why recruiting agents in Java were instructed to reject old hands who wanted to sign on again' (Stecher [1905: 5 and 24], quoted in Breman [1987: 169]).

In the various work regimes in colonial Asia as seen, material practices colluded with discursive practices to sustain the image of the coolie as shifty. First, a coolie was thought to be one who came from elsewhere and regarded his or her stay as temporary. The first half of that claim may have been accepted by most migrant workers, but the second was kept alive by capitalist employers and governments and, under their hegemony, accepted by labourers as well, more often than not to their detriment.[9] Second, new workers were not provided with a period of apprenticeship, but were taught their tasks on the job by old-timers. A formal apprenticeship bestows upon an acquired skill not only a sense of specialisation but also one of depth and permanence; it is an initiatory rite which lends to the job a measure of 'sacredness'. But skills (if they could even be called that) learned on-the-fly, as it were, presumes that they could be shed as easily as they were acquired. Coolies were not recognised as possessing a capacity for 'skill' but were instead expected to possess brute strength. Had such a capacity been granted a coolie, then even had s/he shifted from job to job, s/he would have been treated as a handy man or woman, a bricoleur. A shuffler with only strength and physical endurance for hard labour was a 'coolie', more a 'bricole' than bricoleur, in the sense that s/he was seen as will-less and immobile as the cue ball in billiards, which does not move until poked by a stick, and then only to bounce off the cushion in a trajectory that may be predictable but at times likely to surprise.[10] A coolie embodies that element of chance, like a bricole upon which the billiard cue, as fate or boss, executes its play.

As far as the employer was concerned, the trick was to tame that element of chance. The colonial official and colonial policies in general were only too eager to assist in this. Third, if a coolie could be put to work on arrival, without training, it followed that dismissal could be equally abrupt. As is well known, the high turnover of labour – a notable aspect of most places where coolies were employed in large numbers – was connected to the need to expand or reduce production in accordance with demand, in turn determined by seasonal and economic fluctuations. What was said about the casual labourer in rural England in the early nineteenth century was equally applicable to many early coolies: 'hired and dismissed at will, and lacking even the guarantee, as he set out in a misty dawn, that he would return home that night with any earnings at all' [*Hobsbawm and Rudé*, 1969: 24].

The corollary of the shiftiness attributed to the coolie was of course his or her dispensability, which worked to the employer's convenience. Coolies had to work long days that frequently began while it was still dark and continued until after dusk, but it also happened that they were abandoned to unproductivity for hours, even days. Employers usually blamed the absence of a steady work pace on coolies whom they denounced as lazy and recalcitrant; sometimes, however, they might acknowledge, with perverse paternalism, that the responsibility might well be their own. Thus Goudal [1938: 4] observed that 'the Annamite coolies, although of slight build, usually form excellent workers, ... docile and easy to control, but if good work is to be obtained from them they must be treated with fairness and sympathy'.

The contradictions abound. Once, the immobility of peasant-coolies called for unsettling by the good graces of capitalism. Having been thus unsettled, the potential capriciousness of their mobility called for settling by a series of labour regimes. Once, indigenous nationalists – as in Sri Lanka – demanded the expatriation of Indian coolies on the grounds that they lacked loyalty to their adopted country and were unsettled transients. But the more these coolies became settled patriotic citizens, indigenous nationalists in their turn became unsettled and took steps to re-unsettle and repatriate the coolies. To be more precise, such settlings and unsettlings did not occur in such neat temporal sequences. One may have been amplified at one time and the second at another; but most of the time, they occurred simultaneously. The word that became flesh in order to regulate this contradiction, among employers and colonial officers then and nationalists and many trade union leaders more recently, is 'control'.

The view that coolies are unable to exercise control over their labour power or lives was held not only by employers, but was, and still is, shared by a wider, non-coolie population as well. Where self-control is absent control from without has to take its place. In this regard a close reading of the representations made on behalf of plantation workers by their trade unions is instructive. Trade union leaders did not see themselves representing the expressed wishes and interests of workers as much as being the 'voice of the voiceless', displaying a thoroughly paternalistic attitude.[11] In the same vein, Wesumperuma, the well-meaning director of a Foundation committed to improving the lot of plantation labour in Sri Lanka writes, 'The bulk of the immigrants were from among the most inarticulate, illiterate, and poorest sections of the population of South

India' [1990:11]. And again, in the same paper he writes that the coolie's 'conservatism, poverty, ignorance and illiteracy bound him to his customary way of life in spite of his legal rights to the contrary' [1990: 11]. The coolie lacks agency. The coolie does not move unless moved, is not a bricoleur but a bricole, in short, a victim. But the obverse and ever-present script is that, once moved, the coolie will not stop of his or her own accord: s/he lacks self-control, s/he has to be controlled.

With regard to the coolies' silence, it is worth noting that this same silence, be it imputed or real, voluntary or imposed, is given a different political interpretation by revisionist theory and neo-classical economists (see Brass and Bernstein, this volume). Here, silence is interpreted as assent. Coolies, the argument goes, would not have chosen to stay put on the plantation if it was as bad as depicted, so for this reason such employment must of necessity have been perceived by its subjects as good and beneficial, and could not have been the degrading and exploitative activity associated with the historical image of the plantation. The coolie-as-victim associated with European images is accordingly inverted, making the coolie yet more amenable to exploitation by justifying it in terms of the coolie's own wishes, restoring to him or her a will that had been, for other reasons, taken away.

V. THE FEMALE COOLIE

There is little doubt that the migrant population in the enclaves of colonial capitalist growth consisted predominantly of single males [e.g. *Tinker*, 1974: 201, 204; *Breman*, 1990: 51–2]. The general exclusion of females from labour contingents recruited for destinations overseas was justified, when and if needed, by the claim that such migrants would return as soon as possible [e.g. *Sandhu*, 1969: 98]. As mentioned earlier, such a claim was a corollary to the notion of the coolie as essentially shiftless and immobile, moved only by overwhelming external forces but sooner or later to return, to the point of inert rest. It also implies that until that moment of repose is recovered, the coolie will be restless. What is less evident is how colonial discursive and material practices subtlely and massively contributed to this view through the uses and abuses of female workers. Here, we shall see how the coolie becomes one not only cast adrift from place, skill and purpose, but also from a moral order whose centre was held together by the woman.[12]

From the capitalist producer's point of view, the ideal coolie was one who had abandoned family and home. Women, by and large, could not meet this ideal (without seriously jeopardising their own morality and, by extension, the moral order of coolie identity – a point we return to). Morris [1965: 69n.] summarises the attitude of the managers of Bombay's textile mills thus: 'Even before the passage of the Factories Act of 1891 women in these departments typically came to work later and, except during very busy periods, would leave earlier than the men. They were also customarily allowed to leave mills during the day to feed their children'.[13]

The male worker, on the other hand, came closer to the ideal of the unencumbered labourer; the ideal coolie. Of course, this disregarded the fact that in Asia, a full adult male is a married man and an unmarried man at any age is, in a deep cultural sense, incomplete. It is no accident then that male coolies were quite regularly called 'boys' by their European employers and by non-European foremen. The deprivation in the life of the coolie was such that to have a wife became a symbol of status, security and prosperity in the plantations [*Tinker*, 1974: 202]. But the complete man and the ideal coolie were in tension. If a coolie arrived at a plantation together with his wife, he ran the risk of losing her to a superior, that is, to a foreman or a white employee, thereby reverting to the status of a 'boy', an incomplete man.

In agro-mineral industries in particular, the highly discrepant ratio of men to women encouraged sexual practices described in the colonial literature as 'unnatural'. Young male indentured workers, known as 'calves', were appropriated as lovers by the strongest among the Chinese workers immediately on their arrival on the plantations of Sumatra's east coast. According to the planters, coolies were not interested in women but had a 'natural' preference for their own sex. Yet when several male labourers lived together with one woman, this was not ascribed to force of circumstance but to cultural or racial traits [*Breman*, 1987: 191–2]. In short, the normalising (in Foucault's sense) of capitalism's abnormal conditions made abnormal the labourers' normal adaptive behaviour.

Government reports listed the lack of women as one of the foremost causes of the high rates of suicide wherever indentured labourers were settled [*Tinker*, 1974: 201, 204; *Lal*, 1985: 59–60; *Breman*, 1987: 159–60, 205]. From available statistics Kondapi [1951: 27] found that the rate of suicide among coolie populations in several countries correlated inversely with the proportion of women amongst them. By the middle of the nineteenth century, apprehensive of rising tensions among migrants, the British authorities in India mandated an increased proportion of women in the shiploads of coolies destined for overseas plantations. Capitalists resisted; the mastery exercised by planters went so far that indentured labourers could marry only after obtaining their master's permission [*Lal*, 1985: 64; see also the essay by Behal and Mohapatra in this volume). Casual alliances were frowned upon, and co-habitation of male and female coolies would be brought to an abrupt end whenever one of their superiors found it necessary for economic reasons. Female coolies were even forbidden to bear children [*Breman*, 1987: 192], and abortions were common, for no woman could allow herself the luxury of not working, even temporarily [*De*, 1968]. For plantation managers, mothers with young children represented a double burden: they could not be used to the maximum and they had an extra non-productive mouth to feed. Keeping the price of labour low was, after all, good business practice, the significant side effect of which was to condemn women workers, in the moral sense, to female incompleteness. It was only towards the end of colonialism that capitalist employers became interested in family formation for the sake of building up a more stable work force [e.g. *Stoler*, 1985].

To dwell on the incontrovertible claim that there was a disproportionately high ratio of men to women tends to muffle the more important questions: when, why and how did women form a significant part of the coolie labour force? Indeed, to raise this question is to put the first claim under scrutiny, to read in it effects other than a mere statistical disproportion. If the relative absence of women in a capitalist, predominantly male, work environment made the male coolie a morally blemished man and his world a fragmented moral order, her incomplete presence in this work-world made a woman into an immoral being who helped seal coolie identity in depravity. Even though the numbers of women employed in most capitalist enterprises were relatively small, most were unattached, or at least unaccompanied by parents, children or husband. According to a number of sources, women who sought employment in urban textile mills were mainly widows in needy circumstances, or women judged guilty of illicit sex, whether before or during marriage, or otherwise deserted by their husbands. Even if the female factory worker had not been forced to leave her home for such reasons, she invariably acquired a bad reputation after reaching the mill. The female supervisors (*naikin*) of Bombay's cotton mills, as well as lower-ranking staff members, were notorious for forcing their subordinates to take up prostitution [*Burnett-Hurst*, 1925: 53]. 'Depraved,' Gandhi called the working conditions in the textile mills, and warned men against marrying women mill workers [*Kooiman*, 1978: 51–2]. Equally bleak is the picture of female coolies recruited to work on plantations at home or overseas: 'A very large proportion of the women who now emigrate are persons who have been turned out of the home, or have lost their friends by famine or pestilence; some were Hindu girls who had been forced to become Muslims in some inter-communal quarrel; many were widows' (cited in Tinker [1974: 267]).

To move unaccompanied by a male relative or other household member introduces a new intensity of significance to the notion of the 'coolie' as someone cast adrift. According to some colonial critics, at the end of the nineteenth century some hundreds of young women were shipped each year from Java to Sumatra's east coast where they were forced to prostitute themselves [*Stoler*, 1985: 31; *Breman*, 1987: 108]. This was said of the 'Hindu woman' in Fiji: 'In this country (she) is like a rudderless vessel with its mast broken drifting on the rocks; or like a canoe being whirled down the rapids of a great river without any controlling hand. She passes from one man to another, and has lost even the sense of shame in doing so' (quoted in Lal [1985: 581]).

Women workers on the large plantations of Sumatra were similarly depicted. Employers rejected the idea of wage increases for women workers on the grounds that the greater majority found an easy source of extra income in prostitution. A former resident claimed that even girls of 12 or 14 years of age were no longer virginal when they left Java [*Breman*, 1987: 191–2]. Given the stigma of so grave a fate, it is not surprising that single women reported for emigration to plantations only when no other alternative was available.

Capital's new work-places may have had a lower proportion of women in their labour force, but this did not mean that women's physical and moral workload remained the same or lightened, regardless of where they stayed or moved to. What often goes unnoticed is that women left behind by male migrants had to be even more actively involved in local production processes, agricultural or otherwise, than they were before their men left. Many of those left behind were caught up in seasonal migration to seek work in cottage industries less distant from home, for example in cotton gins: 'The ginning season lasts about eight months, about five of which the hands work from 5 am to 10 pm. and the remaining three months they work days and night. The hands are mostly women who worked day and night for as long as a week at a stretch ... I do not think there is a double set of children anywhere, so they must have worked 23 out of 24 hours' (quoted in Sen [1977: 39]).

Unlike the large-scale capitalist factories, mines and plantations, the work schedules and divisions of labour in such enterprises were more flexible and less standardised. Paradoxically, however, this very flexibility that enabled women to combine their employment with household and family obligations made their overall workload even heavier than that of their sisters on the plantations.

What of the moral burden? Each woman, whether she stayed behind in a village, moved within the vicinity or country of her birth, or migrated to plantations overseas, became an embodiment of the family's, clan's or community's moral centre. Keeping this moral centre above reproach called for relentless diligence, vigilance and work. The women who stayed behind in their villages were, in the watchful eyes of their neighbours, perennially on trial. Oral histories of the first third of the nineteenth century gathered in South Indian villages tell sordid tales of women wronged and gone wrong – women who had been left behind by their husbands, fathers and brothers, who had gone to work in the coffee and, later, tea plantations of Ceylon and Malaya. Reputations tarnished by women who could not uphold the moral burden any longer have left their stigma on generations removed from those unfortunate ancestresses by 75 to 150 years.[14]

Movements of female labour, largely unnoticed and unrecorded by census-takers and statisticians, were common in the countryside, resulting from regional differences in agricultural cycles or crop specialisations. Interpreting the 1931 Indian census, however, Omvedt [1980: 192] notes that: 'of the two major waves of migration from the northern districts of Burma in 1931, women were 1/6th of those who went to the jute mills and factories surrounding Calcutta where there was relatively little work available for them; but they were 40 per cent of the migration which went to help bring in the harvests in rural Bengal where they had specific agricultural tasks'.

If female workers in casual or seasonal employment were invisible in one way, falling in the colonial record keeper's blind spot, women who worked in more structured large-scale industries such as mills, mines and plantations, were made invisible or only partly visible in another way. We already referred to the single woman, cast adrift from the moral centre, debased, debilitated and demoralised. This identity-fractured woman was all too visible in the moral landscape of cooliedom. Moreover, she was more likely to be counted among that low 10 to 20 per cent of female coolies appearing in the census taker's statistics than were those less 'fractured' women who were wives, mothers and sisters of male workers who, with their children, were counted as dependants, not workers.[15] Some men, in their capacity as heads of households, refused to report their female relatives as workers [*Bhattacharya*, 1985: 165–7]. Among their reasons was the conviction that the lower the profile of a female, and the less obligated to work within capitalist enterprises' formal arrangements and contracts, the less vulnerable she would be to the predatory designs of the more powerful.[16] But work she did – work that was acknowledged, if at all, in wages that were but a fraction of a man's (already low) wages. In Sri Lankan and South Indian plantations, the kinds of work such women did were called 'piece-work.' They consisted of a chore here, a job there, called upon to do one thing and interrupted to be told to do another or nothing at all; a couplet in an old Tamil folk song sung by female plantation workers in Sri Lanka goes thus:

> On pieces of paper we received pieces of work
> Piecing and unpiecing us as he wished.

The price paid for keeping intact one's moral identity as a female was to submit to the inclusion of one's labour price in the wage the male worker received. Where payments were based on a contracted job, 'male workers whose families resided on or near the plantations often enlisted the aid of their wives and children in order to meet their daily quotas' [*Murray*, 1980: 292]. In many of these cases, the women and children went unregistered and therefore, in the strict sense, went unpaid. This practice exists to this day in the tea plantations of South India and disappeared from Sri Lanka's plantations only after they were nationalised in the early 1970s.

VI. THE CHILD COOLIE

If male coolies were treated as incomplete men or even as children, and women as fractured or invisible, children were treated as adults. The magnitude of child labour employed in capitalist enterprises will never be known exactly, but we do know that early estimates of children working in the mills of Bombay, for instance, were far too low [*Morris*, 1982: 644; 1965: 70]. As against Morris's six per cent, Newman [1981: 50] offers a revised estimate of more than 50 per cent as a reasonable proportion of workers under seventeen years of age in some textile factories in Bombay before the first World War.[17]

Children's labour was not child's stuff; it was hard, adult labour. Buchanan [1934: 272–3] points out that large numbers of boys and girls between the ages of twelve and sixteen worked in the coal mines of Bihar in the first few decades of the present century and were registered as adult labourers, and even younger children went into the mines with their father or mother and were never reported in the mine's administrative records. According to a 1908 report of the Indian Factory Labour Commission, many children 'not more than 6 or 7 years old are employed for seven or eight hours a day, and the proportion of under age children employed as half-timers probably amounts to 30 to 40 per cent of the total half-timers' (quoted in Buchanan [1966: 308]). Buchanan goes on to observe that, 'the Commission also estimated that 25 per cent of the full time workers under 17 were not yet 14 years of age. These, and doubtless many younger children, worked 10, 11 or more hours per day'. The same Report of 1908 stated that children less than seven years old walked three or four miles to reach the factories of Bengal's jute mills, to start work at the sound of the sirens at 3 a.m. [*Sen*, 1977: 40].

Capitalist employers invented conscience-easing euphemisms. 'Half-timers,' these child workers in the textile mills of Calcutta and Bombay were called, even though prior to 1922 their work day lasted nine hours, six days a week. In 1922, these 'half-timers' worked six hours a day. Over the course of time the definition of a half-timer changed: in 1881 they were seven to eleven-year-olds; in 1891, nine to thirteen-year-olds; from 1922 onwards, it applied to twelve to fifteen year olds. Some employers found the new category of 'half-timers' a boon; it enabled them to use a child in two different factories as a 'half-timer'. Protective legal measures were instituted by the government, but the officials appointed to enforce these laws resorted to the following kind of plea: 'The Special Inspector of Factories ... draws attention to the fact that mills are at present shorthanded; and states that if children are altogether denied admission not only will hardship be entailed on the parents, but that the industry will also be affected adversely, because much female labour will thereby be lost it' (quoted in Gupta [1974: 423]).

In Java, juveniles were called 'half-adults' – again, a label that indicated that they were something other than what they really were [that is, children], and incomplete at that. Here again laws were passed and then circumvented [*Levert*, 1934: 193]. A law introduced in 1925 in Java, prohibiting the employment of children less than twelve, did not apply to those who did 'light fieldwork' on the sugar estates. In 1926, 15 per cent of registered field workers were still 'half-adults', not to mention the unregistered ones. Official admissions of child labour came out sounding like manifestos of benevolent protection. Asian girls were said to mature at an early age; therefore, French employers in Indo-China considered thirteen or fourteen-year-olds 'women'. If they were fit enough to bear children, it was argued, they were fit enough to work for a living [*Goudal*, 1938: 271].

The Act regarding the employment of Indian immigrants, promulgated in colonial Malaysia in 1923, allowed children of older than ten years of age to be employed. Some authors, such as Arasaratnam [1970: 55, 185], maintain that children joined the work force due less to economic need than to parental indifference. The point we wish to stress, however, is that in Europe child-labour had been abolished as an abomination by the end of the nineteenth century. In Europe, a child *qua* child was one who, among other things, had come to be seen as being unsuitable for adult labour; to combine the two, adult labour and childhood, had come to be perceived as cruel and criminal. In the colonial context and the imagination of capitalism in Asia, this double standard in perception was naturalised in the bodies of Asian children. They were children no longer.

CONCLUSION

In the nineteenth century and first half of the twentieth century, with the expansion of European colonisation and a capitalist economy, we find coolies emerging from all over Asia – mainly from India, China, Indonesia and Indo-China, but also from Japan and Oceania. Lydia Potts [1990: 72–3] writes that 'the total number of men, women and children sent abroad as coolies [within a century] could not have been less than 12 million, and an estimate of 37 million or more would not be entirely without foundation'. Their emergence and spread covers almost two centuries and spans all continents except Europe. Any generalisation of 'the coolie,' is therefore, likely to be undermined by exceptions. In this essay, we have tried to steer a middle course, not by working our way through a statistical average but through what we have gathered, from the literature, as constituting strategic perceptions of who or what a coolie was. A single coolie, forced by circumstances into a life of immorality, counted for more in the perceived identity of a coolie than ten virtuous ones.

Rather than renumerize our evidence and interpretation, we end with an admission. We began by saying that a coolie's identity is as much the product of self-definition as it is of the construction of a category by those who did not belong to it. But the identity we have constructed has been drawn mainly from the latter and not from the former. There is a void, a silence: the silence of the coolies themselves. The resulting picture has been one of 'coolie as victim'. The imagery is almost overdone: flotsam and jetsam, being cast adrift, a bricole, demoralised, debased and dejected.

We, as well as the other contributors to this volume, are writing about a period when literacy among coolies was almost non-existent. The documents we have drawn from were all written in the colonisers' languages, languages in which the person (*āl*, in Tamil) was appropriated by the menial payment s/he received (*kūli*), making, thereby, a 'coolie'. Let us return to not too fine an etymological point. The language of its origin, a language spoken by a significant number of coolies who were scattered all over the world by the expanding waves of capitalism, resisted such a conflation. Invoking Tamil here to serve a metonymic function for coolies everywhere, we find that in that language, the suffix '*-āl*' (or 'person') of the Tamil *kūliāl* was never surrendered by those workers whose mother tongue was Tamil. It is as if there was a hidden hyphen that held *kūli* and *āl* together and apart. The struggle of the coolie has been the struggle to recover an identity that is not fractured, not alienated from place, skill, and a moral centre. This has meant, quite literally, severing the *kūli* from the *āl*, and casting away the former into the void of history. There are success stories in this regard as well as stories of continued struggle, portraying the actions and counter-actions of proletarians and not just the suffering of a reserve army written off as *Lumpen* elements.

The collection in this volume, in more ways than one, attempts to let us in on a world where some successes were achieved and struggles were (and still are being) fought. However, we have had to work against a built-in impediment, with a language and in languages that collapsed *āl* into *coolie*. There are sources and resources that wait to be tapped so that clearer accounts of struggle may be written. They are found in the oral histories recalled and retold in the languages of the coolies of Asia. Such work has barely begun; such accounts are, by and large, yet to be written.[18]

NOTES

1. Also see Lydia Potts [1990: 63–4].
2. See Jan Breman [1974] for a discussion of how 'bondage' is conceptualised in patron–client relationships. On the same theme, see Brass [1986] Ramachandran [1990] and Prakash [1990].
3. In E.V. Daniel's fieldwork among contemporary tea plantation workers of South India and Sri Lanka, he finds that these two terms, almost always conceptualized negatively (as a deficiency) constitute the semantic field in which these workers still find themselves. It is a field in which hegemonic interests of class and race (masquerading as ethnicity) find their perverse fulfillment.
4. For the most vivid portrayal of the Indian village community see Maine [1907], also Baden-Powell [1892], Monier-Williams [1891], and L.S.S. O'Malley [1932; 1934]. The last two are 'capsule representations which trace the descent back to Sir Charles Metcalf [1785–1846]', whose 'description of a "village republic" first appeared in a British Parliamentary inquiry of 1890' [*Inden*, 1990]. On the colonial construction of the Asiatic village, see Breman [1988].
5. This is illustrated by the shift in composition of the labour force in Kanpur's textile mills as the numbers from agrarian castes who lost their land gradually increased [Joshi, 1981: 1827].
6. See, for example, Bates [1986: 579]; Yang [1979: 44]; Bhattacharya [1985: 151–4]; Choudhary and Bhowmik [1986: 335]. In all these cases, fluctuations in the agrarian cycle caused agricultural workers to move over considerable distances.
7. Excerpt from oral interview recorded by E.V. Daniel. The speaker is now a member of Sri Lanka's upper middle class and a convert to the hegemonic view of coolie identity

constructed by colonialism and nationalism.

8. Also see Daniel [forthcoming] on how apparently neutral and universal terms such as terms of measurement are ideologically loaded and constitutive of a discursive space that the labourer finds coercive.
9. The plight of Tamil tea estate workers in Sri Lanka being a case in point – most workers who chose to be repatriated to India have fallen into new forms of bondage and destitution in their imagined 'motherland' [*S. Vamadevan*, 1989; *Vidya Sagar et al.*, 1986].
10. 'Bricole', denotes in this sense the rebound of a ball against the wall in squash or against the cushion in billiards, or the move of a horse to avoid an obstacle, and thus connotes 'an indirect unexpected stroke or action' in the words of the dictionary [*The Shorter Oxford English Dictionary* 1956: I, 220, art. 'bricole'; *Dumont*, 1985: 30].
11. See, for example, the well-meaning periodical published by Sarvodaya in Sri Lanka, the title of which is appropriately called, *The Voice of the Voiceless*.
12. For an fascinating and ironic twist in this female custodianship of morality among white women in colonial Asia and their working out of this custiodianship in racist terms, see Stoler [1989].
13. In British India, the 1891 Factory Act reduced the number of hours that women were allowed to work per day. The prohibition on night work for women followed soon thereafter.
14. From oral histories gathered by E.V. Daniel in villages of Tamil Nadu, South India.
15. Long Vinh Ngo, in contrast to the low official figure of 10 to 20 per cent, estimates that women and children may have in fact represented as high as half to two-thirds of the workforce in late-colonial Indo-China [1974: 50–51].
16. The significance of these strategies is underscored by the fact that they are, to this day, an easily recoverable part of oral histories that go back to Ceylon's coffee-years of 1823–50. These oral histories have been gathered by E.V. Daniel.
17. Misra, making his survey of a wider range of cities, estimates child labour as constituting one third of the work force in the late nineteenth century [1975: 218].
18. Examples of such beginnings are to be found in E.V. Daniel [forthcoming]; Tran Tu Binh [1985]; C.V. Velupillai [1970].

REFERENCES

Arasaratnam, S., 1970, *Indians in Malaysia and Singapore*, London Bombay/Kuala Lumpur: Oxford University Press for the Institute of Race Relations.

Baden-Powell, 1892, *The Land Systems of British India*, 3 vols., Oxford: Clarendon Press.

Baden-Powell, B.H., 1896, *The Indian Village Community*, London: Longman, Green.

Bates, C., 1986, 'Tribal Migration in India and Beyond', Paper presented at the Conference on India and World Capitalism, Tufts University, 12–14 Dec. [unpublished].

Bhattacharya, N., 1985, 'Agrarian Change in Punjab: 1880–1940', unpublished Ph.D. thesis, Centre for Historical Studies, School of Social Sciences, Jawaharlal Nehru University, New Delhi.

Boeke, J.H., 1948, *The Interests of the Voiceless Far East; Introduction to Oriental Economics*, Leiden: Universitaire Pers.

Brass, T., 1986, Review of Patnaik and Dingwaney (eds.) [1985], *Journal of Peasant Studies*, Vol.14, No.1, Oct., pp.120–36.

Breman, J. 1974, *Patronage and Exploitation; Changing Agrarian Relations in South Gujarat, India*. Berkeley, CA: California University Press.

Breman, J., 1983, *Control of Land and Labour in Colonial Java: A Case Study of Agrarian Crisis and Reform in the Region of Cirebon during the First Few Decades of the 20th Century*, Verhandelingen van het Koninklijk Instituut voor Taal-, Land- en Volkenkunde No.101, Dordrecht: Foris Publications.

Breman, J., 1987, *Taming the Coolie Beast: Plantation Society and the Colonial Order in Southeast Asia*, Delhi: Oxford University Press.
Breman, J., 1988, *The Shattered Image: Construction and Deconstruction of the Village in Colonial Asia*, Comparative Asian Studies 2, Dordrecht: Foris Publications.
Breman, J., 1990, *Labour Migration and Rural Transformation in Colonial Asia*, Comparative Asian Studies 5, Amsterdam: Free University Press.
Buchanan, D.H., 1934, *The Development of Capitalistic Enterprise in India*, New York: Macmillan (Reprint New York: A.M. Kelly Booksellers' 1966).
Burnett-Hurst, A.R., 1925, *Labour and Housing in Bombay: A Study in the Economic Conditions of the Wage-Earning Classes in Bombay*, London: King & Son.
Carey, P., 1986, 'Waiting for the "Just King": The Agrarian World of South-Central Java, 1755–1825', *Modern Asian Studies*, Vol.20, pp.59–138.
Chakravarty, L., 1978, 'Emergence of Industrial Labour Force in a Dual Economy; British India, 1880–1920', *The Indian Economic and Social History Review*, Vol.15, pp.249–328.
Chaudhuri, P., 1985, 'The Impact of Forced Commerce on the Pattern of Emigration from Orissa, 1901–21', in *Essays on the Commercialization of Indian Agriculture* (eds. K.N. Raj, N. Bhattacharya, S. Guha and S. Padhi), Delhi: Oxford University Press.
Choudhury, N.C. and S.K. Bhowmik, 1986, 'Migration of Chota Nagpur Tribals to West Bengal', in *Studies in Migration: Internal and International Migration in India*, (ed. M.S.A. Rao), Delhi: Manohar.
Cooper, Frederic, 1980, *From Slaves to Squatters: Plantation Labour and Agriculture in Zanzibar and Coastal Kenya, 1890–1925*, New Haven, CT: Yale University Press.
Daniel, E.V. [forthcoming], 'Tea Talk: Violent Measures in the Discursive Practices of Sri Lanka's Estate Tamils', *Comparative Studies in Society and History*.
Das Gupta, R., 1976, 'Factory Labour in Eastern India: Sources of Supply, 1855–1946: Some Preliminary Findings', *The Indian Economic and Social History Review*, Vol.13, pp.277–330.
De, S.S., 1968, 'The Problem of Labour in Assam (1826–1901)', unpublished Ph.D. thesis, University of Gauhati.
Dumont, J.-P., 1985, 'Who Are the Bricoleurs', *American Journal of Semiotics*, Vol.3, No.3.
Emmer, P., 1984, 'The Importation of British Indians into Surinam (Dutch Guiana), 1873–1916', in *International Labour Migration: Historical Perspectives* (eds. S. Marks and P. Richardson), London: Maurice Temple Smith for the Institute of Commonwealth Studies.
Geertz, Clifford, 1980, *Agricultural Involution*, Berkeley, CA: University of California Press.
Goudal, J., 1938, *Labour Conditions in Indo-China*, ILO Studies and Reports, Series B, No.26, Geneva.
Gupta, P.S., 1974, 'Notes on the Origin and Structuring of the Industrial Labour Force in India, 1880 to 1920', in *Indian Society: Historical Probings. In Memory of D.D. Kosambi* (ed. R.S. Sharma), New Delhi: People's Publishing House.
Hobsbawm, E.J. and G. Rudé, 1969, *Captain Swing*, Harmondsworth: Penguin Books.
Inden, R., 1990, *Imagining India*, Oxford: Basil Blackwell.
Joshi, Ch., 1981, 'Kanpur Textile Labour: Some Structural Features of Formative Years', *Economic and Political Weekly*, Vol.16, pp.1823–38.
Kelly, J.D., 1991, *A Politics of Virtue: Hinduism, Sexuality, and Countercolonial Discourse in Fiji*, Chicago, IL: University of Chicago Press.
Kerr, I.J., 1983, 'Constructing Railways in India: An Estimate of the Numbers Employed', *Indian Economic and Social History Review*, Vol.20, pp.317–39.
Kondapi, C., 1951, *Indian Overseas, 1838–1949*, Indian Council of World Affairs, Bombay: Oxford University Press.
Kooiman, D., 1978, 'Koppelbazen, Kommunisten en Ekonomische Krisis: Arbeidsorganisatie in de Textielindustrie van Bombay, 1917–1937' (Jobbers, Communists, and Economic Crisis; Organization of Labour in the Textile Industry of Bombay, 1917–

1937), Ph.D. thesis, Universiteit van Leiden.

Lal, B.V., 1985, 'Kunti's Cry: Indentured Women on Fiji's Plantations', *Indian Economic and Social History Review*, Vol.22, pp.55–71.

Levert, Ph., 1934, 'Inheemsche Arbeid in de Java Suiker-Industrie' (Native Labour in Java's Sugar Industry), Ph.D. thesis, Agricultural University of Wageningen, Wageningen: Veenman & Zn.

Maine, H.S., 1907, *Village Communities in the East and West*, London: J. Murray (first published 1871).

Masselos, J., 1982, 'Jobs and Jobbery: The Sweeper in Bombay under the Raj', *Indian Economic and Social History Review*, Vol.19, pp.101–39.

Meyer Ranneft, J.W., 1916, 'Volksverplaatsing op Java' (Movement of People on Java), *Tijdschrift voor het Binnenlandsch Bestuur*, Vol.49, pp.59–87.

Mintz, Sidney, 1959, 'The Plantation as a socio-cultural type', in *Plantation Systems and the New World* (ed. V Rubin), Washington, DC: Pan American Union.

Mintz, Sidney and Richard Price, 1973, 'An Anthropological Approach to the Study of Afro-American History', New Haven, CT: Yale University Press.

Misra, B., 1975, 'Factory Labour during the Early Years of Industrialization: An Appraisal in the Light of the Indian Factory Commission, 1890', *Indian Economic and Social History Review*, Vol.12, pp.203–28.

Mohapatra, P.P., 1985, 'Coolies and Colliers: A Study of the Agrarian Context of Labour Migration from Chotanagpur, 1880–1920', *Studies in History*, New Series, Vol.1, pp.247–303.

Monier-Williams, Monier, 1891, *Brahmanism and Hinduism or, Religious Thought and Life in India, as Based on the Veda and other Sacred Books of the Hindus*, 4th ed., London: John Murray (first published 1883 as *Religious Thought and Life in India*).

Morris, M.D., 1965, *The Emergence of an Industrial Labor Force in India: A Study of the Bombay Cotton Mills, 1854–1947*, Bombay/Berkeley, CA: Oxford University Press/ University Press of California.

Morris, M.D., 1982, 'The Growth of Large-Scale Industry to 1947', in *The Cambridge Economic History of India, c1757–c1970*, Vol.2 (ed. D. Kumar and M. Desai), Cambridge: Cambridge University Press.

Murray, M.J., 1980, *The Development of Capitalism in Colonial Indo-China (1870–1940)*, Berkeley, CA: University of California Press.

Newman, R., 1981, *Workers and Unions in Bombay: A Study of Organisation in the Cotton Mills*, Australian National University, South Asian History Section, Monographs on South Asia 6, Canberra.

Ngo Vinh Long, 1973, *Before the Revolution: The Vietnamese Peasants under the French*, Cambridge, MA:., MIT Press.

O'Mally, L.S.S., 1932, *Indian Caste Customs*, Cambridge: Cambridge University Press.

O'Malley, L.S.S., 1934, *India's Social Heritage*, Oxford: Clarendon Press.

Omvedt, G., 1980, 'Migration in Colonial India: The Articulation of Feudalism and Capitalism by the Colonial State', *Journal of Peasant Studies*, Vol.7, pp.185–212.

Potts, L., 1990, *The World Labour Market: A History of Migration*, London: Zed Books.

Prakash, G., 1990a, *Bonded Histories; Geneologies of Labor Servitude in Colonial India*, Cambridge/New York, Cambridge University Press, Cambridge South Asian Studies No.44.

Prakash, G., 1990b, 'Writing Post-Orientalist Histories of the Third World: Perspectives from Indian Historiography', *Comparative Studies in Society and History*, Vol.32, No.2.

Ramachaudran, V.K., 1990, *Wage Labour and Unfreedom in Agriculture: An Indian Case Study*, Oxford: Clarendon Press.

Robequain, Ch., 1944, *The Economic Development of French Indo-China*, London: Oxford University Press.

Roy, M.N., 1922, *India in Transition*, Geneva, Reprint Bombay, 1971, Nachiketa Publications.

Saha, P., 1970, *Emigration of Indian Labour (1834–1900)*, Delhi: Popular Publishing House.

Sandhu, K.S., 1969, *Indians in Malaya*, London: Cambridge University Press.
Sen, S., 1977, *Working Class of India: History of Emergence and Movement, 1830–1970*, Calcutta: Bagchi & Co.
Standing, G., 1979, *Migration and Modes of Exploitation: The Social Origins of Mobility and Immobility*, World Employment Programme Research, Working Paper 72, ILO. Geneva; also published in *Journal of Peasant Studies*, 1980, Vol.8, pp.173–211.
Stoler, A.L., 1985, *Capitalism and Confrontation on Sumatra's Plantation Belt, 1870–1979*, New Haven, CT: Yale University Press.
Stoler, A.L., 1986, 'Plantation Politics and Protest on Sumatra's East Coast,' *Journal of Peasant Studies*, Vol.13, No.2.
Stoler, A.L., 1989, 'Rethinking Colonial Categories: European Communities and the Boundaries of Rule', *Comparative Studies in Society and History*, Vol.31, No.1.
Taussig, Michael, 1980, *The Devil and Commodity Fetishism in South America*, Chapel Hill, NC: North Carolina University Press.
Tinker, H., 1974, *A New System of Slavery: The Export of Indian Labour Overseas, 1830–1920*, London: Oxford University Press.
Tran, Tu Binh, 1985, *The Red Earth: A Vietnamese Memoir of Life on a Colonial Plantation as Told to Ha An*, translated by John Spragens Jr., Center for Southeast Asian Studies, Ohio University Center for International Studies.
Vamadevan, M., 1989, *Sri Lankan Repatriates in Tamil Nadu: Rehabilitation and Integration*, Madras: Zen Publishers
Vidya Sagar R., Irudayam, Aloysius and M. Devadoss, 1986, *The Human Cost of Commercial Forestry: A Study of the Conditions of Coupe Labourers in Kodaikanal Hills*, a report commissioned by The Supreme Court Commission in W.P. No.574/86 & S.C. Order 3.4.86 and In W.P. No.633/86 & S.C. Order 15.4.86.
Velupillai, C.V., 1970, *Born to Labour*, Colombo: Gunesena Press.
Volkstelling Nederlandsch-Indië 1930, 1933 (Census of 1930 in Netherlands India), Vol.1, *Native Population in West Java*, Batavia: Landsdrukkerij.
Wesumperuma, D., 1990, 'Plantation Labor System in Colonial Sri Lanka' (Paper presented at the Conference on Capitalist Plantations in Asia), Centre of Asian Studies, Amsterdam.
Wolf, Eric, 1959, 'Specific Aspects of Plantation Systems in the New World: Community Sub-Cultures and Social Classes', in *Plantation Systems of the New World*.
Yang, A.A., 1979, 'Peasants on the Move: A Study of Internal Migration in India', *Journal of Interdisciplinary History*, Vol.10, pp.37–58.

Abstracts

Introduction: Proletarianisation and Deproletarianisation on the Colonial Plantation
TOM BRASS AND HENRY BERNSTEIN

Given continuing debate about the plantation system, this introduction examines some of the theoretical issues surrounding the recruitment, reproduction, and dissolution of its workforce, and in particular the free/unfree character of its social relations of production. It begins by confronting two interrelated revisionist interpretations of production relations on the plantation: on the one hand neoclassical economic theory, and on the other the 'culturalist' arguments derived from moral economy, survival strategies and resistance theory, (re-) interpretations of the plantation regime which involve either a denial or a dilution of its unfree element. By contrast, it is suggested here that the free/unfree dynamic structuring production relations on the plantation, and the specific trajectories of the latter (depeasantisation, peasantisation, proletarianisation, deproletarianisation, reproletarianisation), are determined historically by class struggle (waged from above as well as from below).

'White Gold' or 'White Blood'?: The Rubber Plantations of Colonial Indochina, 1910–40
MARTIN J. MURRAY

The expansion of rubber production in colonial Indochina after 1910 can be understood through the changing patterns in the capitalist world economy. However, the plantation system which evolved in colonial Indochina, including the methods of labour procurement, the closed compounds, the rigorous work regime and harsh disciplining of labourers, the forms of payment as compensation for expenditure of labour, and the strategies designed to speed-up and stretch-out labour-time, was embedded within an historically-specific context, and hence can only be explained by reference to the localised socio-economic conditions in which metropolitan capital and indigenous labour confronted each other.

The Java Sugar Industry as a Capitalist Plantation: A Reappraisal
G.R. KNIGHT

The Java sugar industry rose to the forefront of the world sugar industry in the mid-nineteenth century and at its zenith in 1920s was second only to Cuba as an exporter of cane sugar to world markets. It has been widely assumed, nonetheless, to stand somewhat apart from the mainstream of capitalist plantation development in colonial Asia by virtue of the fact that its workers were notionally 'peasants' and its raw material produced on rural farmland in rotation with other 'peasant' crops. The present paper suggests the need for a major re-appraisal of this 'sugar-with-peasants' image and argues that in many important respects Java sugar production in the late colonial era did not differ significantly from the other main sectors of capitalist plantation production carried on elsewhere in Asia in the nineteenth and early twentieth centuries.

Labour Control and Labour Resistance in the Plantations of Colonial Malaya
P. RAMASAMY

This essay is about control of and resistance by plantation labourers in pre-war colonial Malaya. Despite differences in the way Indian and Chinese labourers were recruited and retained, the system of control was not fundamentally different. When the (Indian) *kangani* and (Chinese) contract systems replaced indenture, control was maintained through more indirect means. Under indenture resistance took the form of individual resistance, desertion being a common manifestation. With improvements in labour conditions, labour began to develop a more assertive role; collective resistance complemented individual forms of resistance.

Sugar Barons: Formation of a Native Planter Class in the Colonial Philippines
ALFRED W. McCOY

While Europeans owned the plantations elsewhere in colonial Asia, in the Philippines local planters developed sugar *haciendas* and eventually acquired the milling factories. Indigenous ownership gave Philippine plantations a depth of social and political penetration not found elsewhere in tropical Asia. Controlling the economy's advanced sector, Filipino planters, particularly those from Negros Island, were able to influence the US colonial regime and shape the emerging Philippine state to suit the imperatives of their production system. Since independence in 1946, the Negros planters have used their political influence to assure the survival of their quasi-colonial *hacienda* regime, thereby preserving both an inefficient industry and an inequitable society.

'Tea and Money versus Human Life': The Rise and Fall of the Indenture System in the Assam Tea Plantations 1840–1908
RANA P. BEHAL AND PRABHU P. MOHAPATRA

This article examines the processes involved in the emergence, operation and the eventual decline of the indentured labour system in the Assam tea plantations in India in the colonial period. As against the recent revisionist view which justifies the existence of indentured system of supply and employment of labour, it is argued here that the indenture system designed to supply low cost labour to the expanding plantations was wasteful and inimical to the physical reproduction of the labour force. The decline of the indenture system is attributed to the sharpening internal contradictions of the system rather than humanitarian actions on the part of the colonial state.

Plantation Labour in Colonial India
RANAJIT DAS GUPTA

The article attempts an analysis of the nature of plantation enterprises in colonial India. The focus is on the nature of capital–labour relations in the plantations, the methods of mobilisation and control of labour, the production and reproduction of labour power and the process of proletarianisation. It argues that capital–labour relations were marked by unfreedom, which is analysed with reference to class struggle.

'Enclave' Plantations, 'Hemmed-in' Villages and Dualistic Representations in Colonial Ceylon
ERIC MEYER

This contribution intends first to question the dualistic model that denies the extent of the interaction between plantations and villages in the case of colonial Ceylon, then to understand dualistic mythical representations (the plantation as enclave, the village as hemmed-in) as elements of colonial and nationalist discourses. It is argued that the estate/village dynamic is crucial to an understanding of the rural history of Sri Lanka, and that enclave and hemmed-in metaphors were and still are a key component in mythical representations of the rural elaborated by non-peasant groups (European planters, Kandyan aristocracy, Low Country elite) to foster their own interests.

Strategies of Labour Mobilisation in Colonial India: The Recruitment of Indentured Workers for Mauritius
MARINA CARTER

Studies of indentured migration have assumed either that recruits were obtained by deception or that the mobile unemployed voluntarily engaged, and have not adequately addressed the issue of how the dynamism of recruitment for a bonded labour system was maintained. This article explores the means by which an Indian indentured workforce was recruited for Mauritius and in particular the use of time-expired labourers (returnees) to encourage family settlement and to sustain the flow of immigrants despite poor material benefits offered.

'Coolie' as a Labour Commodity: Race, Sex and European Dignity in Colonial Fiji
JOHN D. KELLY

Asian capitalist plantations were organised by conceptions of racial and moral difference between Europeans and others. In Fiji, an enacted European vision of 'coolies' gave shape to the indenture system, its form of labour commodity and its penal disciplinary mechanisms. The efforts to make 'coolie' status real generated contradictions in Indian social life and exploitation of Indian women. Indians resisted 'coolie' status and found direct and effective ways to attack European claims to dignity. We can understand better the history of Fiji's plantations if we do not presume that the descriptive tools provided by political-economic discourse can grasp colonial capitalist realities.

Conclusion: The Making of a Coolie
JAN BREMAN AND E. VALENTINE DANIEL

This essay attempts to trace some of the more significant discursive and material practices of a gendered and racist colonial Asia of the late nineteenth and early twentieth centuries that went towards constituting the category of 'the coolie'. Especially dear to the discourse in question was the notion that a 'coolie' was essentially a peasant, given to a generalised disposition of immobility. The persistence of this fictive construction, we argue, served to obstruct and/or deny the proletarianisation of migrant coolie labour, spawning a set of contradictions that in turn revealed the contradictions inherent in the material practices of capitalism in Asia. Chief among these contradictions was that of a peasant-coolie who was both immobile and unstable, hyper-settled and unsettled. These contradictions infected nationalist as much as colonialist literature, and persist to this day, not only in the rhetoric of emerging nationalisms of certain nation-states where 'coolies' are immigrants, but also, to one degree or another, in the discourses of labour unions and 'coolies' themselves. The resultant identity of a 'coolie' is best described as a fractured one, whose most poignant representations are to be found in the fractured moral identities of the female coolie and child coolie.